THE
AGE OF
SCIENCE

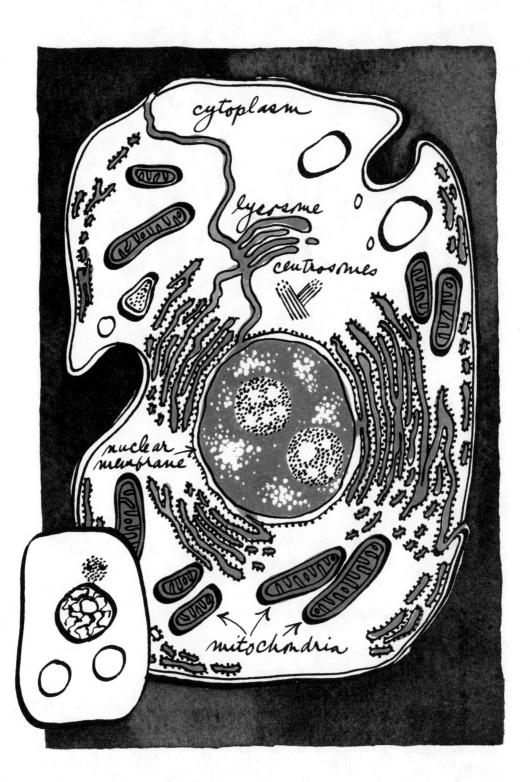

THE AGE OF SCIENCE

WHAT SCIENTISTS LEARNED IN THE 20TH CENTURY

GERARD PIEL

WITH ILLUSTRATIONS BY
PETER BRADFORD

A CORNELIA AND
MICHAEL BESSIE BOOK

BASIC
BOOKS

A MEMBER OF THE
PERSEUS BOOKS GROUP

Frontispiece: **Anatomy of the eukaryotic cell,** *a nearly empty blob of pro-toplasm in 1900, [see page 268] organizes its large molecules for the ordered, sequential, interacting cycles of reaction that are the life of the cell. Infolding of the membrane brings the interior of the cell into contact with the surrounding medium, from which it incessantly draws substance. Nucleus, containing chromosomes, is the seat of heredity; centrosomes govern the sep-aration of replicated chromosomes at cell division; mitochondria generate cell's energy [see page 285]; the lysosome is the cell's center for recycling molecular piece parts for reuse.*

Published by Basic Books,
A Member of the Perseus Books Group

Book design by Joshua Passe

Piel, Gerard.
    The age of science: what scientists learned in the 20th century / Gerard Piel; with illustrations by Peter Bradford.—1st ed.
        p. cm.
    Includes bibliographical references and index.
    ISBN 0-465-05755-1 (hc)
        1. Science—History—20th century. I. title.

Q125.P598 2001
509.04—dc21   2001043178

First Edition

01 02 03 04 / 10 9 8 7 6 5 4 3 2 1

TO PHILIP MORRISON

CONTENTS AND ILLUSTRATIONS

.

ILLUSTRATIONS

About This Book
and Myself

In this book, I have undertaken to tell my readers what I have learned about what scientists learned in the 20th century, in sum: The work of science is converging on seamless comprehension of the world around us and the identity of ourselves in it. Not since revelation explained everything so long ago has the picture been so complete. Within and beyond the world around us, the work has opened new immensities to understanding.

This time, understanding is anchored in verifiable human experience. The experience I have to report here transpired principally in the 20th century. In the last decades of the century, work in diverse disciplines converged on the same ultimate questions. Re-framed in the light of objective knowledge, those ancient questions assume immediate relevance to life. The work goes on, ever open to enlargement and amendment by experience.

The objectivity of the new knowledge has had public demonstration in technologies of immense subtlety and reach. They are enabling individuals and society with power to any purpose. The rate, scale and irreversibility of attendant economic, social and environmental change compel re-examination of the received premises from which human purpose has hitherto been drawn.

To such re-examination, the scientific enterprise offers the example of reason, tolerance and individual autonomy.

The picture I have put together here, I must state at the outset, is nowhere assembled in the formal literature of science. The authors of the work say little about relevance. Colleagues in a field have ready appreciation of relevance or they would not be engaged in the work.

Not a scientist, like most people, I have put the picture together myself. My readers, I hope, will find here the material and the encouragement to assemble her and his own picture.

As writer, editor and publisher, I have been a spectator at the inquiry after objective knowledge as it has unfolded in the second half of the 20th century. My employment has been to report on work in each field of science to the wider audience of people who want to follow developments on the whole advancing front of the scientific enterprise.

The challenge set for me and my colleagues by that audience was to meet the interest, in the first place, that almost all scientists have in fields outside their own. They begin their lives wanting to learn all about the world around them—every child is a natural scientist. Not far into the experience, they discover that they must narrow their interests and specialize. Laymen in fields outside their own, they look for ways into those fields with the same innocence and readiness to be surprised as not-elsewhere-classified interested laymen. If we could not measure up to their interest, we had no business talking to anyone else.

I stumbled onto the need of this community in the public-at-large during the 1940s in my service as science editor of the then recently established mass-circulation picture magazine, *Life.* For readers whom I came to know during that apprenticeship, I set out to create a magazine of science. Enlisting partners and staff and rounding up the necessary venture capital, I observed a precedent set by my employer. Time Inc. had purchased the name *Life,* all that was left of a defunct funny magazine. For our enterprise, I bought the 102-year-old name of the then moribund SCIENTIFIC AMERICAN. We published the first issue of the new magazine under the old name in May 1948.

For this lifework, I had the most unlikely preparation. My A.B. in history from Harvard College in 1937 was a certificate, *magna cum laude,* of illiteracy in science. For me, as for most pupils in U.S. elementary and secondary schools even now, recoil from mathematics and science started in earliest grade school. Arithmetic, taught by drill, had to be learned by rote. That invited resistance. Stubborn mistakes in arithmetic brought humiliation in algebra and mathematics from then on.

Science was then (and is still) taught from books. High school and college texts were contained in modest octavos, unlike the four-color quartos—some of the best of them published by W. H. Freeman and Company, the science-textbook subsidiary of SCIENTIFIC AMERICAN—that burden the backpacks of students today. Physics, my teacher gave me to understand, was finite. Like Euclid, it was all there in a little green book, and one of its authors was Charles Elwood Dull. A typical laboratory exercise was to measure the boiling point of water, which I knew to be 212 degrees Fahrenheit. I was instructed to find it at 100 degrees Celsius (C). My measurements missed that target and averaged, I recall to my prolonged confusion, 98.6 degrees C.

The sociology of science

At Harvard, my closest approach to science was a tutorial in sociology, to which my attention was riveted by a graduate student named Robert K. Merton. On the principle of total immersion, he set me to reading the works of the giants: R. H. Tawney, Emile Durkheim, Karl Marx, Werner Sombart, Max Weber and even Vilfredo Pareto, who had just then been translated from the Italian. From this humbling experience I turned to history as my field of concentration.

With the ambition to cover the Second World War that history told me was on the verge, I went to work at *Life* in 1938. My illiteracy in science emerged suddenly, however, as a qualification surpassing my preparation in history. John Shaw Billings had been managing editor of *Time.* Drafted to responsibility for the company's considerable investment in *Life,* he determined to make it the photographic edition of the company's flagship, devoted to the same earnest mission of public instruction in all matters. (Admittedly, this has not been to be believed from the appearance now of the magazines bearing those names at the supermarket checkout counter.) He appointed me to cover science with the thought that ignorance is safer than a little knowledge, which is so very well known to be a dangerous thing.

As I learned from the moment I undertook my assignment, there was no magazine of science. My colleagues—in domestic and foreign news, sports, Hollywood, the arts, the theater—had numerous sources, starting with *The New York Times,* to tell them what to put in the next issue of *Life.* Science, I discovered, is served by its own press, with more publications than any other human enterprise. Its narrowly specialized journals serve internal communication in each narrow line of work and are unintelligible to the outsider, even to scientists in other fields. I found little that I was able to read about science anywhere, not even in *The New York Times,* which covers science so well today.

For good reason, in those days, scientists were wary of the press: they were too often embarrassed by overheated, improbable accounts of their work. It took some doing to persuade my first two collaborator-scientists to go to work with me and a photographer on a story about his and then her work. Once I had succeeded in getting my first few stories into the pages of *Life,* however, I found it happily easier to persuade the next scientist I approached to work with me.

That was because the camera told the story. Displaying the thing itself in its own inherent interest, the photograph relieved the prose of the task of exciting the reader's interest. In other mass publications, it was the carnival barking for the public's attention that occasioned the embarrassment of scientists in their experience with the press. When the photographs were successful, moreover, they were self-explanatory, to the relief of the prose again. The captions and the story that I wrote for the pictures were certified to be accurate, finally, because my scientific collaborators insisted upon vetting my copy. They had to be conceded that right because they put as much trouble into setting up our stories as our photographer and I did.

For the first time, science had responsible, authentic coverage in a mass-circulation magazine. By the summer of 1944, I found myself welcomed by the scientists I called on for help. I was able to enlist even the remote and proud Rockefeller Institute in collaboration on a story. That was the first isolation and chemical disassembly of a virus, a feat for which Wendell Stanley, a member of the Institute, was to receive the first postwar Nobel prize in chemistry in 1946. With Stanley's help and the loan of a pint of his tobacco mosaic virus, Fritz Goro, my principal collaborating photographer, was able to demonstrate the distinctive geometry of the virus in four colors stirred by a swimming guppy.

By this time, I was aware that my science department had isolated its own warmly interested audience inside the then 4-million circulation of *Life*. The keen interest they took in the pages I filled in the magazine told me that those readers, so well informed in their own fields, had the same need for a magazine of science that I did.

I learn about the "secret"

Meanwhile I had been learning at my work much that I could not report. This was convincing me that a magazine of science would have to stand on its own independent editorial legs. I would not propose the idea to my employer. For all my poor preparation in science, I had learned the "secret" of the Atom Bomb.

My principal source of intelligence was, naturally, the process of censorship itself. Early in 1942, it became the duty of John Billings to share with me a telegram from the wartime Office of Censorship in Washington. The Office was invoking the voluntary censorship agreement to which the U.S. press had subscribed. The telegram proscribed a long list of topics that started with "atomic energy" and went on, as I recall it now, to put "radium," "uranium," "isotopes," "critical mass" and "atomic fission" on its *Index expurgatorius*.

The recognition of atomic fission by Otto Hahn in Germany, by his collaborator Lise Meitner, in exile from Hitler in Sweden, and then in 1939 at Columbia University had made a minor sensation in the press before I began my assignment. The Office of Censorship telegram made it clear to any reader that this phenomenon now had a place in the war effort. The topics it cited gave me a reading list.

There was already more than enough in the open literature, I found, to give anyone who could read it an idea of how an atomic weapon would work and how it could be made. I could not, of course, be counted as practiced reader of any part of it. With such motivating questions in mind, however, I could tease out some meaning even from the paper by E. Fermi and J. A. Wheeler [see page 151]. Outside the primary literature, to my relief, I found a college textbook, published in 1939 before the physicists in the U.S., France and Britain had suppressed publication of their development of the breakthrough. It con-

cluded with a chapter on the promise of energy for free from the nucleus of the atom.

My studies put me on the alert to learn how the atom bomb project was progressing whenever I encountered the censors. Thus, in the midst of our photographing an elegant demonstration of the mass spectrometer at the University of Illinois, Professor A. O. Nier had to call our collaboration to a halt for reasons he could not discuss. With that story, I was eager to show my readers something I had just learned: physicists could tell apart atoms that escape discrimination by chemists. It could be important to tell which was which because, along with differences in weight, "isotopes" of the same chemical identity could exhibit other significantly different physical properties—such as tendency to fission in a chain reaction. By their interference, the censors told me that mass spectroscopy was being taken seriously as a method for separating the rare and fissionable uranium isotope of atomic weight 235 from the much more abundant and stable isotope 238.

In this context, I learned that the scientists who had been involved in the uranium-fission work at Columbia had all gone to Chicago. Then I was told that I could not interview Professor Nils

Bohr, in exile from Denmark, and admonished that I was not "cleared" even to know that he was in the country. I first heard the name of the Manhattan Project from a procurement agent for the Canadian government when, at night, in the middle of a deserted Manhattan street, he asked in a whisper what I knew about it. The war-effort "priority" of the Canadian government could not compete for supplies with a giant Manhattan Project undertaking in Tennessee. I got an idea of what the enterprise in the hills down there might be about when I overheard at lunch at Eastman Kodak, that one of its bright young physicists was to be saved from the draft by "going to Tennessee."

All at once, in the late summer of 1944, I had my gathering surmises confirmed by R. W. Wood, professor emeritus of physics at Johns Hopkins. He was helping us to do a story on the "shaped charge" that had made German antitank shells so lethal in the desert battles of North Africa in 1941–1942. The shaped charge had been belatedly adopted in U.S. ordnance and was no secret because the enemy had it first.

Wood had learned of the phenomenon early in the century, in a technical bulletin published by the U.S. Coast Guard. One of its ordnance experts had discovered that armor plate could be punctured by the proper shaping of an explosive charge set upon its surface; for his report, he had fashioned shaped charges to puncture "USA" into a sample of naval armor plate. From his reading, Wood had been able to explain a safecracking to the Baltimore police. They had found no metal shavings but a clean hole in the door of the safe and congealed droplets of steel inside. Wood had not, however, been able to interest the ordnance officers of the U.S. Army in the phenomenon.

In the battle of North Africa, the Germans had demonstrated that a shaped charge in an antitank shell would deliver most of the energy of its explosion in one direction, forward, instead dissipating it in all directions [opposite]. With Wood's help, our camera showed that much of the force of the directed-energy beam was supplied by the ultrahigh-velocity vaporized heavy atoms of the metal liner that the German armorers fitted into the conical depression in the charge. The forward-facing conical nose of the shell was a windscreen and fuse carrier; a higher order of physics did the armor piercing.

The atoms of vaporized metal made themselves known in our pictures through Wood's own plastic replica of a Johns Hopkins diffraction grating, said to be the most perfect object made by the human hand [see page 34]. The replica served as a filter on the lens of our camera. A graduate student apprentice to Henry Augustus Rowland, who made the first of those perfect gratings, Wood had succeeded him in his professorship. At the outbreak of the Second World War, with Wood in charge, the laboratory found itself serving a huge new market in the U.S. war effort. Wood had devised his plastic replica of the Rowland diffraction grating to meet the demand. At much lower cost, it reproduced the perfect grating with its own perfection. For *Life's* readers, the plastic grating clearly delineated the lines of light at the wavelengths of the vaporized metals in the skyward detonation of the charge.

One of the customers for which Wood had been manufacturing his replicas in quantity was known to him only as the Manhattan Project. Because the security agents refused him clearance to know what the Manhattan Project was using his gratings for—knowledge by which he might have improved the performance of his replicas—he had given the question a good deal of thought. Until he had my ear, Wood had managed to bottle up his indignation at the censors' affront to his age and eminence. Because they had denied him the mystical security clearance, he was free to talk about what he had deduced. The Manhattan Project wanted his gratings, of course, for use in the precise monitoring by spectroscopy of the purity of all materials involved in the project, including the fissionable U-235, present in nature in minute quantity, to be captured on industrial scale from its more abundant stable isotope U-238. From the timing and the volume of orders for his gratings, he could calculate the progress of the enterprise, which, at that time, he estimated at several critical masses of U-235.

Our shaped-charge story had its own relevance to the Atom Bomb. Wood explained that shaped charges offered the surest way, by implosion, to assemble and hold the critical mass of fissionable material in place for the split instant of its explosion. A hollow sphere of high explosive lined with uranium-235 or plutonium would serve the purpose. In working on the shaped-charge story with me, Wood was playing a high-level joke on the censors. Julius and Ethel Rosenberg were nonetheless to be executed in 1953 for betraying this secret.

The demonstration of the atomic weapon, Wood thought, was imminent. He held its manufacture to be a reckless and criminal enterprise. In particular, he had anxieties about the long-term consequences of radioactive fallout. He wanted me to know about the bomb, he declared, because the press should do something about it as soon as it could, if possible, keep it from being used.

The magazine of science

Putting nothing in writing, I reported to John Billings what I had learned, in line of duty, from R. W. Wood. My boss and anyone else with responsibility at Time Inc. would be ready to respond when the time came. Thereupon, I determined that I could lose no time on the start-up of the magazine of science for which responsible citizens, scientists and nonscientists, would now have such urgent need.

The well-timed publication in June 1942 of a "Profile" of Henry J. Kaiser, under my byline, had made him the Paul Bunyan of our country's industrial war effort. He did, in fact, build one-third of all the cargo- and troop-carrying ships launched by our country during the war. Kaiser's 50 convoy-escort carriers, reluctantly accepted by the Navy on orders from the Commander-in-Chief, did critical service as ships of the line in the Battle of the Philippines that saw the destruction of the Japanese navy. Employment as his "personal assistant" through the year 1945 taught me enough about business to secure the launching in May 1948 of the new SCIENTIFIC AMERICAN.

Like my *Life* science department, SCIENTIFIC AMERICAN was the product of the collaboration of scientist and editor. We enlisted the scientists who were authors of work that caught our interest to write the articles about that work. In their concern to reach the wider audience interested in science, they accepted our invitation and then our editing. They collaborated, as well, in producing the illustrations that lifted so much of the burden from our prose.

The magazine did not solve the still outstanding problem of public understanding of science. For the audience to which we addressed it, we were assured "it fills a real need very satisfactorily." We had this assurance from our reader and author Albert Einstein, who volunteered it in writing before the end of our second year of publication. Objective evidence from the market confirmed his judgment: the circulation of our magazine in the English language increased steadily over the years to more than 600,000. Uniquely assembled in this circulation was the U.S. community of science, broadly defined to include scientists and the people most interested in what they were doing.

By 1986, SCIENTIFIC AMERICAN had found a worldwide circulation of more than 1 million in English and in nine translated editions—in Italian, Japanese, Spanish, French, German, Chinese, Russian, Hungarian and Arabic, in the order of their launching. In each case, the start-up of the new edition was on the initiative of speakers of that language; they were seeking to fill, so "very satisfactorily," the need of their fellow citizens. SCIENTIFIC AMERICAN had assembled in its circulation the world community of science.

As my end notes suggest [see page 445], this book is my synthesis of my necessarily close study, then and now, of the articles we published over four decades in SCIENTIFIC AMERICAN. What were accounts of current events, from issue to issue, I have read again as history of a glorious progress in human understanding.

In the first chapter, we will see that scientists do what we all could do if we would clear our heads of preconception and bring our questions in reach of experience. The next three chapters show the world around us reconstructed in continuity with the immensities now opened up within it and outside it. The last three chapters tell the story of life from its beginning, soon after Earth cooled down to hold water in the liquid state, on to the understanding that the people must soon find inside their own heads the purpose of human existence.

1

Science Is What Scientists Do

What science ought *to be is what*
the ablest scientists really want to do.
Warren Weaver

How did something come from nothing? How did something come to think about nothing?

Everyone has asked these two questions in one way or another, at one time or another. Stories of creation have answered in every language. Those stories never fail to remark the special creation of man and woman. Whatever else they have meant to people, accounts of genesis have set cornerstones in the literatures of their languages and exhibited the power and glory of the imagination.

Now, at the beginning of the 21st century, scientific inquiry is approaching still other answers to these two first questions. This is remarkable, because scientific inquiry is constrained by three rules, self-imposed by the mutual agreement of its practitioners. A scientist can admit to rational consideration nothing but the physical reality that is accessible to experience. On the experience in question and what it may mean, a scientist can recognize no authority but his or her own judgment and must, at all times, hold that authority in suspicion. The experience must yield evidence open to inspection by others.

The "reality" of physical reality remains an open question in the philosophy of knowledge. Recent work in the sciences has warmed the interest of philosophers in the question. Is reality independent of human perception, they ask, or an invention of it? Scientists go about their work, however, with little or no interest in the matter. Over the 20th century, by stubborn practice of natural philosophy, they have extended the boundaries of the universe accessible to experience.

Astronomical instruments carry perception a million times farther out into the universe. On the inner frontier, the tools of experiment and observation reach 10 billion times deeper into the substance the universe is made of. At the extremes of dimension now in reach, scientists have discovered entirely new orders of physical reality. Human consciousness must strain and stretch to comprehend the universe in which our little world of everyday experience now floats.

Thus far, the effort sustains the conviction that order and simplicity will be found to underlie and unify the still unfolding complexity of the extended universe. The observed simplicity collapses hitherto separate compartments of physical reality, however, and the order discerned is proving to be of an entirely unanticipated kind.

In the new grand scale of the universe, time and space—so separate in sensory experience—are fused in a four-dimensional continuum. Space and time are no longer the theater of cosmic events; they are "spacetime," engaged in events along with matter and energy. Light from the most distant reaches of the observed universe shows us earlier times in the life of the cosmos. In those times, space was correspondingly smaller. How, in the expansion of spacetime, the universe came from next to nothing has begun to yield to understanding from close study of its constituent particles.

Matter and energy have been found, likewise, to manifest a single entity, transforming interchangeably, one into the other. Here at the root of physical reality, the matter-energy interchange is governed, if that is the word, by chance. Chance is, however, constrained to discrete possibilities: Energy transforms to matter not just anywhere along a continuous scale but at fixed points of intensity—energy is "quantized." In the realization of probabilities, thus constrained by possibility, chance gives rise to order. At appropriate intervals on the scale, matter-energy self-organizes in particles; particles in atoms, and atoms in molecules. Thereupon, especially in that part of the universe nearest at hand, the diversity of the natural world unfolds. In their autonomy, living organisms exhibit the protean potential of the self-organization of matter-energy.

Evidence from the oldest sedimentary rock indicates that life must have begun on Earth almost as soon as the planet could hold liquid water. This suggests that life inevitably arises wherever there is water. The liquid state of water specifies, however, a narrow range of temperature: only 100 degrees, perhaps midway on the Kelvin scale between the hundreds of million degrees in the observed universe and the equal number of fractions of 1 degree above unreachable absolute zero. Given water, by weight its principal substance, and other elements from the atmosphere and crust of the Earth, life animates itself with energy captured from the narrow band of light in the spectrum of radiation broadcast by the Sun. Light, in all its colors, extends but an octave in the electromagnetic spectrum that reaches from the kilocycles of radio to the trillion times a trillion vibrations per second at which the transformations of matter and energy occur [see illustration, pages 6–7].

Special as these circumstances are, they suggest that life may be a relatively common occurrence. Within the solar system, the possibility of life on one of the terrestrial planets, or on a terrestrial satellite of one of the gassy giant Jovian planets, has not been excluded. Human observers want to know, of course, whether life elsewhere in the universe has given rise to creatures like themselves. That possibility is small enough. Whether the human species will live long enough to learn of it is largely in human hands.

The human brain is the most highly organized assembly of matter-energy known. Its evolution began, as learned late in the 20th century, 2.5 million years ago when the first toolmakers framed conscious purpose in their heads. They were 90-pound primates otherwise indistinguishable from cousins that did not form the toolmaking habit. Homo *sapiens* made his-her appearance only in the last few hundred thousand years of the 4.5-billion-year history of the Earth.

Only in the 20th century—in the last microsecond of Earth history—have people attained the capability to pursue the question whether they are alone in the universe. The possibility of communication with life elsewhere plainly depends upon the chance that this split second on

Earth coincides with the corresponding phase in the history of life near some other star. This possibility will improve with every century that the capability is prolonged on Earth. In increase of the stock of objective knowledge will be found the surest ways to that end.

Powers of ten

At the start of the 20th century, the observed universe reached outward 50,000 light-years to the center of what is now recognized to be a local congregation of 100 billion stars, called the Milky Way or the Galaxy. A light-year is the distance that light travels in one year at its velocity of 300,000 (more exactly, at latest measurement, 299,800) kilometers per second. That velocity multiplied by the number of seconds in a year gives the light-year a round-number length of 10,000,000,000,000 kilometers. That gives the presently observed universe a radius of 100,000,000,000,000,000,000,000 kilometers.

These cosmic numbers are unintelligible, as well as incomprehensible. People have been learning, however, to live with ever larger numbers. Within a lifetime the gross domestic product of the United States has increased, in current dollars, from $100 billion to nearly $10 trillion. This is less comfortably read as from $100,000,000,000 to $10,000,000,000,000.

To manage the large and small numbers that describe physical reality, scientists employ the notation of powers of 10, or orders of magnitude. In this notation the expansion of the U.S. gross national product reads much more easily as an increase from 10^{11} ($100 billion) to 10^{13} ($10 trillion). The exponent counts the zeros needed to write out the number. A light-year is, accordingly, a more legible, if still inconceivable, 10^{13} kilometers. The 10^{10} light-years to the edge of the observed universe is 10^{23} kilometers, or 10^{26} meters.

Powers-of-ten notation works for ever smaller numbers, too. Particle accelerators have extended human perception inward to the dimension of the quarks—the particles of the particles of the nucleus of the atom. That is one ten-thousandth of one-trillionth of one meter, or 1/10,000,000,000,000,000 meter. This fraction is more easily written and read as 10^{-16} meter, the minus sign signifying that the exponent is the denominator of a fraction. The observed universe thus reaches across 43 powers of 10: from 10^{-16} to 10^{26}. Just off center, at the boundary between the reach to the large and descent to the small, is the 1-meter (call it 10^0) order of magnitude closest to every-day human experience [see illustration pages 6–7].

Now within observation, therefore, is a cosmos vaster than the infinities that once bemused the mind. The powers-of-10 notation can help somewhat to bring the observed incomprehensible differences in scale in the reach of the imagination. The expressions 10^2 and 10^9 tell, by subtraction of exponents, that 10^9 is 10^7 (10,000,000) times larger than 10^2. This means that a billion (10^9) contains 10 million (10^7) hundreds (10^2s). Conversely, 10^{-2} is 10^7 (10,000,000) times larger than 10^{-9}; there is room inside 10^{-2} for 10 million (10^7) billionths (10^{-9}ths).

The universe brought into the consciousness of most human beings by their unaided senses reaches from the starry sky at 1,000 light-years or 10^{19} meters down to 0.1 millimeter or 10^{-4} meter, the limit of close inspection. That is 24 powers of 10 (counting the 10^0 power). It must not be thought, however, that our 24-order-of-magnitude naked-eye world comprehends more than half of the 46-order-of-magnitude universe. The deep interior is grossly out of reach, and there is room for 10^{22} naked-eye worlds in the observable universe.

This book will employ exponents to manage magnitudes outside of common experience. Even without practice, this is easier than counting boxcars of zeroes. There are, for now, no common terms for the big numbers beyond the "trillion" that the GDP only recently brought into general discourse. International convention has designated names, adapted from the Greek, for still bigger numbers in scientific discourse, along with a corresponding set of names for magnitudes in the micro world. Trillion is the biggest magnitude with a name in this book. "Billion" is used with apology for it excites misunderstanding between speakers of English and American: in one language it means 10^{12}; in the other, 10^9. In this book, it is American: 10^9.

By 1900, optical telescopes and microscopes had extended the reach of the naked eye by not much more than an order of magnitude outward and inward. Improvements in optical instruments have since

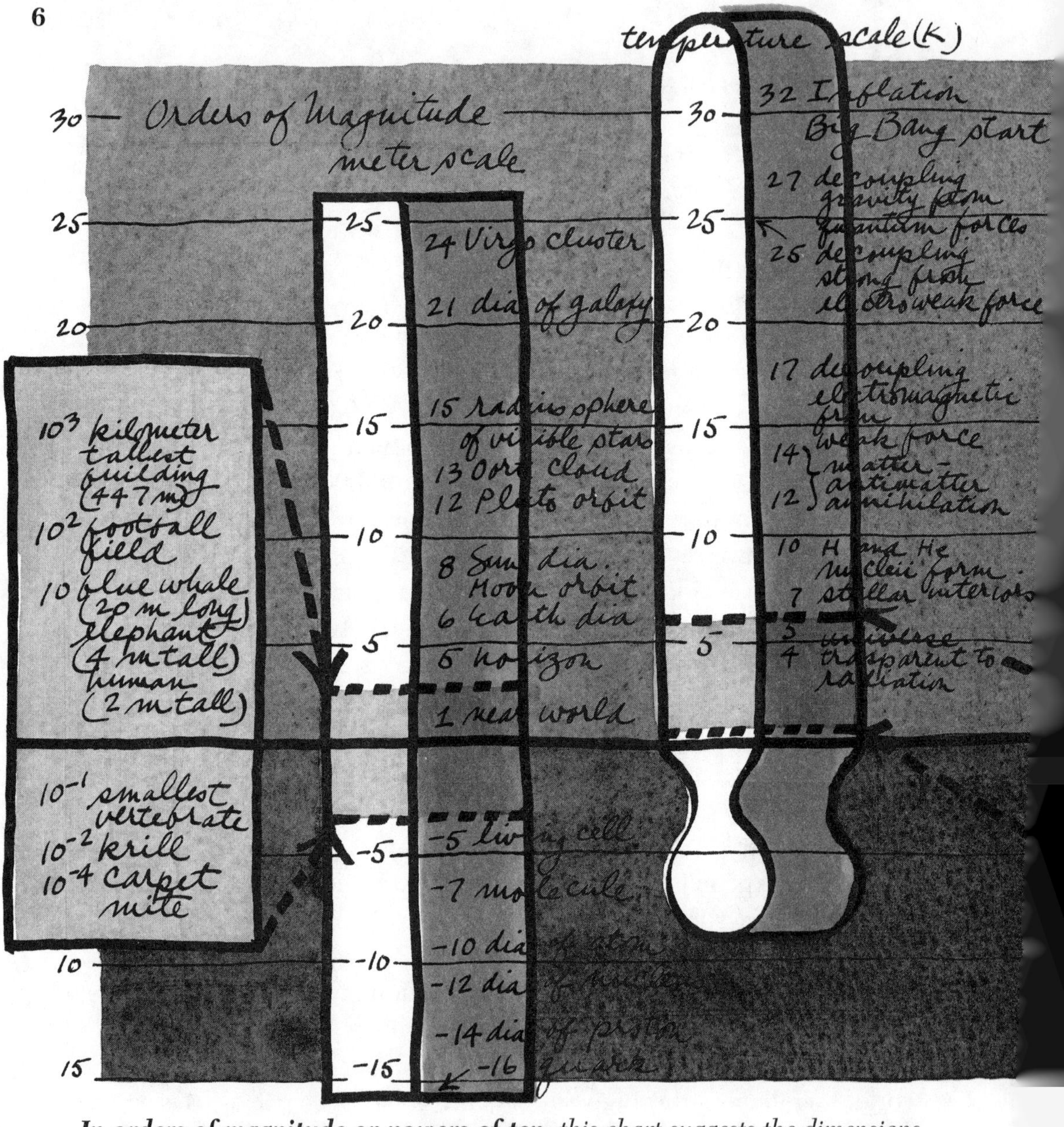

6

temperature scale (K)

Orders of magnitude
meter scale

30 — 30
32 Inflation
Big Bang start
27 decoupling gravity from quantum forces
26 decoupling strong from electroweak force

25 — 25 — 25
24 Virgo cluster
21 dia of galaxy

20 — 20 — 20
17 decoupling electromagnetic from weak force

10^3 kilometer tallest building (447 m)
10^2 football field
10 blue whale (20 m long)
elephant (4 m tall)
human (2 m tall)

15 — 15 radius sphere of visible stars
13 Oort cloud
12 Pluto orbit
15
14 } matter-antimatter annihilation
12 }

10 — 8 Sun dia. Moon orbit
6 earth dia
10 — 10 H and He nuclei form. stellar interiors
7 stellar interiors

5 — 6 horizon
5 — 5 universe transparent to radiation
4

1 real world

10^{-1} smallest vertebrate
10^{-2} krill
10^{-4} carpet mite

-5 — -5 living cell
-7 molecule
10 — -10 — -10 dia of atom
-12 dia of nucleus
-14 dia of proton
15 — -15 — -16 quarks

In orders of magnitude or powers of ten, this chart suggests the dimensions of what scientists learned in the 20th century. On the scale of distance, they increased the radius of the observed universe by nearly 20 orders of magnitude, nine orders of magnitude outward to the quasar region and era of the universe and its history and 10 orders of magnitude inward to the quarks [see page 161]. On temperature and energy scales, accelerators approach 10^{13} electron volts (eV) and the corresponding temperature of 10^{17} K, at which particles of matter

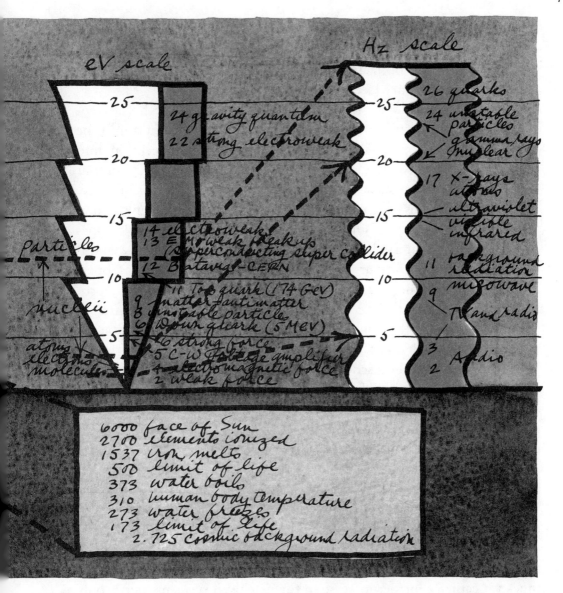

took their rest mass from the background energy of the universe [see page 202], and so approach the first 10^{-12} second of universal history [see page 232]. The "limits of life" temperatures are spanned by prokaryotes; eukaryotic life is confined to a narrow range between the boiling and freezing points of water. From the corresponding spectrum of radiant energy, here scaled in frequency, Hz, we know most of what we "know for certain," by spectroscopy on frequencies up to gamma rays and the new spectroscopies of the accelerators [see page 34].

expanded the radius of the observed universe perhaps another four orders of magnitude. The further reach of observation, outward and inward by 14 orders of magnitude, is owing to technologies developed during the 20th century [see illustration, pages 6–7].

These technologies have brought long reaches of the electromagnetic spectrum on both sides of the visible light spectrum into the service of perception. They have extended human perception into realms to which perception is blind. Radio telescopes see objects in our local galaxy like proto-stars not yet ignited to radiate in the visible spectrum, as well as quasars, galaxies undergoing gravitational implosion at the edge of the universe. On the 21-centimeter band, these telescopes have mapped the interstellar distribution of hydrogen, the most abundant of the elements, in our local Galaxy. Gamma-ray telescopes, operating in the 10^{-10} meter band aboard Earth satellites outside the protecting atmosphere, take the measure of distant galactic cataclysms.

No less unsettling, in the scale of human experience, must be the expansion of time that has accompanied the expansion of space. In the 17th century, the biblical scholarship of Bishop Ussher set the moment of creation in 4004 B.C. The geologists of the early 19th century were already in need of a universe tens of millions of years old to accommodate their observations. Today, radio telescopes have set the lifetime-to-date of the expanding universe at 10 to 15 billion years.

Space and time, it has only recently been realized, go together toward the smaller and the earlier as well. Accelerators that reach into the 10^{-16}–meter realm approach conditions at around the same fraction of the first second of time. Pivotal questions respecting the unification of the forces of nature, the origin of matter, and the beginning of the universe now tantalize inquiry out of reach at dimensions still tinier and times yet earlier. Further increase in the negative exponents needs only the ingenuity to design still more powerful instruments and the willingness of society to support the building of them.

It is easier to imagine the exponent increasing in the positive direction. There is room for human beings inside the universe described by those big numbers. The rules by which scientists have brought the universe under observation across 46 orders of magnitude must here be recalled: experience alone merits consideration; evidence of the experience must be open to inspection. By no verifiable experience has the universe been observed from the outside. There is, then, in accordance with the rules, no space or time outside the expanding universe.

Observed from inside, the universe has a radius of the order of 10^{10} light-years and is expanding. During the last 50 years, the telescopes that have been tracking the expansion of spacetime into the past have, in effect, reversed the expansion. They have brought into perception and understanding a universe that was then correspondingly smaller, denser and hotter. There and then, 13 or 14 billion years ago, they have caught galaxies in stupendous convulsion. Still deeper into the past, instruments have measured, at 3,000 degrees Kelvin (K), the temperature of the universe in a late phase of its natal explosion. Sustained part of the way by the experiments of particle physics, theory has brought the universe into view at the age of 10^{-35} second and 10^{-6}–meter radius. This is an impossibly small universe. The picture of it in the mind's eye prompts the reminder that the universe can be seen only from inside, there being no time or space outside. The mighty nucleus is all there was. There were no galaxies, much less people, inside; only the matter-energy from which they were to materialize.

By extrapolation on strongly constructed theory, partially sustained by experiment, spacetime has been imploded further, to approach a geometrical point and infinity of matter-energy density. This is as close to nothing as experience can hope to reach.

The ground of knowledge

Uncertainty surely attends stretching the evidence beyond the 43 orders of magnitude reached by the instrumentation of science. An antecedent uncertainty inheres, however, in the very procedure of scientific inquiry. By induction, scientists seek the general law in the observed particular. In violation of at least one general law drawn from observation, it is well known that the nth crow may turn out to be white.

As a way to the truth, induction has been held inferior to deductive logic and the certainty with which logic derives the particular from the general. Nothing new issues from deduction, however, for the particulars are already there, implicit in the general. Moreover, it is now established that deduction does not always yield the truth. One of the great mathematical discoveries of the century has shown that, under sufficient logical torture, any finite set of premises can be made to yield a self-contradiction or paradox. As Kurt Gödel, the author of that discovery, allowed, the paradox may be cured by enlargement of the premises, but at the cost of setting up a new paradox.

The pragmatic method of science employs both approaches to the truth. By "free invention," Albert Einstein's happy phrase—or by agonized second thought—a scientist chooses a set of premises from which to pose the question. By deduction, she derives a proposition that can be confronted by experience. The experiment or observation thus prescribed assesses, by induction, the validity of the premises.

The uncertainty of the procedure is compounded by the fallibility of the human being who is the self-selected authority on the subject at hand. Against error from this quarter, scientific inquiry proceeds as a self-correcting social enterprise.

The work of a scientist has no existence until it is published to other scientists. If they find it of sufficient interest, they will take the trouble to verify the evidence to their satisfaction by repeating the observation or experiment. The small community of scientists most interested in a line of inquiry forms a consensus about the validity of the work and its meaning. That a finding has the consensus of such a community is the best that can be said for the certainty of science. Scientists engaged in establishing a truth are likely to find more interest, meanwhile, in the questions that a finding asks than in the one that it answers.

Scientific revolutions

The recognition that science is the work of fallible human beings has disconcerted serious scholars of the scientific enterprise, among them Thomas Kuhn, author of *The Structure of Scientific Revolutions*. Kuhn found a consistent scenario in these revolutions: Workers in a field refer their premises and fit their findings to the prevailing "paradigm," or general theory of a field of inquiry. They continue to do so even as the accumulation of findings exceeds the strength of the paradigm to contain

them. The crisis yields to a "paradigm shift" that resolves the contra-
dictions. To the new general theory, workers in the field thereafter refer
their questions and findings until the pressure brings on the next crisis.

The foregoing is a faithful summary of the revolutions in understanding to be encountered in these pages. The circulation of "paradigm shift" as a common coin in current discourse testifies to the wide
readership of Kuhn's treatise. Kuhn is not to be held responsible for
such misunderstanding of his reading of history as the notion that the
Einstein revolution, in our time, displaced the universe of Isaac Newton.
In the end, however, he lost his epistemological nerve. Kuhn concluded: "We may, to be more precise, have to relinquish the notion, explicit
or implicit, that changes of paradigm carry scientists and those who learn
from them closer and closer to the truth."

Kuhn supplied no definition of "the truth" he so despaired of.
The truth sought by scientific inquiry has its plain statement in the pragmatic philosophy of C. S. Peirce, as restated by his admirer William
James: "The truth of an idea is not a stagnant property in it. Truth happens to an idea. It becomes true, is made true by events. Its verity is,
in fact, an event, a process: the process namely … of its verification."

Someone does the verification. Others verify the verification.
The verification remains subject to verification.

From evidence available in the 3rd century B.C., Aristarchus of
Samos found, to his own satisfaction, that the Sun is a great deal larger
than the Earth, and the Moon a great deal smaller; that the Earth rotates
on its axis and, with the Moon revolving about it, revolves around the
stationary Sun; that the Earth's axis is inclined to the plane of its orbit
and that this inclination brings the change in seasons. For lack of interest in such evidence—and the prepotency of paradigms resting on other
grounds—people lived for nearly two millennia with quite different pictures of the cosmos. Verification of observations by Aristarchus and his
contemporaries, renewed in the 16th and 17th centuries by Tycho,
Copernicus, Galileo and Kepler, set matters straight. People today see
the picture much as Aristarchus did.

If the shift from Newton's paradigm to Einstein's did not carry
closer to the truth Thomas Kuhn had in mind, it surely brought unity to

an assembly of truths verified to the Peirce-James standard. As the next chapter will develop for its lift to appreciation of General Relativity, the Newtonian equations persist in the larger system and work satisfactorily in the nearer world. From Aristarchus to Einstein, the rule holds: a firm grip on any corner of physical reality is a grasp upon the whole.

In the end the work of science finds its public verification in technology. The firm grip on any corner of physical reality soon tightens to control in that corner. Each "technology" is first a scientific experiment designed to test a proposition about physical reality. If successful, the experiment is scaled up and retooled by engineers for reliable repetition or continuous operation. Whatever the military or political objectives served by Alamogordo, Hiroshima, Nagasaki and the succession of above- and underground weapons tests that followed, they redundantly demonstrated to the world that matter transforms to energy; that $m = E/c^2$. The equation is more familiar in its inverse statement: $E = mc^2$. In English, this translates to the mysterious statement that Energy equals Mass multiplied by the Velocity of Light, squared. So redundantly demonstrated, this is but one outcome of Einstein's Special Relativity that collapsed space and time, and matter and energy.

The history leading to this outcome is recounted in the next chapter. The third chapter will show how Special Relativity, together with Max Planck's quantum, arrived at the theory of quantum electrodynamics. It was the work of the most extraordinary concert of intellect in history. This incomparable invention of the mind supplies the ultimate reference for everything that is known and yet to be learned about the corner of physical reality closest to human concern and existence.

Quantum electrodynamics has public verification in the surge of industrial revolution that got under way in the second half of the 20th century. Producer and consumer goods and services derived from the new understanding at quantum scale in time and space are aggregating to 25 percent of the gross domestic product of the industrialized countries. The new technologies exploit principally the theory's grasp on the interaction of light and matter and on the nature of matter in the solid state. They draw on "flea-power," not horsepower; they mechanize the capacity of the nervous, rather than the muscular, system. On the now familiar computer chip, programmed reversals of tiny voltages open and shut millions of switches in ultramicroseconds to permit and to stop the flow of tiny bursts of electric current. The totally abstract logical operations performed thereby serve real-world functions from aircraft flight control to accessing the Library of Congress.

Meaningful inquiry into the nature of the physical world commenced with the experimental work of Galileo Galilei. More celebrated as astronomer than as laboratory physicist, Galileo was the first to observe the sky through a telescope. With it he saw mountains on the Moon casting shadows in the light of the Sun. He observed Jupiter eclipsing its moons on their encircling orbits and, no less important, the changing, moonlike phases of Venus.

Galileo's inclined plane

Galileo developed the significance of these observations in his *Dialogue on Two World Systems.* In 1632, he addressed this work not in Latin, the language of learning, but in Italian to the wider audience. The systems under discussion are the Sun-centered cosmology proposed by Copernicus and Ptolemy's Earth-centered model [see illustration, page 38]. *The Dialogue* argues equally, as the historian of science Owen Gingerich observed, the weight of the book of nature against that of the book of scriptures. To the character called Simplicio, Galileo gave the defense of the losing book and system. Pope Urban VIII, it seems, found echoes of his own arguments with Galileo (they were fellow members of the Lincei, the world's oldest scientific society) in Simplicio's. The Inquisition put Galileo under house arrest for the rest of his life.

Quite apart from Galileo's casting of the characters in the *Dialogue,* the heresy advanced by Nicolaus Copernicus in his *De revolutionibus orbium coelestium* was of sufficient consequence. For centuries, on the authority of Aristotle, the Ptolemaic cosmos had placed the Earth at the center of the sphere of fixed stars [see illustration, following page]. From outside, heaven shone through the stars. Inside, the Sun, the Moon and the planets could be seen wheeling in the sky, each fixed in its celestial sphere concentric with the Earth. Deep within the Earth, whether taken to be flat or round, was the netherworld of the damned.

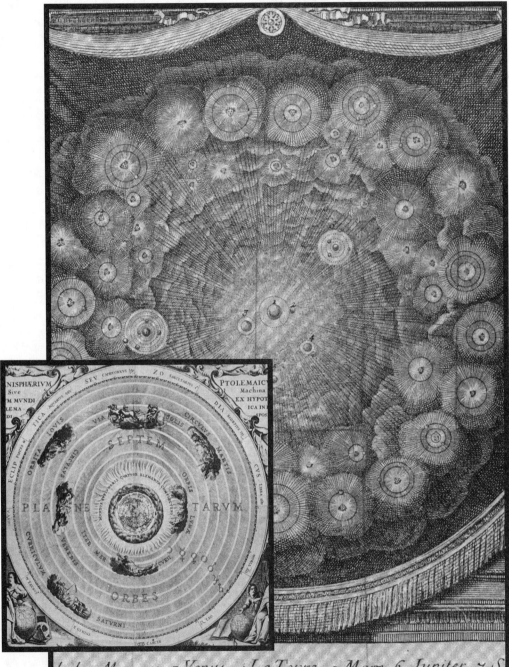

...leil. 2 Mercure. 3 Venus. 4 La Terre. 5 Mars. 6 Jupiter. 7 S...

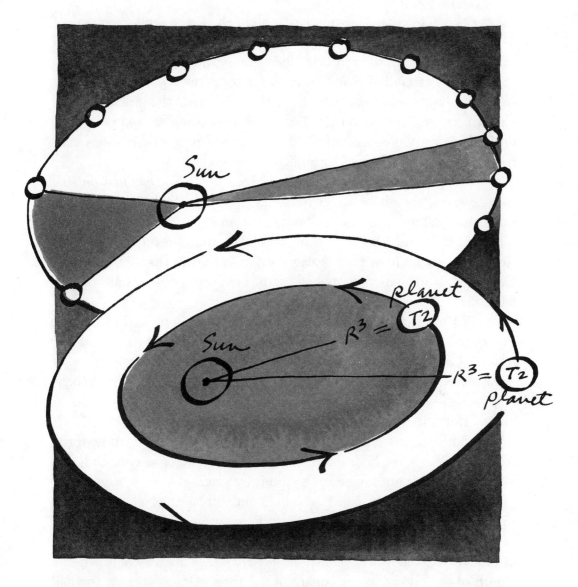

Copernican revolution [opposite] displaced Earth from the center of Ptolemy's cosmos [opposite, inset]. In this 17th-century celebration, Earth is at 45-degree angle to right below Sun at center; stars in the surrounding firmament have planets. Kepler's first law [see page 48] puts planets on elliptical orbits, here exaggerated. Faster travel of planet when nearest Sun, conserving angular momentum, sweeps out same area as when most distant. Third law finds cube of planet's average distance from Sun (R^3) proportional to square of length of its year (T^2).

The Ptolemaic-Aristotelian system had the endorsement of Thomas Aquinas. To upset that system in an impious dialogue, with Man displaced from the center of the cosmos and pitched into improbable motion around the Sun, was to strike at the foundations of the authority and submissive faith that contained the social order. Copernicus did not go on the *Index librorum prohibitorum,* however, until Galileo's popularization of his own confirming observations captured the interest of the wider audience.

At age 70, under house arrest, Galileo wrote his *Discourse on Two New Sciences,* the dialogues in Italian again for the wider audience but the formal statement of the new physics from experiments he had done 30 years earlier in Latin. This time Simplicio expounds the notion, from Aristotle and Aquinas, that heavier bodies fall faster than lighter ones. Galileo's historic experiments had established that bodies of all weights fall under the same constant acceleration.

To measure the rate of acceleration Galileo contrived an elegant experiment. No clock then could time a vertical fall, even from a tall tower. Galileo slowed the fall by "diluting," as he said, the force of "heaviness" that brought the body to the ground. He let bronze balls "fall"—roll—down an inclined plane: a wooden board with a polished groove down its center, marked off at equally spaced *punti* (points).

But how did Galileo time the roll of the ball? From study of the minutes he kept on his experiments, the historian of science Stillman Drake found out not long ago: he did it by singing!

Galileo sang, over and over, the same ditty as the ball rolled down the inclined plane. Along the groove he set frets at or in between the *punti* where the rolling ball clicked in time with the ditty. Measuring the distances between frets, Galileo found that they increased down the length of the groove as the square of the time. (If, by chance, the ball clicked at the first *punto* on the first beat, then it must have clicked at the fourth on the second and at the ninth on the third.) Inclining the board at steeper angles, he found that the intervals between frets lengthened but still increased down the groove as the square of the time [see illustration, page 18]. Plainly the ball was moving under constant acceleration at each angle of inclination.

To verify the counting by his ditty, Galileo weighed the water that ran from a pail through a spout soldered to its bottom, "in a fine thread … which was collected in a small glass all the time the ball rolled down the groove … [the] weight differences and proportions giving us the differences and proportions of the lengths of time."

Drake repeated Galileo's experiments in the 1970s at the University of Toronto. He found that Galileo's measurements come remarkably close to the 980.7 centimeters (or about 32 feet) per-second per-second acceleration of gravity at Padua, where Galileo did his work.

Galileo's parabola

Galileo next addressed a more general question: that of motion itself. This question posed itself in the trajectory of the ball after it came to the bottom of the incline and rolled off the table to the floor. Measurement and calculation confirmed Galileo's hunch that the trajectory of the ball, at every distance, lay on a parabola. For this finding Galileo won a certain fame. It improved the sighting of artillery pieces. Here, utility hardly reflects significance. Galileo's parabola tracks the trajectory of starlight in Einstein's General Relativity.

Contrary to intuition, a ball in flight does not "lose momentum"—except that lost to air resistance. Galileo saw that a ball in flight traces, in the parabola, two motions at once. One is its accelerating fall in the vertical plane. The other is the forward horizontal velocity of the roll down the inclined plane. Galileo found that the distance from the table at which the ball hit the floor increased in proportion with the increase in velocity imparted by the steeper tilt of the incline. Having shown that the ball falls vertically each time at the same accelerating velocity, he saw that the ball must retain its initial forward velocity as a constant component of its compound motion throughout its flight (i.e., the ball did not "lose its momentum"). The two velocities fitted the motion to the parabolic trajectory that would bring the ball to the floor at the point it reached at each initial forward velocity [see illustration, page 18]. Galileo perceived further that the parabola traced the path of least time and distance at each velocity between the bottom of the inclined plane and the point at which each flight ended.

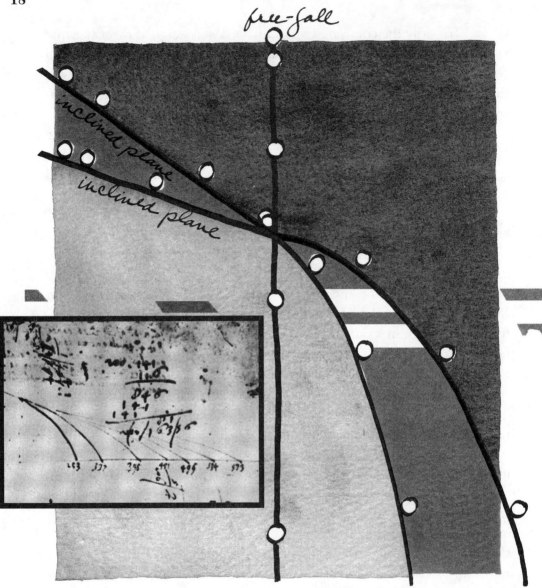

free-fall

inclined plane

inclined plane

Galileo's inclined plane [inset] "diluted" the force of gravity and permitted him to fix the acceleration of a body in free fall at roughly 9.8 meters per second per second [see page 17]. He saw that a ball retained the forward velocity of its roll down the plane. The constant forward velocity and gravitational acceleration of its fall carried the ball on parabolic, shortest-time-and-distance trajectories at each velocity. Galileo proposed, therefore, that constant motion "would be perpetual," as Newton's first law later affirmed.

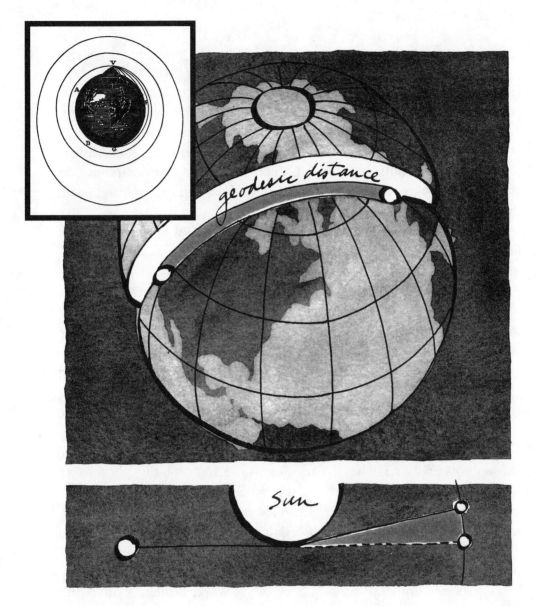

Newton's first law of motion *says a body continues forever in constant motion subject to deflection by another force, such as gravity [see page 47]. Newton saw that increase in the velocity of a body in Earth's gravitational field would stretch Galileo's parabola to orbits around the Earth [inset]. A plane struck through Earth's center describes "geodesic" of shortest time and distance on surface. At the "unpassable" velocity, Einstein showed [see page 105] starlight traces geodesic in the warped gravitational field of the Sun.*

In his *Two New Sciences* Galileo later drew this general conclusion: "I mentally conceive of some moveable [body] projected on a horizontal plane, all impediments being put aside … equable motion on this plane would be perpetual if the plane were of infinite extent." This generalization reads today in textbooks as Newton's first law of motion: "A body remains at rest, or moves in a straight line at constant speed unless acted upon by an outside force." Galileo was the first to isolate motion as the natural state of matter. It is change of motion, not motion itself, that is impelled by the force applied.

Galileo's experiments show that the first law of motion, for all its first-impression improbability, is true. One can repeat those experiments—at the very least repeat them in one's head—and see the ball travel on its parabolic trajectory.

Newton stretched the parabola, on higher velocities, all the way to the velocity that would set an object on orbit around the Earth [see illustration, page 19]. At around 11 kilometers per second, the velocity of contemporary Earth satellites, the constant component of the motion of Galileo's brass ball offsets the acceleration of gravity. At correspondingly higher velocities, Newton set the Moon on its orbit around the Earth and the Earth-Moon system on orbit around the Sun.

At the velocity of light, in the General Relativity of Albert Einstein, Galileo's parabola stretches to the least-time track of the photons of starlight. As will be seen in the next chapter, the curvature in the Sun's gravitational field traced by the star track explains the apparent displacement of a star in a solar eclipse observation made by Arthur Eddington in 1919. That observation supplied confirmation not only of Einstein's theory of General Relativity but also of a conjecture of Newton's.

For the immense consequence of his work, Galileo is often and rightly cited as the Founder of modern science. He surely showed how science should be done. In the trajectory of a moving body, Galileo isolated for consideration the vertical and the horizontal, the accelerated and the constant. He proceeded then to isolate those components of motion physically in the artless simplicity of his experimental apparatus.

Seldom today does the design of an experiment, with black-box instrumentation, reach its object with the simple elegance of Galileo's inclined plane. Working in realms far outside the reach of the unaided senses, physicists today pose questions about hypothetical particles and forces drawn from labyrinths of mathematical abstraction. They confront them with experience in accelerator experiments that yield results in computer readouts at second and third remove from the event. The Galilean example remains the ideal, if nostalgic, model.

No higher authority

To Galileo is owed also the governing demonstration of the ethic that enjoins a scientist to recognize no higher authority for the truth than his own experience: "In questions of science, the authority of a thousand is not worth the humble reasoning of a single individual."

Galileo did accede to the secular authority of the Inquisition by recanting the thesis of his *Two World Systems.* He retired from that contest to write his *Two New Sciences.* Reaching the places of learning in Europe, this book started the avalanche of events that established the autonomy of the individual in the realm of objective knowledge.

The intensely individual experience of work in science was well described by Percy Bridgman, who pioneered the study of matter under high pressure at Harvard University in the 1920s and 1930s, when no one else was interested. His lonely enterprise won a Nobel prize in 1946. Today his work helps to explain the churning in the Earth's interior that moves the continents; machines modeled on his experimental apparatus manufacture industrial diamonds. Bridgman wrote:

> The process I want to call scientific is a process that involves the continual apprehension of meaning, the constant appraisal of significance, accompanied by the running act of checking to be sure I am doing what I want to do, and of judging correctness or incorrectness. This checking and judging and accepting, that together constitute understanding, are done by me and can be done for me by no one else.... They are as private as my toothache, and without them science is dead.

Upon publication, the work enters the public, social process of science. Members of the community who are interested will address it in their individual responsibility. They are a democracy of warring sovereigns. If science is not dead, they will root out frailty in the design of the experiment and error in the data. They will challenge the premises on which the work was undertaken and the meaning the author has found in it and, perhaps, argue for their own. Debate will be unsparing in the common cause of consensus.

The small number of inquirers actively at work on each question form an invisible college. Engaged in competition with one another, their shared uncertainty commits them to collaboration. They are in continual communication on many closed as well as open channels. The word that one scientist has visited another will be enough to tell a third what they talked about, perhaps a notion that none would argue at an open meeting. The airplane ticket was said, in a time of more generous public support of science, to be the universal scientific instrument. The Internet today facilitates continual communication within the invisible colleges. It is the rare scientist who goes to press without the attestation and encouragement of some prior informal peer review. Much publication, especially in Big Science, goes to press under multiple authorship.

A work in science has not happened until it is published for the interested community; the breadth of that community is itself a measure of the prospective significance of the work. The Royal Society of London began in 1662 to publish the correspondence that collaborating adversaries were then exchanging with one another. Today, thousands of scientific journals give the world's librarians their worst financial and administrative problems.

The ethic of objective knowledge

The organized advance of objective knowledge, in the four centuries of its history, has evoked the world's first truly international community. It is recruiting members in countries that look to histories and cultures more ancient than the West. Wherever they hail from, members come to share closer interests and to communicate more easily with one another than they do with most of their fellow citizens at home. They share the languages of reason and mathematics. From habit and practice, this world community self-governs under its own common law.

On his election to the first chair in molecular biology in the

Collège de France in 1967, the molecular biologist Jacques Monod undertook his personal statement of the communal ethic:

> The sole end, the sovereign good, the supreme value in the ethic of knowledge—let us acknowledge it—is not the happiness of man, much less his comfort and security—it is objective knowledge itself. I believe it is necessary to state clearly and to systematize this ethic and, regardless of the moral, social and political consequences, to teach and spread it abroad; for, creator of the modern world, this is the only ethic consistent with life in this world.
>
> This, it must not be concealed, is a harsh and constraining ethic; while it looks to man to advance knowledge, it declares a value superior to man himself.
>
> It is an ethic of conquest, a will to power, but to power solely in the noösphere [the sphere of knowledge]. It is, in consequence, an ethic that teaches the evil of violence and of temporal domination.
>
> It is an ethic of personal and political liberty, because to contest, to criticize, to constantly put in question is not only a right therein but a duty.
>
> It is a social ethic, because objective knowledge cannot be cherished except in a society that respects its norms.

With the austerity of his Huguenot heritage, Monod described a society in which almost everyone would want to live. For citizenship in it scientists hold no automatic qualification. On trial in the social process of science, the greatest have betrayed their merely human selves.

Robert K. Merton, at Columbia University, the founder of the sociology of science, declared "communism" to be the central ethic governing the behavior of scientists. Pursuant to it, they consign "the substantive product of their collaboration to the ownership of the community. The scientist's claim to 'his' intellectual 'property' is limited to that of the recognition and esteem which, if the institution functions with a modicum of efficiency, is roughly commensurate with the significance of the increments brought to the common fund of knowledge."

This is a prize without price. In consequence, "concern with scientific priority becomes a 'normal' response." Contests over priority recur throughout the history of science. Galileo had to endure the prior publication by Bonaventura Cavalieri, a student of a student of his, of a proof of the parabolic trajectory. Drake cites Galileo's indignant protest that "the proof was the fruit of studies he had begun 40 years earlier and that the least he deserved was the courtesy of first publication."

Merton found the most distressing case in the prolonged contest between Isaac Newton and Robert Hooke. For a decade they quarreled over which of them first showed that the colors of things represent the partial reflection by a surface of the gamut of color perceived by the human eye. Newton protested: "There is nothing wch I desire to avoyde in matters of Philosophy more than contention, nor any kind of contention more than one in print." He delayed publication of his *Treatise on Opticks* until after Hooke, an older man, had died. At the threat of prior claim by Hooke on the inverse-square law of gravity, Newton hastened publication of his *Principia.*

While the iconography of Newton counts more than a dozen authenticated contemporary portraits of himself, not a single such image of Hooke survives. Suspicion points to Newton in his authority as president of the Royal Society.

Charles Darwin, in his turn, protested: "I rather hate the idea of writing for priority; yet I certainly should be vexed if anyone should publish my doctrines before me." He was spared that vexation down to the last minute of the two decades he gave—partly from intimidation by the prospect of the controversies that were sure to follow—to the writing of his great work. Then, in 1858, he received in the mail an elegant précis of the principle of natural selection written by a young naturalist named Alfred Russel Wallace. Writing from the faraway Molucca islands, Wallace was seeking the great naturalist's critique of his idea.

Friends arranged for simultaneous publication of Wallace's précis and a brief statement by Darwin. The fuller statement, in *The Origin of Species,* was hastened from the press in 1859. Wallace was a Darwin champion in the controversies that did ensue.

Whatever these episodes tell about people, they testify to the objectivity of the knowledge sought by scientific inquiry. Simultaneity of discovery is more the rule than the exception in the history of science.

Work that would otherwise be ratified by repetition acquires instant ratification, so to speak, by simultaneity of discovery.

In favor of people, the record shows that quarrels about priority are rarer than the occasions for them. Scientists in each field are likely to be working on the same important question wherever they are. Simultaneity of discovery—as reflected, for example, in the number of Nobel prizes that must be shared—testifies to the collaboration that engages scientists, if only implicitly, in every field.

The collaboration is evident in citations appended to every published paper. They document references to and acknowledge the priority of the work of others. Searches of the literature often more fruitfully proceed upstream through the cascade of citations than down the conventional author-title route. Such

	1	2	3	4	5
1 AGRICULTURE AND FISHERIES	10.86	15.70	2.16	0.02	0.19
2 FOOD AND KINDRED PRODUCTS	2.38	5.75	0.06	0.01	*
3 TEXTILE MILL PRODUCTS	0.06	*	1.30	3.88	*
4 APPAREL	0.04	0.20	—	1.96	—
5 LUMBER AND WOOD PRODUCTS	0.15	0.10	0.02	*	1.09
6 FURNITURE AND FIXTURES	—	—	0.01	—	—
7 PAPER AND ALLIED PRODUCTS	*	0.52	0.08	0.02	*
8 PRINTING AND PUBLISHING	—	0.04	*	—	—
9 CHEMICALS	0.83	1.48	0.80	0.14	0.03
10 PRODUCTS OF PETROLEUM AND COAL	0.46	0.06	0.03	*	0.07
11 RUBBER PRODUCTS	0.12	0.01	0.01	0.02	0.01
12 LEATHER AND LEATHER PRODUCTS	—	—	*	0.05	*

Column headers (diagonal): 1 AGRICULTURE AND FISHERIES, 2 FOOD AND KINDRED PRODUCTS, 3 TEXTILE MILL PRODUCTS, 4 APPAREL, 5 LUMBER AND WOOD PRODUCTS, 6 FURNITURE AND FIXTURES, 7 PAPER AND ALLIED PRODUCTS, 8 PRINTING AND PUBLISHING, 9 CHEMICALS

searches are facilitated by the indexes to citations offered by publishers to which the nonprofit library system long ago surrendered the demanding librarianship of science. These indexes hold riches for historians and sociologists of science. The economist Wassily W. Leontief—author of the input-output matrix [above] that displays the inter-industry flow of goods and services—found in his matrix a way to map the interdisciplinary flow and exchange of understanding among fields of science.

Scientists watch these indexes to keep track of their competitive ranking. Steven Weinberg declared pride, greater than he ever acknowledged in his Nobel prize, in the finding by one index that his 1967 paper on the "electroweak theory" "was the most frequently cited article on elementary particle physics of the previous half century."

The genealogy of ideas, as might be expected, has live family trees in successive generations of teachers and students or, better said, masters and apprentices. After 100 years, it is the rare Nobel prize, in physics especially, that goes to a first-generation worker; those prizes go preponderantly to students of prize-winners.

Bridgman's line of work and his not-to-be-distracted engagement in it did not draw many graduate students, but he had in John Bardeen—one of the authors of the transistor, now the computer chip— a laureled successor in the physics of condensed matter. Two-generation and three-generation successions are common. One four-generation succession links Wolfgang Pauli, a founder of quantum mechanics, to I. I. Rabi, who brought his learning home in the 1930s to establish Columbia University as one of the first centers of the new physics in the United States, to Julian Schwinger, who helped to bring quantum mechanics to its mature statement in quantum electrodynamics, to Sheldon Glashow, who has essayed to bring all but the cosmic realm of gravitation within the embrace of quantum physics.

An unacknowledged caste system stratifies the democracy of science. Work in the sciences may be spread across a spectrum from those still engaged in identifying, classifying and describing their subject matter to those offering fully quantitated, mathematized descriptions of their realm of physical reality.

Explanation and verification

The question "why" asked about any solidly established observation in one field finds its answer in the field that underpins the observation. To insist upon knowing why anything is the way it is leads the inquirer ineluctably to physics. One must finally engage that branch of physics, called particle physics, that seeks to understand the ultimate nature of matter-energy. The last century's work established those fixed points on the energy scale that are the sites of principal events in the interaction of matter and energy. Physics has no answer yet for the question: why those points on the scale? For what it describes there is no explanation, only questions. Pursuit of those questions has brought particle physics into collaboration with the spacetime physics of cosmology. The answer to the question of why it is has come now to be sought in the question of how it came to be the way it is.

In the community of science the logical hierarchy engenders questions of social hierarchy. The identification of the genetic molecule and the decoding of the gene evoked the new discipline of molecular biology, but broke up departments of biology. Only now are young biologists of "the whole organism"—zoologists and botanists—acquiring the tools of that discipline. To the loss of the whole community questions of social hierarchy have created political division at a time of declining public investment in human and social capital. Condensed-matter physicists in 1992 testified to the Senate and House in opposition to the $8-billion appropriation for the particle physicists' dream of the Superconducting SuperCollider.

The logical hierarchy draws from some quarters the protest of "Reductionism!" In language reminiscent of John Keats's dismay that natural philosophy would "unweave a rainbow," scientists will avow that the wonders they encounter in their field cannot be reduced to electrons

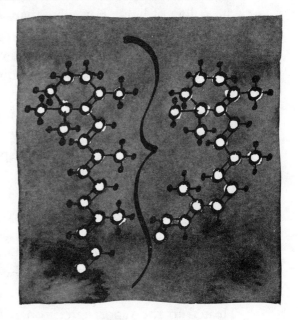

and photons. To allay such anxiety, T. C. Schneirla, curator of animal behavior at the American Museum of Natural History, advanced the idea of "levels of integration." Schneirla's subject was the Central American army ants. Strictly at the level of behavior, he developed a total explanation of "the most complex instance of organized mass behavior occurring regularly outside the homesite in any insect or, for that matter, in any infra-human animal." For neurophysiologist and molecular biologist, at the next levels of integration, Schneirla's work asks numerous questions and shows them just where to look. Their work is, of course, underpinned by physical chemistry and at the deeper level by quantum mechanics.

At Harvard, George Wald showed at that deeper level how light excites vision. A quantum of light delivers just the right energy to release energy stored in the visual pigment by metabolism. The three-dimensional retinene molecule [inset, preceding page] twists abruptly from one of two configurations to the other, releasing the energy to the nerve impulse. That stored energy came also from the Sun, first captured by photosynthesis in a leaf. Wald urged anxious whole-animal colleagues to see that "physics and chemistry have grown up to be biology."

Truth and light

Verification of every proposition about the nature of things arrives at last at physics. That is because physics secures the first ground for all science. This is the level of integration at which we know most of what we "know for certain."

For such certainty, physicists find much of the hard evidence in the spectrum of electromagnetic radiation. The prism that spreads out the colors of light shows less than an octave of the frequencies of that radiation [see illustration, pages 6–7]. Given the constancy of the velocity of light, frequency is inverse to wavelength: 300,000 kilometers (3×10^8 meters, 3×10^{10} centimeters) divided by 10^6 frequencies per second yields a longer wavelength than does division by 10^{12} frequencies per second. From the low-frequency red end of the visible spectrum, electromagnetic radiation is experienced first in the warmth of the infrared frequencies and then, with the help of electronic technology, as radar, TV, wireless telephony, FM and then AM and arcane uses beyond. From the blue high-frequency end, ultraviolet radiation is first encountered, then X rays, then gamma rays, and on out to the frequencies of electrons and other atomic particles.

Every event in the matter-energy interchange shows up in that spectrum somewhere. An iron stove radiates a continuous spectrum at the red end of the visible that grades out of vision into its "warmth." Heated to the vapor state, iron radiates a discontinuous spectrum, in narrow bands of light, at its signature wavelengths and frequencies. The corresponding "lines" in the discontinuous spectra radiated by the incandescent vapor of each of the other 91 naturally occurring elements declare their presence, whether in distant stars or in a compound in the laboratory [see illustration, page 30]. Close to the wick of a candle can be seen the blue emission of hydrogen glowing inside the yellow flame of incandescent soot particles. A pinch of salt dropped in the flame of a Bunsen burner will give out flashes of yellow at the characteristic emission line of sodium, familiar now in much street lighting.

With the spectroscope, the "queen of scientific instruments," physicists have catalogued millions of lines in the spectrum of visible light and in the octaves of radiation in the ultraviolet and infrared. Spectroscopic evidence shows the universe out to its farthest reaches to be composed of the same chemical elements, if not in the same proportions, as on Earth. The increasing shift toward the red end of the spectrum of the lines of radiation from sources outside the Galaxy shows the universe to be expanding and gives measure of its age.

The spectral lines of chemical elements here on Earth have disclosed the boundaries of the metaphorical shells in which negatively charged electrons spin around the positively charged nucleus of their atoms and the energy state of the nucleus inside. The infrared region of the spectrum opens up molecules to inspection; the lines give a measure of the strength and angle of the chemical bonds between component atoms and show the stretching, twisting and warping of those bonds under the stress of rising temperature.

By the end of the 19th century, spectroscopy had shown that radiation at shorter wavelength and higher frequency carries higher energy. That discovery brought on the crisis that launched the new physics of Special Relativity and the quantum that fuses matter and energy, space and time. The catalogue of the spectral lines of the element hydrogen in its different energy states supplied the first confirming evidence for the quantized electron-shell structure of the atom. An observed shift of the lines of solar radiation toward the red end of the spectrum accorded with a crucial finding in the General Relativity of Albert Einstein that frames the new observational cosmology.

The precision of spectroscopy depends ultimately upon the precision of some human being's hand and eye. In a research spectroscope, a diffraction grating, rather than the familiar prism, disperses the radiant energy of light. This is a flat or spherically curved slab of mirrored glass

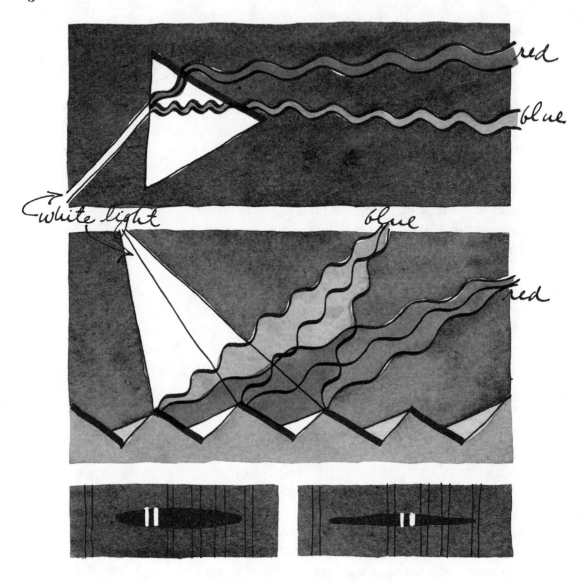

Prism separates colors in visible "white" light. *Blue light is more strongly refracted than red; light at shorter wavelengths loses more velocity in passage through prism. Newton attributed this difference to "weight" differences among particles of light and so to gravitational attraction. From a diffraction grating [see page 32], in cross section, much enlarged [middle] blue light reflects at sharper angle than red. Two bright lines in calcium spectrum [bottom] are re-redshifted in faint spectra from cosmic sources in proportion to their distance.*

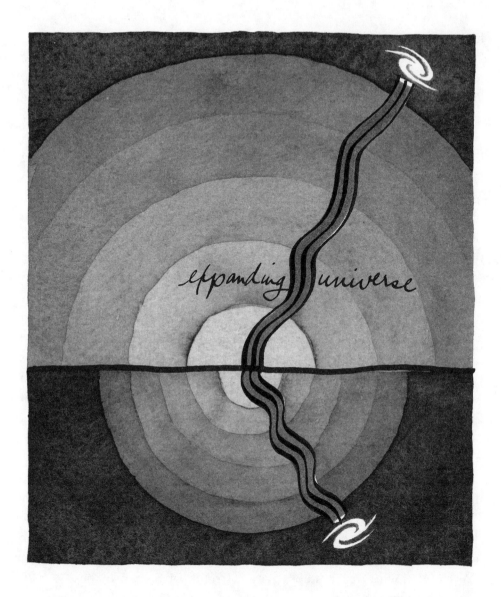

Redshift measures expansion of universe. *Expansion of spacetime stretches the wavelengths of light traveling across it, shifting lines in spectra from distant galaxies to longer red wavelengths; the larger this redshift, the more distant the galaxy [see pages 181 and 189]. The rate of expansion appears to increase with distance because short and long distances in the earlier, smaller universe have increased by same multiple. The pair of bright lines in the calcium spectrum [opposite] registers in the faintest spectra from the most distant galaxies.*

from which the light is diffracted by parallel lines—hence the term *grating*—scratched on the reflecting surface [see illustration, page 30]. The resolution of a grating—the narrowness and faintness of the spectral lines it captures—is a function of the number of lines per inch and the exactitude with which they are spaced in the scratching.

Until the advent of electronic feedback-control technology following the Second World War, the perfection of spectroscopy was owing to the pleasure that Henry Augustus Rowland found in contending with physical reality, after the manner of Galileo, with bare hands. His tangible legacy is 1,000 of the most perfect diffraction gratings and a million or more precisely established spectral lines in the catalogue of spectroscopy recorded from them.

The founding professor of physics at Johns Hopkins University, at age 27 in 1875, Rowland acted on his determination to get the foundation work of spectroscopy done quickly and precisely. He settled on a standard of 14,400 2-inch lines of identical width and depth, to be plowed in the mirrored surface exactly parallel to one another across each inch and at exactly the same spacing from one another across a width of 5 or 6 inches. To eliminate the focusing lens and the loss of light it would entail, he hit on the self-focusing spherical surface.

14,400 lines to the inch

Henry Augustus Rowland once boasted that he never saw a machine that he did not immediately understand [right, portrait by Eakins]. He recognized that to make the perfect "ruling engine" to rule the perfect grating required perfection of the critical engine part: the lead screw that would advance a diamond stylus exactly 1.44×10^{-4} (or $1/14.400$) inch at each step across a stretch of 5 or 6 inches. At 20 threads to the inch, each full rotation of the screw would advance the diamond stylus of his ruling engine one-twentieth of an inch; each of the 720 settings in each rotation advanced it 1.44×10^{-4} inch.

To grind the perfect thread in the 9-inch lead screw, Rowland invented a new machine tool. This was an 11-inch-long grinding or lapping nut with the same threading, which he split lengthwise in four sections. With the segments reassembled around the screw, he could adjust the tightness of the nut through the course of grinding with emery powder and oil and finally the rouge used to polish telescope mirrors.

"Now grind the screw in the nut," Rowland wrote, "making the nut pass backwards and forwards over the whole screw ... turn the nut end for end every ten minutes and continue for two weeks." In that time the high spots on the threading of the nut and the screw found one another and wore themselves away.

With his own perfect grating in hand, Rowland turned to mapping the solar spectrum. He was "able to do as much in an hour as has been done hitherto in three years." Rowland's "preliminary table" of solar-spectrum wavelengths, published in 1898, served until 1928 as the world standard. Its revision then at Mount Wilson Observatory found the largest necessary correction at 3×10^{-4} (or one part in 30,000). Rowland also undertook to catalogue all of the known elements, establishing the spectral signature of each. Comparing these with the solar spectrum, he was challenged by the number of "important lines [in the solar spectrum] not accounted for." Some corresponded to lines of unknown, not yet isolated elements, including helium. At his untimely death in 1901, Rowland left his successors a substantial agenda and the tool with which to

accomplish it. His gratings continue in use in laboratories and observatories around the world, and the Johns Hopkins physics department continues to supply the world community with gratings made on the descendants of his ruling engine.

For still higher precision, astrophysicists now turn to interferometry. The interferometer, perfected by Rowland's contemporary A. A.

Michelson, established the constancy of the speed of light [see illustration, page 91]. Its perfection is now given reach into depths of space by electron multipliers that respond to light by the photoelectric effect [see illustration, pages 92–93]. They "see" cosmic sources far below visibility, equivalent to the arrival of one quantum of light at a time.

As scientists discovered toward the end of the 19th century, the spectrum of electromagnetic radiation stretches across many octaves of higher and lower frequency on either side of the 30 or so octaves dispersed by the diffraction grating. Technologies now exploit these regions of the spectrum in a host of ways, not least in the service of spectroscopy. On radio frequencies, organic molecules disclose their presence in interstellar space. In the same low-energy range, by the spectroscopy of nuclear magnetic resonance (the phenomenon employed in medical imaging), chemical reactions may be observed under way in the living cell. On the other, high-frequency, side of the diffraction-grating spectrum, X-ray and gamma-ray spectroscopy tune in on violent events in the lives of stars and galaxies (signified, for example, by a line in the X-ray spectrum emitted by neon that turns up in the debris of supernovae).

High-energy spectroscopy

The particle accelerators of the high-energy physicists may be understood best as spectroscopes. That is the metaphor of Victor F. Weiskopf of the Massachusetts Institute of Technology, who served a term as director of CERN (Centre Européenne de Recherche Nucléaire), the international accelerator laboratory near Geneva. On the energy scale of particle physics—stated in "electron volts" or eV—classical spectroscopy studies the more or less intact atom at energies at the very bottom of that scale. In its lowest-energy or ground state, the single electron in the hydrogen atom is bound to the single proton in the nucleus with an energy of 13.6 eV. Interatomic reactions, the combustion of carbon and oxygen, for example, typically yield energy in the range of one or a few electron volts per molecule of reaction product, of carbon dioxide, for example.

Accelerators with power in the lower million-electron-volt (MeV) range take spectroscopy into the interior of stars. This is the spectroscopy of the excited atomic nucleus stripped of its electrons. A sufficiently excited nucleus emits not only quanta of high-energy radiation but particles—electrons, neutrinos and even neutrons—as well.

At energies of 800 MeV and beyond, accelerators bring the particles that compose the nucleus of the atom, the protons and neutrons, into their excited states. These states are manifested in the multiplicity of unstable, increasingly massive, speedily decaying particles that took the community by surprise in the 1950s.

The confusion found resolution during the 1970s in the recognition that the proton and neutron, in their familiar and excited states, are composed of the still more elementary "quarks." By then the power of accelerators had reached the GeV—G for "giga" in Greek or billion in American parlance—range. At 2,000 GeV, the present limit of accelerator power, the sixth quark made its momentary appearance in 1994.

For the present, the high-energy spectroscopies of the particle accelerators and the lore of their findings hold the interest of physicists principally, and those who share their interest in the why and how and whence of physical reality. As such understanding comes into wider circulation it will illuminate other realms of inquiry. For example, physicists' understanding of the electron compelled the chemists to accept from Linus Pauling, a third-generation quantum physicist, the complete revision of their chemical bond. Chemists put away their cookbooks thereafter and began to design molecules to specification by the function they are to serve. The diffraction of X rays by inorganic crystals gave physicists clear pictures of the ordered ranks of atoms in those structures. Perfected in that use, X-ray diffraction was turned to the resolution of the architecture, in three dimensions, of the large molecules of the living cell. Resolution of the famous double helix of DNA launched the new discipline of molecular biology. By feedback, other fields present new questions to physics. The reliability of life processes confronts quantum physics with many such questions.

Little is made of this cross-disciplinary exchange in the much that is made of the fractionation of science into ever more highly specialized disciplines. The tools, the skills, the conceptual apparatus, the established knowledge, the new questions that arise all go, indeed, to compel specialization. The drawbacks of specialization bring on the preaching of "holism." The preachers tend to overlook the necessary virtues of specialization and forget that, as disciplines proceed with their work, they converge again. Cytology, the study of the anatomy and physiology of the living cell, and biochemistry are, for example, converging in molecular cell biology—fulfilling the ambition voiced 30 years ago by Monod "to bring into biology the complete conceptual unity that still fails it." The convergence of particle physics with cosmology has taken inquiry into the nature of matter back to the genesis of the universe.

Constructing a cosmology

No one but specialists on the frontier need be intimidated by the much-touted knowledge explosion, and then only that in their own field. The comprehending concepts in each field reduce that volume to containment in the human head. Understanding that relates knowledge from each discipline of science to the emerging picture of the whole is within the reach, in proportion to motivating interest, of everyone.

To find one's bearing in the new world requires some grasp on the whole. From the diverse compartments of objective knowledge—variously subject to inherent uncertainties, it must be kept in mind—each is free to assemble one's own comprehensive cosmos. The cosmologist is constrained only to consider external reality accessible to experience and the consensus of the scientists most concerned with each element in the whole. Inevitably, there will remain as much as and more outside, around the 360-degree horizon of each personal cosmos, than within. The expanding sphere of knowledge, it has been said, puts us everywhere in touch with the unknown.

In this book I have set out my cosmology, with due respect not only for the unknown but for the much that is known and is beyond my grasp. I was lucky to stumble into the occupation that invited, indeed compelled, a lifelong education in the sciences.

It was then a tide taken at the flood that launched SCIENTIFIC AMERICAN. With the end of the Second World War came public realization of the role that science had played in the conduct of the war. The atomic bomb gave awesome reality to the mathematical abstraction of the new physics.   Penicillin and vacuum–freeze-dried blood plasma, which secured life for 95 percent of the wounded who reached battalion aid stations, promised miracles from the life sciences.   Science had become an urgent public enterprise.

Franklin D. Roosevelt in 1944 invited Vannevar Bush, chairman of the Office of Scientific Research and Development and so chief of the wartime mobilization of university scientists, to design an agency through which the federal government might sustain the work of science in the nation's universities after the impending victory.   The Bush Report, *Science, the Endless Frontier,* called for the creation of a foundation under governance by an extragovernmental board of trustees appointed by the president.   It would disburse federal funds in project grants.   Legislation of this proposal reached the desk of Harry Truman in 1947.   The president administered a harsh veto: "The proposed [foundation] would be divorced from control by the people to an extent that implies a distinct lack of faith in democratic processes."

Legislation in 1948 at last established the National Science Foundation under governance of a director appointed by and responsible to the president and counseled by an advisory board.   By 1953, appropriations in support of the Foundation's mission reached $4 million.   In that year, the Department of Defense and the paramilitary Atomic Energy Commission laid out more than $100 million for the support of science in the universities, principally the physical sciences.   From the health department of the federal government, the life sciences in the universities, principally in their medical schools, received another $50 million.

Purchase of the product

Since 1945, the outlay on the promise of product from science— weapons and pharmaceuticals, in particular—has mounted to a cumulative total approaching $500 billion in 1990 dollars.   In the first flush of postwar enthusiasm and rising appropriations, the granting agencies of the federal military and para-military and health departments construed their missions broadly to cover the most remotely relevant enterprises in fundamental research.   Since 1970, funding from those sources has been going to ever more narrowly construed "mission-oriented" projects.   It has gone to support projects, not scientists, and for the short term, not for the long term of sustained scientific inquiry.

Until the 1980s, the National Science Foundation disbursed less than 7 percent of the annual federal expenditure for science. Its budget climbed then into the billion-dollar range after it was charged to install "institutes" for the promotion of U.S. "industrial competitiveness" in the universities. The agency otherwise has managed to grant half the applications approved by peer review a quarter of the funding requested.

Public expenditure on product from science meanwhile got its money's full worth. U.S. industrial supremacy is owing, in no small respect, to the high technology purchased incidental to the country's expenditure of $18.7 trillion (1996 dollars) on its military establishment over the long course of the Cold War arms race.

In the settling of federal "science policy," the universities and the scientific community have their complicity. They did not seize the opportunity for public education in science represented by the proper argument for the outlay of $500 billion of public funding. They went along with the case for utility and its ready appeal to Congress.

Freely motivated inquiry

Federal funding has been directed, in consequence, to less than the full spectrum of freely motivated scientific enterprise. In this respect, it is notable that the plant sciences—not of interest to the defense or health department—withered in all but the few universities where botany was well and restrictedly endowed.

On the sheer volume of the funding, on the other hand, science in the U.S. has flourished. So the Nobel prize scoreboard testifies year after year. Yet, after half a century, the federal commitment to the support of science amounts to no more than a small fraction of the government's annual expenditures for the purchase of science. Presently this funding, along with the rest of the country's investment in human and material resources, is in decline.

That decline has been in part offset, to the greater peril of freely motivated open scientific inquiry, by market-motivated industrial financing. The intrusion of the market compromises work in the life sciences especially. Whole university medical-school departments now operate as subsidiaries of pharmaceutical companies. They are generously funded under agreements that induce or compel restraint on publication and the open communication that is the life of competitive collaboration in science. Intellectual property formerly deeded to the community is now private property. The choice of question to be investigated turns on the movement of NASDAQ exchange. The ablest scientists—in particular young scientists framing their first venture—find it increasingly difficult to do what they really want to do.

SCIENTIFIC AMERICAN had much to report over those decades on the new technologies that supply public verification of the advance of science. From the first demonstration of the "transistor" at Bell Telephone Laboratories, reported in our pages in 1948, the magazine tracked the solid-state revolution in electronics and its ramifications into every economic activity. The latest editions of the computer chip put millions of transistors and equivalent circuits to work. Now, in accordance with predictions made in 1960 from the data placed in reach of computer analysis by Leontief interindustry tables, ever more powerful computers are downsizing the clerical and middle-management payroll.

The computer is only part of the story. With elegant electronic sensors on the input side and electromechanical actuators on the output side of the computer, automatic production increasingly replaces people in the "process" industries. In 1990, the petroleum refining industry employed half as many "production" or blue-collar workers as it did in 1950, while multiplying its output by three. The steel industry reduced its production payroll by half while producing the same ingot tonnage. The percentage of the U.S. labor force employed in production functions declined from more than one-third in 1950 to less than 20 percent in 1990. White-collar professionals replaced them.

The flat production of steel between 1950 and 1990—in an economy that multiplied its total output four times—signified another impact of the new physics. Understanding of the structure of matter from the inside displaced and replaced materials in their traditional end uses. By 1960, organic plastics overtook steel in bulk; a decade later, in tonnage. Ceramics reinforced by microscopic carbon fibers—on the ancient model of bricks made with straw—stood up to fiercer heat in turbine blades. The optical-glass fiber—another triumph of the late Bell Telephone Laboratories—began to take over from copper in the communication systems, piping photons in place of electrons.

The habit of consensus

In 1952, a single-topic issue of SCIENTIFIC AMERICAN reported on the economic and social consequences of this impending revolution. The reconstruction of the labor force and the declining compensation paid to labor, in white as well as blue collars, now excite political concern for the well-being of the "middle class," which now means higher-paid wage-earners. This country and the rest of the industrialized world face questions not yet articulated about purpose, value and equity in securing the blessings of the workless economy.

On the questions of the arms race and its control, so much at the center of public concern over all those years, the country is indebted to the consensus-forming habit of the scientific community. This is belied by the mass-media coverage of these issues. The media give equal time to the consensus and to the dissenting maverick and kook, often a spokesperson for an economic interest in the issue. By way of balance, we reserved the pages of SCIENTIFIC AMERICAN for the consensus.

The lay public, including its representatives in Congress, had independent counsel, therefore, on the arcane technical issues of the arms race and arms control from authorities as fully informed as the official security-cleared "defense intellectuals." Our authors made public record of the ominous transformation of our country's "nuclear deterrent" from retaliatory to first-strike weaponry during the 1960s. They exposed in hard numbers the lunatic stockpiling of tons of nuclear explosives that perpetuate, in the now-prevailing international anarchy, the peril to civilization laid by mutual assured destruction. The publication of these articles in our Russian-language edition, V MIRE NAUKI, shows the consensus-forming habit of the community to be international.

Beginning with an article on the Amazon frontier in its second issue, the new SCIENTIFIC AMERICAN kept the interlocking determinants of the world future—population, environment and development—

under surveillance. I still hope to see recognition in U.S. public policy of the concept, first published in the magazine in 1955, of the "demographic transition," of the transit of the population:

a) from near-zero growth at high death rates and high birth rates and life expectancy of less than 30 years
b) through the population explosion
c) to near-zero growth again at low death rates and low birth rates and life expectancy exceeding 70 years

The 1.25-billion population of the industrialized countries has already arrived at zero growth. If all goes well, every indicator says that the rest of the world population may complete the transition to bring population growth to a halt by the end of the 21st century. The population explosion sustains in public understanding, however, the Malthusian vision of population growth to self-extermination in the war of all against all. That vision continues to determine the foreign policies of nations, including our own.

International conventions now recognize what SCIENTIFIC AMERICAN authors began telling their readers 30 years ago: combustion of fossil fuels exacts the principal cost and peril laid to the environment by industrial civilization. The fourfold multiplication of energy consumption since 1950 has increased the carbon dioxide input from human activity to more than 25 percent of the planetary atmospheric turnover. The next fourfold increase in energy supply necessary to carry the rest of the world population through the demographic transition cannot conceivably be secured from fossil fuels. Alternative primary energy sources—including photovoltaic conversion of solar energy and extraction of solar energy stored in the ocean and not excluding nuclear power—were all appraised for our readers in time to have allayed the present international anxiety. In the words of the British molecular biologist P. B. Medawar, "Problems caused by technology must, by definition, be cured by technology."

Warren Weaver, in the September 1953 issue of SCIENTIFIC AMERICAN, declared the hope that "the citizens of a free democracy, understanding and prizing the work of science, will provide the support and terms of support that will cause science to prosper and bring its benefits, power and beauty to the service of all the people." As bursar for the sciences at the Rockefeller Foundation from 1932, Weaver had administered the principal fund—a few million dollars a year—that supplemented university science department budgets before the Second World War. That September issue was devoted to "Fundamental Questions in Science." Weaver's concern, shared by the editors, was with the terms on which the large and growing expenditures by the federal government were then flowing to support the work of science in the country's universities. That work, reported by the scientists engaged in it, principally filled the pages of SCIENTIFIC AMERICAN and brought readers to those pages. To those authors I owe the understanding I hope to share with the readers of this book.

What scientists learned

The painting on the cover of the May 1948 issue of SCIENTIFIC AMERICAN displayed, with magic realism, the apparatus employed in an experiment by C. J. Davisson and L. H. Germer at Bell Telephone Laboratories in 1925 [opposite]. They showed that the electron has wavelength, that a particle also has the nature of a wave. Their experiment completed a symmetry: the photoelectric effect, explained by Albert Einstein in his paper on Planck's quantum in 1905, shows that waves of radiant energy have the nature of particles.

SCIENTIFIC AMERICAN, with that May issue, began its coverage of the first of the consequential developments in understanding reported over the next four decades. That is the ramification of quantum electrodynamics into understanding of the physical world and its history from the beginning. It is said that quantum electrodynamics defies understanding even by some who give their full time to it. So intimidated were the SCIENTIFIC AMERICAN editors, that in our second issue, we allowed an author to say so. The difficulty is not in the theory. Percy Bridgman had the right diagnosis: "Hundreds of years of attempting to find inside our own heads the necessary pattern of the external world has proved a dismal failure.... We can only take experience as it comes and must try to get our thinking into conformity with it."

By the mid-20th century, quantum electrodynamics had comprehended the physics of the tiny region of the cosmos—the nearby "natural world"—where atoms are intact, or nearly so, and form chemical compounds including those enchanted with the properties of life. The second half of the century saw quantum electrodynamics comprehend all of
physical reality except that in the embrace of gravity and the theory of
General Relativity.   Together,
quantum theory and the general
theory of relativity bring into nearly complete unity of understanding
all of physical reality now in reach
of observation and experiment.

THE ELECTRON: PARTICLE AND WAVE (PAGE 50)   FIFTY CENTS

May 1948

In the next three chapters,
therefore, it will be seen that
inquiry guided by these powerful
constructs supplies the current
answer to the first of the two
questions that began this chapter.

The chapters that follow
address the second, warmer, question of how we came to be and
who we are. The resolution of the
double helix of DNA, deoxyribonucleic acid, recounted by
Francis H. C. Crick in the
October 1954 issue of SCIENTIFIC
AMERICAN, undoubtedly won
more public attention than any other single finding in the second half of
the century. This was the materialization, in an elaborately structured
molecule, of the abstract unit of heredity postulated by Gregor Mendel
a century before. The way was open then to understanding of the
molecular anatomy and physiology of the living cell and, thereafter,
understanding of the origin of life, the branching of the evolutionary
family tree, the genesis of form in embryonic development and the generation of consciousness in the brain.

Central to understanding the place of life on Earth and in the cosmos is the convergence of the knowledge obtained from mapping the ocean bottom and exploring the solar system. Continental drift or plate tectonics—the understanding that this planet remains geologically alive, with convective turnover of its plastic interior constantly remaking its crustal map—helps explain the success of life on this planet. For nearly 3 billion years life on Earth was confined to the ocean and to expression in single-celled organisms. Early in those eons, such organisms engaged the crust of the Earth along with the hydrosphere and the atmosphere in generation of the biosphere and thereafter exerted major geological force. In diversity of organisms, life is thought to have arrived now at its all-time peak—the very era in which the human species has stumbled into custody of evolution.

Inquiry into the identity of humankind, thus set apart within the natural world, must proceed now from the discovery that toolmaking began 2.5 million years ago as the adaptive strategy of an infrahuman primate. For thousands of years—while, as Bridgman said, we were looking "inside our own heads" for the necessary pattern of the external world—we looked to the external world, to the sky and beyond, for the reason and purpose of our existence. The new understanding of our evolution places the natural locus of purpose inside the human head. It can be seen now that human purpose has been undergoing evolution ever since purpose dawned in the heads of the toolmaking primates. That evolution is presently under pressure to accelerate in response to fundamental changes in the human condition brought on by the last century's advance in the quest for objective knowledge that was started by those toolmakers.

2

Revolution at the Ground of Knowledge

There exists a passion for comprehension.
Albert Einstein

Albert Einstein, in 1950, his 70th year, in a retrospective essay on his life and work, wrote: "Without this passion there would be neither mathematics nor natural science. Time and again, the passion for understanding has led to the illusion that man is able to comprehend the objective world rationally, by pure thought, without any empirical foundations—in short, by metaphysics. I believe that every true theorist is a kind of tamed metaphysicist, no matter how pure a 'positivist' he may fancy himself.

"The metaphysicist believes that the logically simple is also the real. The tamed metaphysicist believes that not all that is logically simple is embodied in experienced reality, but that the totality of all sensory experience can be comprehended on the basis of a conceptual system built on premises of great simplicity."

In another context, Einstein cited the Newtonian world system as one built on such premises:

Essentially there are two laws:
1. the law of motion 2. the expression for force or potential energy

Einstein did not mention that Newton rested these premises on the further premise that time and space are absolute. That might seem to go without saying, as it does in every-day experience. But Newton spelled it out:

> Absolute space in its own nature, without relation to anything external remains always similar and immovable and by another name is called extension.
>
> Absolute, true, and mathematical time, of itself and from its own nature, flows equably without relation to anything external, and by another name is called duration.

The Einstein world system sets these two propositions aside at the outset. That is what gives people their first difficulty when they seek admission there. Absolute time allows simultaneity, even across time zones. Absolute space permits absolute motion, the kind of motion most people experience when going from here to there. Time and space are when and where things happen. These two absolutes stand outside Newtonian physics as well; they are the theater of its events. In the cosmos of Einstein, time and space are closely linked physical variables. "From preordained perfection high above the battles of matter and energy," in the words of the cosmologist John A. Wheeler, Einstein transformed space and time to "spacetime … a new dynamic entity participating actively in the contest."

Only in extreme situations can a person experience the suspension of the nearby familiar world in the four-dimensional spacetime matter-energy continuum comprehended by Einstein's Special Relativity. The fourth variable comes into play, for example, with the 40,000-fold increase in mass acquired by electrons when boosted by physicists to 99.9999… percent of the velocity of light in high-energy particle accelerators. A motorist can ignore, on the other hand, the increase by a whole 1.5×10^{-8} gram in the mass of her automobile accelerated to 100 kilometers per hour. The corrections to Newton's laws called for by Special Relativity are negligible for almost all purposes but the understanding that makes the world so much more interesting than it was at the beginning of the 20th century.

Absolute space made it possible for Newton to posit the state of absolute rest. That established necessity for the "expression for force"—

to get bodies from rest into motion. Behind that necessity was another premise so evident that he did not state it: effects have causes.

Quantum electrodynamics also sets certain premises of classical physics aside, this one in particular. The new physics gets on without necessary cause and effect. It shows the physical world happening spontaneously. Events are the outcome of statistical trends; probabilities are conditioned by what precedes, never without the play of chance.

Einstein, having made a pivotal contribution to the founding of quantum physics, could not accept this outcome. He held steadfast to his conviction that the universe is governed by law that relates the past to the present and future and the parts to the whole, and that the mission of science is discovery of such law. This showed him to be at heart a classical physicist and, perhaps, a not fully "tamed metaphysicist." In his last exchange of views with his successors, Einstein could sympathize with Newton, to whom he addressed this apology:

> Newton, forgive me; you found the only way which, in your age, was just about possible for a man of highest thought- and creative power. The concepts, which you created, are even today still guiding our thinking in physics, although we know that they will have to be replaced by others far removed from the sphere of immediate experience, if we aim at a profounder understanding of relationships.

The laws of motion

The first of Newton's three laws of motion—which Einstein cited as "the law of motion"—begins with a body at rest—at rest until it is acted upon by a force. When and so long as it is acted upon by the force, the body goes into and continues in accelerated motion. Thereafter it persists in constant motion in the direction and at the velocity to which it was accelerated, until it is acted upon by another force.

Newton's second law of motion supplies what Einstein called "the expression for force." The force is equal to the mass of the body multiplied by the acceleration that gets the body from rest into constant motion or deflects it from that motion.

The third law gives the expression for the "inertia" of the body

at rest and in motion. It is equal and opposite to the force that accelerates it, and it increases with the acceleration imparted by the force. For
motion in absolute space, absolute time measures the rate of motion and
the rate of change in it and so the magnitudes of force and inertia.

On these minimal principles, Newton secured the unification of
terrestrial and celestial mechanics. For his own generation, with some
debate even then about the absoluteness of space and time, and for generations after, his "conceptual system built upon premises of great simplicity" comprehended the physical world. The same laws of motion
governed the famous fall of the apple and the orbiting of the Earth and
the then-known five other planets around the Sun.   Even now
Newtonian mechanics underpins industrial civilization in all technologies
involving "ponderable" masses, such as pile drivers and steam turbines,
pendulum clocks and spring-wound watches. Over the years, it came to
embrace, as Einstein said, "areas which apparently had nothing to do
with mechanics," such as the theory of heat elaborated in thermodynamics. It budded subbranches, such as aerodynamics, to manage contingencies that Newton did not anticipate. In the celestial sphere, it governs the solar system satisfactorily with only minor corrections required
until people actually venture out into it.

Isaac Newton was born in 1642, as no one fails to mention, in
the year of Galileo's death. For his ambitious undertaking, Newton had
Galileo's demonstration of accelerated and constant motion. Newton
was endowed also with the work of Johannes Kepler. Contemporaneous
with Galileo, who remained unpersuaded, Kepler had found in the
naked-eye, high-precision observations of the Danish astronomer Tycho
Brahe three laws of planetary motion. The first put Earth and the then-
known five other planets on their elliptical orbits. Kepler therewith displaced the Sun from its momentary eminence at the center of the new
Copernican universe to an uncertain location at one of the two always
moving foci of the planetary ellipses. His second law states that a line
drawn from an orbiting planet to the Sun "sweeps out equal areas in
equal times." This is because the planets travel faster on closer approach
to the Sun and slower on the far side of their orbits [see illustration, page
14]. Newton found here strong correlation of celestial to terrestrial

mechanics. Kepler's second law demonstrates the conservation of angular momentum—celebrated in the quickening spin of a figure skater as she brings her arms to her sides—in the motion of the planets.

Newton must have had these predecessors in mind when he wrote to Robert Hooke: "If I have seen further, it is by standing on the shoulders of giants." Hooke, his *bête noir* [see page 50], was a dwarf.

In his lonely rural childhood, in the care of his grandmother, Newton had the solitude to think about matters such as the seasonal migration on the horizon of the sunrise and the sunset. He tracked it by ruling the shadow lines from the windows on the floors and walls of his grandmother's house. In Trinity College at Cambridge University, at age 19, Newton attracted the attention of Isaac Barrow, the Lucasian professor of mathematics. Barrow engaged his promising pupil in work on the new world system and in optics.

"... to answer pretty nearly"

When the Black Death that had been terrifying London found its way to Cambridge in 1665, Newton took refuge in his childhood village. There, in solitude again for 18 months, he accomplished most of his life-work. He invented the algorithms of differential and integral calculus that make it possible to calculate increments in the continuous change of a variable such as motion. Using these procedures, he "compared the force requisite to keep the Moon in her orb [*sic*] with the force of gravity at the surface of the Earth and found them to answer pretty nearly." He found time to think about the problem in optics of color aberration, which blurs with faint rainbows the edges of objects viewed through a less than perfect lens. This, he surmised, must be caused by the difference in angle at which light of different colors is refracted through a lens. To cure the problem for astronomy he invented the reflecting telescope.

Upon his return to Cambridge, Isaac Barrow retired from his professorship so that Newton might succeed to the chair. At age 26, Newton held a secure academic appointment that invited him to get on with his work. Not long afterward, he presented a model of his reflecting telescope to the Royal Society together with his first publication: "a new theory of light and colors." Newton was at once elected Fellow and into the company of contemporaries who would have greatest interest in his work. The world's big optical telescopes have, ever since, been reflectors with a "Newtonian" light path to the eye or camera.

The contest on priority with respect to his work in optics in which he was soon embroiled with Robert Hooke and others made Newton a pugnaciously diffident author. Not until the same Hooke had published his calculations showing that the force of gravity weakens as the square of the distance between two bodies did Newton turn to the writing of his monumental *Philosophiae naturalis principia mathematica* (Latin was still the *lingua franca* of learning), published in 1687. Here, Newton set out the results of those 18 months in refuge: his three laws of motion and the law of universal gravitation. He also propounded the law of conservation of momentum, guaranteed by his third law of motion: the total mass times the velocity of two bodies in collision is conserved after the collision, though perhaps shared by them differently.

The law of gravity

Newton conceived of gravity as a force acting among all bodies in the universe, between earthly objects and the Earth, between the Moon and Earth, between the Earth and Sun. It was the force that deflected the planets from the constant motion of his first law and kept them falling toward the Sun. The orbits of the planets were thus the resultant of two motions. Their constant motion just offset the acceleration of gravity. (Toward the Sun, 150 million kilometers away, the Earth falls on its orbit about 2 million kilometers per day away from the straight-line path of its constant motion.)

Kepler's third law declares that the square of the time of travel of each planet around the Sun is proportional to the cube of the planet's average distance from the Sun. Newton's Principia gave the strength, in each case, of the force that held them in common thrall to the Sun. It is equal to the product of the multiplication of the masses of the Sun and of the planet in question divided by the square of that planet's distance from the Sun. The mass of a body thus acquired two functions. In Newton's third law of motion, mass measures a body's inertia or resistance to a force. In the law of gravity, mass measures the force by which a body attracts other bodies and is attracted by them. To the limit of the precision he could secure in experiments with pendulums, Newton determined that inertial mass and gravitational mass are close to equal or "equivalent." The equivalence only made the duality more puzzling.

Convinced as he was of gravity as a real force, and proud of its conception, Newton could not explain it in terms of any familiar experience of force, as with the inertial force: the action of one body on another, with momentum conserved in the outcome. For gravitational force there is no such evidence of "action"; in Newton's words:

> That gravity should be innate, inherent and essential to matter, so that one body may act upon another at a distance through a vacuum, without the mediation of anything else, by and through which their action and force may be conveyed from one to another, is to me so great an absurdity, that I believe no man who has in philosophical matters a competent faculty of thinking, can ever fall into it.

For the absurdity of action at a distance Newton had no explanation. He gave thought nonetheless to the possibility of determining the magnitude of the force of gravity per unit of mass. His calculations showed that this "gravitational constant" must be immeasurably small.

That conclusion laid a challenge to successors over the course of the next century. Henry Cavendish came closest to establishing the now-accepted value in 1798. He measured the deflection of a torsion balance by the presence of two large lead balls placed near two small lead balls suspended from the balance [see illustration, page 190]. The force proved tiny enough: 6.754×10^{-8} dyne—an infinitesimal fraction of the negligible force required to accelerate 1 gram to a velocity of 1 centimeter per second in one second. Measurement in 1994 set the value at $6.672 \times 10^{-8} \pm .001$ dyne.

Philosophers among Newton's contemporaries took issue with his absolute space and time. Bishop Berkeley held them to be elements in the subjectivity of human perception. Gottfried Leibniz, Newton's rival in the invention of the calculus, saw space and time as mere functions, respectively, of the distribution of matter and the sequence of events, with no objective existence independent of physical reality.

Newton had to acknowledge what Galileo had noted: within a moving body's own frame of reference, constant motion is indistinguishable from the state of rest. In the first law of motion was concealed Galilean relativity. It could cut all three laws of motion loose from their mooring in absolute space.

To put that unsettling notion down, Newton sought to establish a physical locus of absolute rest to which motion in the cosmos could be referred. It could not be the Sun. He recognized that the planets, reciprocating the Sun's gravitational attraction, must pull the Sun this way and that. By calculation, Newton placed the immovable center of the cosmos

just outside the Sun, but close enough to lie inside when the planets pulled the Sun across it.

Newton went on wrestling with the locus of absolute rest in the universe for many years. Near the end of his life, Newton added an appendix to the *Principia*. Here he conflated absolute space and time with the all-pervading presence of the Supreme God.

Over the next century, the *Principia* became classical physics. Newton had supplied the scheme by which successors could account for all they proceeded to learn about the physical world. The laws of motion describe the order governing the succession of events, ensuring that effect follows cause. With the laws in hand and the algorithms of the calculus to apply them, the art of mechanics entered on a period of accelerating progress through the next two centuries.

The great mechanicians who succeeded Newton found especially useful the concept of energy—the capacity to do work—drawn from Newton's second law. In acceleration from rest into motion by action of a force, a body acquires "kinetic" energy; this is equal to the product of one-half its inertial mass multiplied by the square of the velocity to which the body is accelerated. Even at rest, a body in a gravitational field possesses energy. This is "potential" energy, proportional to its inertial mass. When the body falls or is otherwise impelled into motion, the potential energy becomes kinetic energy. Upon impact or redirection of its motion in other ways, as by the turning of a crankshaft, some portion of the kinetic energy is conserved in the motion imparted to another body. This, in mechanics, is the "work done." The rest of the energy is transformed to heat at impact or by friction and thereby dissipated into the wide world, as extension of the laws of motion to thermodynamics was to show in the 19th century.

Experiencing the loss of energy to friction in every system, the mechanicians came early to have deep respect for the law of conservation of energy. The law interdicts the fantasy of the perpetual motion machine. Some external source must supply the energy lost to the heat of friction. In rebuke of the wish to get work done for nothing, the law requires that the energy expended must always exceed the work done.

The mechanicians soon recognized that the transformation of energy might be made, in part, to work the other way, from heat to motion. With the fly-ball governor [opposite], James Watt employed the principle of the conservation of angular momentum to make the steam engine self-regulating. Machines afforded scale-model approximations of the deterministic design of the cosmos. Assurance of celestial and terrestrial order found its grandest statement in the five immense volumes of *Mécanique céleste*—meaning nothing less than the celestial *machine*—published between 1799 and 1825 by Pierre-Simon Laplace.

We must then regard the present state of the universe as the effect of its anterior state and as the cause of the one which is to follow. Given for one instant an intelligence which could comprehend all the forces by which nature is animated and the respective situation of the beings which compose it ...—it would embrace in the same formula the movements of the greatest bodies in the universe and those of the lightest atom; for it, nothing would be uncertain and the future, as the past, would be present to its eyes.

In all its serenity, the cosmos of Laplace and Newton might seem a barren place. By his other great contribution to understanding, Newton showed his successors how the laws governing the mechanical cosmos might be brought to encompass the rich world of common experience. This was the "Opticks," which explained how the gamut of color conveyed by light may be separated by refraction or selectively reflected to fill the eye with color. Into a beam of sunlight he interposed a glass prism, thereby dispersing the colors of what he called its "spectrum." He showed that a single color passes through a second prism without further dispersion. He then showed that the dispersed colors reassemble into white light upon reverse passage through a second prism.

For the dispersion of the colors in their passage through the prism Newton had an explanation consistent with his mechanics. He conceived of light as the streaming of tiny material particles. In evidence, he cited the straight-line travel of a beam of light and the sharp edges of the shadows it casts. Given mass, a beam of starlight would be attracted by the gravitational mass of the Sun. In proof, Newton predicted that the apparent position of a star near the Sun must be displaced by a half-second of arc. This prediction was confirmed—incidental to the test of the corresponding proposal from General Relativity.

In affirmation of his mechanistic model of light, Newton had the pleasure of hearing of the first attempt to measure the velocity of light that produced a credible result. The Dutch astronomer Olauf Roemer compared the intervals between eclipses by Jupiter of one of its satellites when that planet was on the far side of the Sun and when it was nearer to Earth on the same side of the Sun [see illustration, page 90]. The duration of the eclipse, whether Jupiter was near or far, was the same. On the surmise that longer intervals between the far-side eclipses must be attributed to the longer time it took for light to bring the news across the solar system, Roemer undertook an estimate of the speed of light; he got the respectable result of approximately 200,000 kilometers per second. With all the then uncertainties in timekeeping, it was not far from the now accepted velocity of 300,000 kilometers per second. A measured finite velocity plainly distinguished light from the force of gravity and its instantaneous action at a distance.

On the precedent of the founder's invocation of corpuscles of light, physicists in the 18th and 19th centuries called upon a variety of hypothetical substances to take the principles of mechanics into other realms of inquiry. Phlogiston was the substance carried off in flames until oxygen took over in the understanding of combustion. Caloric served for a time as the downhill running substance of heat. Two different fluids—"vitreous," associated with glass rubbed by silk, and "resinous," with amber rubbed by flannel—were called in to explain the attraction and repulsion of static charges of electricity. Benjamin Franklin's single electric fluid flows today from the "positive" to the "negative" pole even though what flows is negative electrons that ought to be repelled by the negative pole. Such scaffolding serves useful purposes until it can be kicked away or reduced to service as metaphor.

Phlogiston and chemistry

The first of these hypothetical substances to yield to understanding was phlogiston. The substance of a brief-lived, 18th-century paradigm, phlogiston burned "out of" wood, leaving ash. Heated out of cinnabar, it left mercury. Flames died in enclosed chambers because air locally had finite capacity to absorb phlogiston.

Joseph Priestley in 1774 discovered that candles burned more brightly in the phlogiston released from cinnabar. He had isolated oxygen. Faithful to the paradigm, he called it "dephlogisticated air."

Antoine Lavoisier gave oxygen its status as a substance and its name. He identified combustion as the energetic combining of oxygen with carbon or hydrogen. The less energetic combination of oxygen with mercury, he saw, yields cinnabar [see illustration, page 57]. The phlogiston paradigm gave way to the new science of chemistry.

Before the end of the 18th century, chemists showed that irreducible substances, which they called elements, combine to make compounds. Their instrument was the horizontal balance, a pan hung at each end and contrived to make ever finer measurements of weight. In 1799, Joseph Louis Proust showed that elements combine in "definite proportions" [see illustration, page 58].

John Dalton found that they combine in definite ratios of weight.

Joseph Louis Guy-Lussac showed that gases, such as oxygen and hydrogen, combine in fixed ratios of volume. This evidence persuaded chemists that elements reduced to atoms, irreducible particles of themselves; if they did not, they would blend in varying proportions.

Dalton's definite ratios of weight and Guy-Lussac's definite ratios of volume prompted Amadeo Avogadro to the bold proposal, in 1811, that equal volumes of gas—at the same pressure and temperature—must contain equal numbers of particles, atoms or molecules. Determined a century later, the number of atoms in the standard 22.4 liters of any gas at a standard pressure and temperature proved to be 6.02×10^{23}. It is properly called Avogadro's number.

Dalton's table of atomic weights, expanded by others, prompted William Prout, in 1815, to a still bolder proposal. Scaled around oxygen, set at the arbitrarily round number of 16.00, the weights varied from around 1.00 for hydrogen all the way to 195 or so for platinum. The weights of all were close to whole-number multiples of the weight of hydrogen. Prout proposed accordingly that the atoms of all elements were made of larger or smaller numbers of hydrogen atoms.

Prout's hydrogen atoms are today's protons and neutrons, the particles of the atomic nucleus. In the division of labor that divided chemists and physicists during the 19th century, they were divided as well by the "atomic hypothesis." Working with substances, chemists had use for atoms. Working with forces, physicists continued to hold atoms in suspicion until J. J. Thomson demonstrated the existence of the first subatomic particle, the electron, at the end of the century.

Early in the 19th century, the particulate nature of light came under serious challenge. Thomas Young demonstrated that light has properties of waves. He shone a beam of light through two narrow slits. On the screen beyond, the beam cast a pattern of alternate bright and dark line-images of the slits [see illustration, page 114]. Geometry shows that wave trains, traveling slightly different distances to their encounters on the same plane, must alternately reinforce or cancel one another depending on whether they arrive in or out of phase at each line.

In keeping with mechanistic principles, light waves were first envisioned as longitudinal waves, like the compression waves of sound in

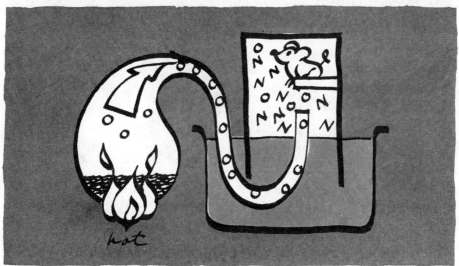

Discovery of oxygen by Lavoisier began modern chemistry. Cinnabar, an oxide of mercury, in flask [top] takes up additional oxygen from the air when it is warmed; mouse suffocates in fanciful experiment. Heat liberates weakly linked oxygen from cinnabar [bottom]. In an actual experiment, Lavoisier demonstrated that the liberated gas sustains animal respiration, as here the mouse. Lavoisier demonstrated further that the gas, which he called oxygen, sustains the combustion of wood, combining with carbon in CO_2 [see page 55].

Law of definite proportions, *along with other basic principles in the new science of chemistry, was established by the precise weighing of elements before and of the compounds after reaction that combined them [see page 56]. If, in the proportion of two to one, two elements produced a compound of weight three, then in ratio of four to two they produced a compound of weight six. Combination in different proportion yielded a surplus of one of the elements. Early chemists were thereby convinced of the atom's reality.*

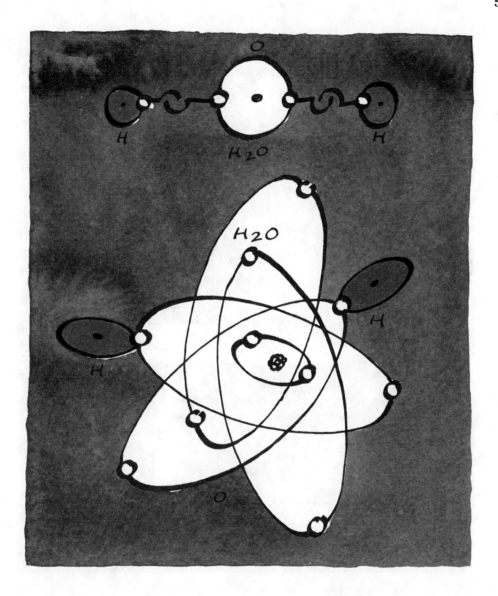

A "valence" joined atoms in compounds in the mind's eye of early chemists
[top]. From the combination of hydrogen and oxygen in ratio of two volumes to
one they saw that hydrogen had one valence and oxygen, two [see page 70].
Discovery of the electron gave the valence the physical reality of the electric
force [see pages 71 and 123]. The oxygen atom, with eight electrons, has places
on its outer orbits for two more. The single electron on each of two hydrogens
fills those places when the two elements join in their familiar compound, water.

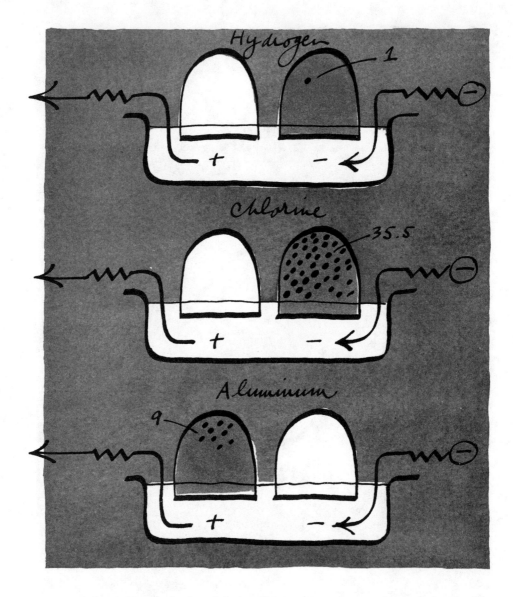

Laws of electrolysis, *the breakdown of a compound to its elements by the electric force, were established by Faraday. He found that the flow of 96,500 coulombs of electric current (a "farad") yields 1 gram of hydrogen from the breakdown of water and 193,000 coulombs, 2 grams. A farad yields 35.5 grams of chlorine, an atom 35.5 times heavier than hydrogen. The yield is divided, however, by the number of valences; a farad yields 9 grams of aluminum, with a weight of 27 but a valence of three [see page 70].*

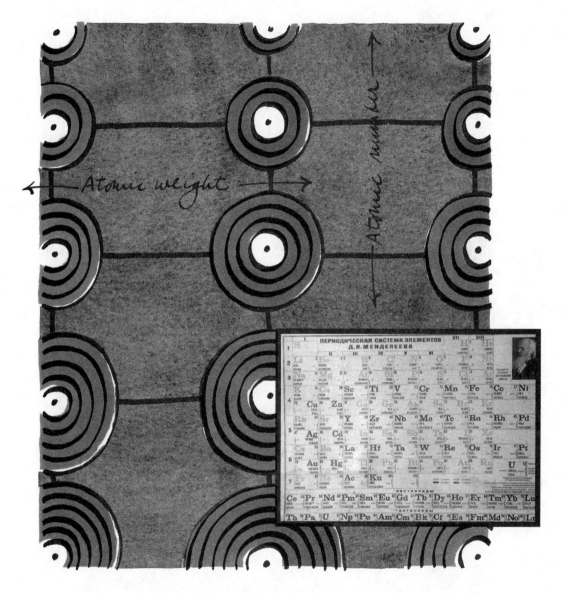

Periodic table of the elements *arrays them along the rows in ascending order of their "atomic weight" (the total number of protons and neutrons in their nuclei) and in the columns by the similarity of their chemical behavior, established by valence, which is related to the number of protons in their nuclei, their "atomic number." As published by Mendeleyev [see pages 71–72] in 1869, the table had open cells awaiting discovery of elements with the indicated properties. The celebratory Russian edition of the table [inset] bears his portrait.*

air. Certain crystals were observed, however, to polarize light—that is, to allow its passage in one plane and block it in others. This effect, familiar to wearers of polarizing sunglasses today, proved that light waves are transverse or perpendicular to their line of travel. They can be pictured as waves whipped into a lariat or a skipping rope [below], except that light waves usually oscillate in all planes around their line of travel.

Empty absolute space suited gravity; action at a distance offered no explanation of its instantaneous conveyance and so asked for no medium to intervene between two gravitational masses. The transverse

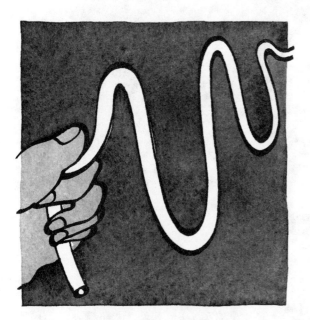

waves of light traveling, as shown, at finite speed had recognizable mechanical properties. They called for a medium to carry them. Newton's successors were compelled to make hypotheses about such a medium.

To carry light waves they conceived the "aether," a perfectly transparent, incompressible, solid medium, as dictated by the laws of mechanics for a carrier of transverse waves of light. Yet aether had to offer no resistance to the motion of bodies on Earth, much less to the motion of planets

on their orbits. Late in the century, Lord Kelvin proposed that a "plastic" quality would answer; plastilene yielded to steady pressure, yet had the rigidity to fracture under sufficiently abrupt force.

Endowed with these and other features, the aether stands as the most elaborate of the hypothetical substances devised to convey the forces of classical mechanics. It kept the Newtonian picture intact. Its ultimate nature was a question to be settled someday by people not preoccupied by more immediate questions posed by classical mechanics.

These abounded: in thermodynamics, in the kinetic theory of gases, in the atomic theory of matter and in the novelties of magnetism and electricity. From the soul of classical physics, thermodynamics has arrived intact in the 21st century, a complete and enduring comprehension of its universal content. Along with the conservation of momentum and energy and matter, its laws govern throughout the contemporary cosmos. "It is the only physical theory of universal content," Albert Einstein said, "of which I am convinced that, within the framework and applicability of its basic concepts, it will never be overthrown."

The imponderable caloric fluid carried thermodynamics to some of its more profound insights. Caloric was heat. There was more of it in a hot body than a cold one; from the hot body, it ran downhill to the cold. Early in the 19th century Sadi Carnot made good use of this recognizably mechanistic concept of heat to explain how to get the most out of a steam engine: give caloric the longest possible downhill run. His law states that an engine's power to do work increases as the difference between the temperature of the "working fluid"—be it steam or the gases of combustion—at the face of the piston or turbine blade and at the exhaust [see illustration, page 65]. Caloric sufficed to yield the first statement of the second law of thermodynamics: Heat flows always from the hot body to the cold, never the other way.

Benjamin Thompson, the dashing Tory renegade from New England known as Count Rumford, demonstrated the simpler, still mechanical and now established explanation of heat. In the armory of the king of Bavaria in the 1790s he arranged the boring of a brass cannon under water. A gallery of onlookers saw the heat of friction between the cutting tool and the brass bring the water to a boil. Thompson's weighing of the metal shavings when hot and then cold showed that caloric must be imponderable. Heat, he declared, could be nothing more than what the lathe supplied to the metal: "namely, motion!"

James Prescott Joule established the equivalence of work and heat more precisely in 1849. Through a gear train, he let a falling weight spin an array of paddles in a water tank, stirring its motion into the water. Measuring the "foot-pounds" of fall and the rise of temperature in the water, Joule came close to establishing the value of the modern "joule" [see illustration, page 65]. Henry Augustus Rowland resolved to fix that value upon installing his laboratory at Johns Hopkins University in

1872. With a steam engine, thermometer and calorimeter designed for the occasion, he determined it at 4.19 joules = 1 calorie. A calorie is in turn the unit of heat that raises one cubic centimeter of water one degree centigrade in temperature. (The dietary "grand" calorie, correctly written Calorie, is 1,000 of Joule's "small" calories.) Joule's experiment established the first law of thermodynamics: energy can be transformed— for example from motion to heat—but not created or destroyed.

Physicists interested in heat necessarily embraced the atomic hypothesis. Because the motion of particles could be more readily visualized in a gas than in Joule's water tank, they first considered gases. In a quantity of gas, as in a bicycle pump, the increase in pressure forced by sudden decrease in volume can be sensed as an increase in the temperature of the barrel of the pump. This could be seen as a consequence of increase in the velocity of the molecules of air and in the frequency and force of their collision with one another and the walls of the container.

In the incessant exchange by collision of their individual energy of motion, it was thought, the atoms and molecules shared the total heat energy of the gas equally. To this principle of "equipartition of energy," James Clerk Maxwell—author of the equations of electromagnetism— offered a significant amendment, his second major contribution to understanding. It was not to be thought that each and every one of Avogadro's 6×10^{23} atoms or molecules moves at exactly the same velocity. Calculation set the average velocity in air at 500 meters per second and the average distance between energy-exchanging collisions, or "mean free path," at .00001 centimeter. Around such averages, the Maxwell velocity-distribution curve plots the most probable partitioning of the energy of motion among the molecules at different temperatures and pressures [see illustration, page 66].

The "statistical mechanics"—for disordered motion of molecules assembled in Avogadro numbers—of Maxwell and of Ludwig Boltzmann in Germany introduced chance and probability into the deterministic world of classical physics. Hindsight would show this to be a first portent of the revolution to come at the end of the century. For those who cared, Laplace's all-seeing intelligence could continue to keep book on the conservation of momentum at each collision. Given the reliability of

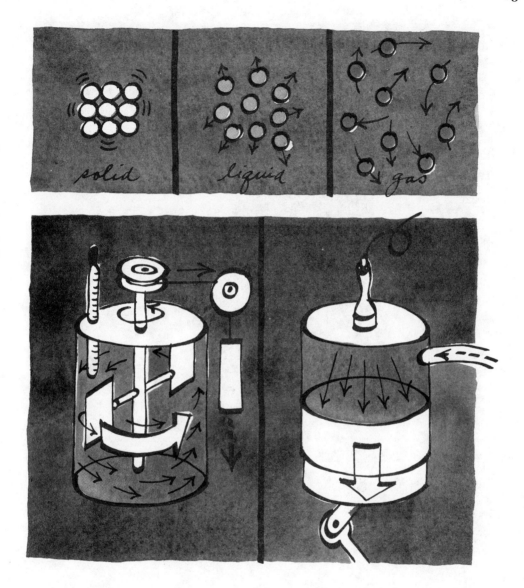

Heat is motion of particles of matter, increasing from solid, through liquid to gas phase [top]. Stirring motion into water with a falling weight, Joule showed the transformation of motion to heat, measuring the temperature increase by "foot-pound" of the weight's fall [see page 63]. Imagining heat as "caloric," an insubstantial fluid [see page 63], Carnot correctly found that the reverse transformation of heat to motion by a heat engine increases by the difference between heat of the "working fluid" at the face of piston and in the exhaust.

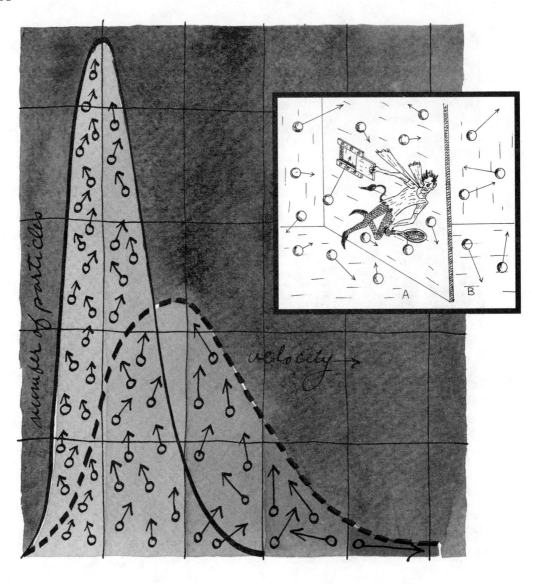

Distribution of heat energy *among particles changes with temperature, as shown by a plot of the number of particles (vertical scale) against velocity (horizontal) at different temperatures. The number particles being constant, the area under curves is the same. Average velocity increases as the square root of absolute temperature. "Statistical mechanics" thus qualified the classical assumption of "equipartition" of energy among the particles [see page 64]. Maxwell's "demon" [inset, by George Gamow] separates hot from cold particles.*

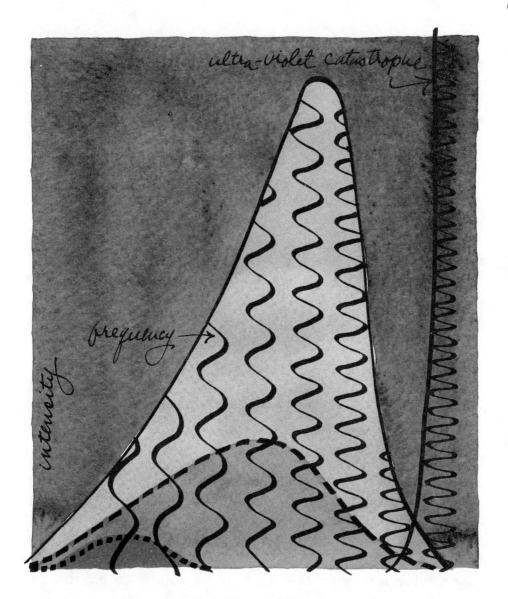

"Ultraviolet catastrophe" *followed from the "equipartition" of energy among waves of radiation from hot bodies: wavelets in hugest number at high frequencies, at right, would carry off all the energy [see pages 84–85]. Measured intensity of radiation across the spectrum corresponds to distribution of energy among the particles [opposite]. Planck showed each wavelet carries same energy "quantum":* 6.77×10^{-27} *erg [see page 85]. Einstein showed the quantum or photon carries the mechanical force of momentum [see pages 87 and 95].*

prediction by statistical mechanics, however, the next generation of physicists could think of order so determined to be possible but not necessary. The generation after came to think it improbable.

The second law of thermodynamics acquired new depth with recognition that the dissipation of energy from a system represents increase in the disorder, or "entropy," of the particles in motion in a system. All forms of energy transform ultimately to heat, and heat dissipates into the wide world. Within a closed system, entropy may decrease—for example in the synthesis of large molecules by a living organism. This increase in order does not happen "for free"; the energy demand must be met by the intake of energy from outside the system and its dissipation to the outside thereafter as the increase of entropy there.

Maxwell's "demon"—"being," as Maxwell himself more dispassionately called him—makes the connection of heat to motion direct:

> Now let us suppose that a vessel is divided into two portions, A and B, by a partition in which there is a small hole and that a being, who can see the individual molecules, opens and closes this hole, so as to allow only the swifter molecules to pass from A to B, and only the slower ones to pass from B to A. He will thus, without expenditure of work, raise the temperature of B and lower that of A, in contradiction to the second law of thermodynamics.

Maxwell's demon mocks human subjection to the second law: "as we can deal with bodies only in the mass and have no power of perceiving or handling the separate molecules of which they are made up." Now, with such power in hand, an entirely different order of mechanics, quantum mechanics, has been shown to rule in the domain of the demon. The new understanding makes it possible to perceive and manipulate molecules and atoms in small numbers and even individually. To reach down there requires, however, inversely huge expenditure of energy—in a particle accelerator, for example. As ever, the second law of thermodynamics exacts the price of increase in entropy.

In its simplicity and universality, the second law exemplifies the rule that truth is beauty, that elegance warrants the power of an idea. It is a secondary law, however, in the sense that it does not tell why systems obey it. The answer to such questions must be sought in the primary laws that describe the nature of forces and particles and their interactions. The equations of those laws are symmetrical; in principle, they may run forward or backward, as would a motion picture of an event they describe. Entropy breaks that symmetry by pointing the inexorable direction of time; it is, Arthur Eddington said, "time's arrow."

Electricity and magnetism

The effort to bring electromagnetism into the comprehension of classical mechanics persisted into the last quarter of the 19th century. As pervasive as gravity in the life of the industrialized countries today, the electromagnetic force first came into human experience as an occasional curiosity. Magnetism was known in the mysterious power of the lodestone and had found use in the compass needle. That force found no association, however, with the spark at the fingertip in the discharge of static electricity, and that spark had none with lightning. Static electrical charges, as gathered from a carpet and discharged at a doorknob, were monopole, either positive or negative. Magnets, however, came in dipoles, called north and south from their use in the mariner's compass. No one can chop the north and south poles of a magnet apart [see illustration, page 74]. Tabletop experiments showed that the two forces were related in that like charges and like poles repelled each other and unlike attracted. The two forces could thus be strongly distinguished from the exclusively attractive force of gravity. With reassuring resemblance to that force, however, it was shown by Charles-Augustin de Coulomb in the late 18th century that the electrostatic and magnetic forces diminish as the square of the distance between the bodies exerting them. Coulomb is honored in the "coulomb," the standard unit of the electric charge, of the force exerted by an electrified body.

Magnetism and electricity at first engaged separate schools of inquiry. Electricity held wider popular interest, attracted by the exploits of Benjamin Franklin, the New World's first international celebrity. A portrait of Franklin, circulating in Paris during his embassy there, bore the inscription *"Fulgum eripuit coelo, sceptrumque tyrannis!"* ("He snatched the lightning from the sky and the scepter from the tyrant!")

Franklin handled lightning with more circumspection. From the observation that lightning tends to strike the highest point, he mounted the first of his lightning rods on the silken kite that he tugged aloft into promising thunderclouds. The pointed rod, as he anticipated, drew electric charges from the cloud that sparked across the gap from the kite string to a brass key that he dangled near it on a hempen cord. This was the first clue to the present understanding that the turnover of the atmosphere maintains a varying electric potential between the sky and the ground that averages half a million volts around the world. Franklin's lightning rods today draw charges from the sky before they accumulate to jump as lightning bolts. His experiments and his lightning rods—the first practical application of knowledge of electricity—showed that this formidable force is ubiquitous in nature.

Franklin's electric fluid

Franklin held that his electric "fluid" was conserved in its transfer from one body to another. A neutral body contained a "normal" amount of it; a positive charge signified an excess of the fluid and a negative charge, a deficiency. "The electrical matter," as this Newtonian conceived it, "consists of particles extremely subtile, since it can permeate common matter, even the densest metals." By the first decade of the 19th century, Alessandro Volta had the electrical fluid flowing steadily from piles of alternate zinc and copper sheets insulated from one another, bathed in brine and connected by a conducting wire. Volta is memorialized in the volt, the term for electric potential energy.

Volta's experiments demonstrated the presence of the electric force in the structure of matter. The breakdown of water by electric current, with hydrogen collecting at the negative pole and oxygen at the positive [see illustration, page 60], begins the history of modern chemistry in many textbooks. Michael Faraday, succeeding Sir Humphry Davy as director of the Royal Institution, established the laws of electrochemistry early in the 19th century.

To Davy, Faraday owed the development of electrolysis, by which Davy had broken down salt to sodium and chlorine. Faraday's first law declares that the mass of an element liberated from a compound is directly proportional to the quantity of electricity employed in the electrolysis. He had found that 96,500 coulombs would liberate 1 gram of hydrogen, the lightest of the elements, and that 193,000 coulombs would liberate 2 grams of hydrogen. For a heavier element such as chlorine, shown by the chemists to weigh 35.5 times as much as hydrogen, he discovered that 96,500 coulombs liberate 35.5 grams. Such a quantity of a given element—the weight of all of them is scaled to hydrogen = 1—is now known as its "gram molecular weight" or "mole." In a mole is an Avogadro's number of atoms: 6.02×10^{23}; it is the mole of a gaseous element that sets the standard volume of gas at 22.4 liters.

Avogadro's number here invites a moment's contemplation. From its recognition early in the history of modern science, it remains one of the hugest significant numbers describing the world in common experience. The attempt to visualize 6.02×10^{23} of anything is assuredly futile. Spread out one atom per square millimeter, a gram atom of any element would envelop the surface of the Earth 12 times. To accept such numbers as in the nature of immediate reality takes practice.

For elements like oxygen, which forms compounds with two volumes of another element such as hydrogen, Faraday made a still more interesting discovery. He found that 96,500 coulombs will liberate only half, 8 grams, of its mole of 16. By that time chemists had invented the term "valence" for the combining power of the various elements; hydrogen and chlorine were credited with a valence of one; oxygen, with a valence of two. This led Faraday to articulate his second law of electrochemistry: The mass of an element liberated by a given quantity of electricity is proportional to the atomic weight of that element divided by its valence-number [see illustration, page 60].

This final bit of evidence established for many chemists the atomic basis of matter. In the mental image of an atom of oxygen "valenced" to two atoms of hydrogen they could see two gases combining to yield water, a liquid [see illustration, page 59]. Faraday gave the real-world presence of the electric charge to their hypothetical valence. The memory of this contribution to understanding is held in the farad, the 96,500 coulombs that liberate one mole of an element.

In the physical reality of the valence, Dmitry Mendeleyev found the hook on which to hang the periodic table of the elements that adorns every chemistry classroom and laboratory today. Other chemists had observed that qualities such as higher and lower density, melting point and heat conductivity, hardness and softness, reactivity and affinity with other elements recurred from element to element along the scale of increasing atomic weight. They had classified them in families such as metals and nonmetals, halogens and alkalis. Mendeleyev gave these per-

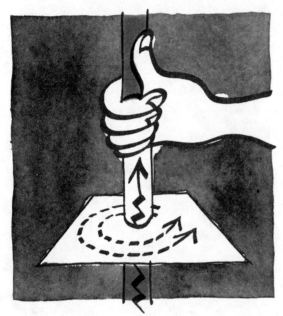

ceptions the force of law: "The properties of the elements are in periodic dependence on their atomic weights." In the layout of his table, valence is the periodically recurring property that governs. The elements increase in weight across the rows; the columns are arrayed by valence, from one to as many as seven, the cycle repeating from row to row.

The physicist Faraday did not embrace the chemists' atomic hypothesis. He was satisfied that the separation of elements in electrolytic solutions had established a defined and universal unit of electric charge: 9.65×10^4 coulombs. By 1833, Faraday was engaged in the experiments that were to demonstrate the induction of electricity from the magnetic field.

Hans Christian Ørsted, in Denmark, made the first connection between electricity and magnetism in 1820. He ran a current through a wire aligned north and south above a north-pointing compass needle. The needle swung abruptly to point east. When Ørsted reversed the direction of the current, the needle spun around to point west. Was this evidence of still another force? It was quite unlike gravity or even the magnetic and electric forces, each of which runs straight from one body to the other—not at a right angle! [see illustration, page 74].

Meanwhile, in Paris, André-Marie Ampère had been trying to

make sense of his discovery that parallel wires carrying electric currents in the same direction attract one another and repel one another when the currents flow in opposite directions. Upon learning of Ørsted's experiment, he correctly collapsed the two and possible three forces into a single electromagnetic force. The flow of electric current sets up a magnetic force that, Ampère found, follows a "right hand rule" [opposite].

The hazard laid to the universal sway of mechanics by these discoveries was immediately appreciated by Carl Friedrich Gauss. Among "pure" mathematicians he is known as "the divine" Gauss. In life, Gauss took his inspiration from the real world and applied his mathematics to the physics of his day. He could not reconcile with the laws of mechanics a force that depends upon motion as well as distance. "Two elements of electricity in a state of relative motion attract and repel one another," he protested in astonishment, "but not in the same way as when they are in a state of relative rest!" His mathematical statement of the situation showed it to be in violation of the law of conservation of energy.

The Galileo of electromagnetism

In 1831, a decade after Ørsted had discovered the induction of the magnetic force by the flow of electric current, Michael Faraday showed the relationship to be symmetrical: he demonstrated the induction of an electric current by the magnetic force. He did so simply by moving a conductor between the poles of a magnet and, alternatively, moving a bar magnet through a coil of conducting wire. The motion of one relative to the other excited the electric current in the conductor so long as they were in relative motion.

Faraday soon improved on this rudimentary experiment and secured the sustained conversion of motion to electricity—of mechanical energy to electrical energy—with the first crude approximation of a spinning dynamo [see illustration, page 75]. He showed further that the strength of the current depends upon the amount of mechanical energy expended to generate it. Faraday's dynamo, scaled up, multiplied by the thousands and operated continuously, now lights the world.

Rightly called the Galileo of electromagnetism, Faraday found his first employment as apprentice to a bookbinder. He read the books he

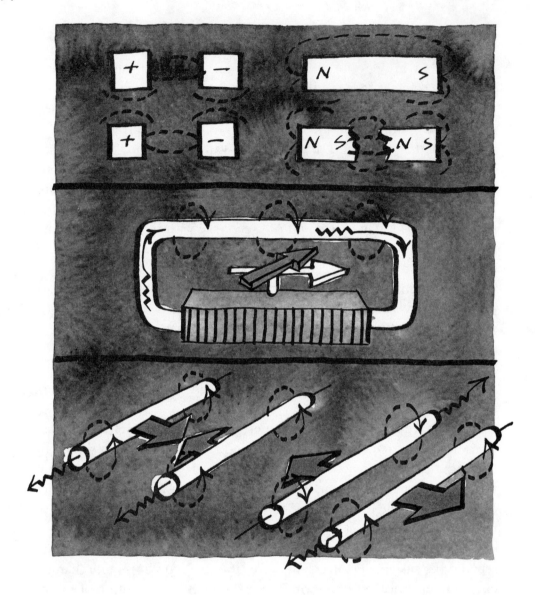

Electricity and magnetism *were long ago recognized as alike but different: like charges (plus or minus) or poles (north or south) repel, and unlike attract one another; electric charges, however, are one or the other and magnets, bipolar. Ørsted made first connection between them, observing the deflection of a compass needle by an electric current [see page 72]. Ampère showed that electric currents induce magnetic fields; fields around currents flowing in same direction attract; around currents in opposite directions, fields repel [see page 73].*

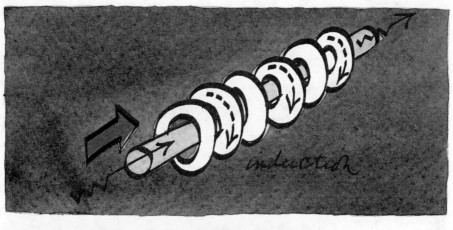

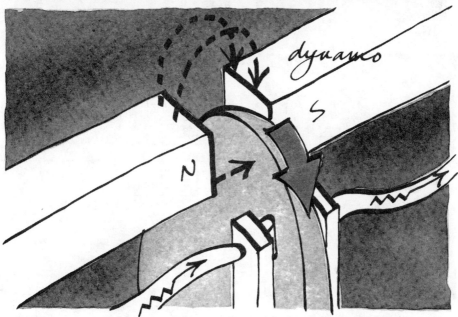

Induction of electric current *was demonstrated by Faraday in a simple exper-*
iment. He moved a conductor in a magnetic field induced, as Ampère had shown
[opposite], by the flow of current in a coil of wire. Faraday noted that fast
motion of the conductor in the magnetic field produces stronger electric current
for a shorter time. By spinning a metal wheel in a magnetic field [bottom] he
induced continuous flow of electric current [see page 73]. That experiment now
supplies electric energy from spinning dynamos throughout the world.

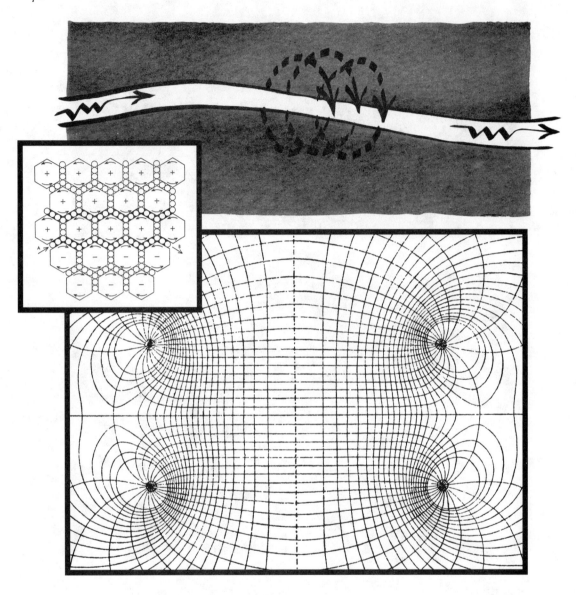

Faraday's "field" of electromagnetic force *opened the way to Einstein's field theories. Maxwell, whose equations developed the full force of Faraday's concept [see page 79], at first constructed mental mechanical models in the attempt to bring it into accord with classical physics [inset]. The cross section of two circular conductors fixes the four centers in this contour map of a magnetic field designed by Maxwell. Currents flowing in the same direction induce the field; spacing of the contour lines expresses field strength.*

Electromagnetic wave generated by the oscillation of an electric charge travels at the velocity of light as a mutually induced disturbance in electric and magnetic fields, as Faraday conjectured and Maxwell's equations showed [see pages 78 and 82]. The electric force, shown in vertical plane, alternates from plus to minus. The magnetic force, shown in horizontal plane, switches correspondingly from north to south. Upon arrival at a conductor, perhaps an antenna, the wave induces the flow of electric current activating a telephone or television.

was learning to bind and discovered special interest in those from the laboratory of the newly established Royal Institution. On the strength of a précis he wrote of one work, he persuaded the director, Sir Humphry Davy, to hire him in 1813 as laboratory assistant. In evidence of the open lines of social mobility in the British establishment, Faraday was director-emeritus of the Institution and had refused a knighthood when he died in 1867. Davy conceded that his choice of apprentice was his own greatest contribution to understanding.

Faraday's gifts as a natural scientist more than offset his lack of formal education. The way iron filings arrange themselves around a magnet fired his visual imagination. He saw "lines of force" crowding and merging in a "field" around a magnet—his terms. A conductor moving in the magnetic field "cuts" the lines of force. "If a wire moves slowly [across the lines of force], a feeble current is produced in it," Faraday wrote in his laboratory notebook, "if it moves across the same lines quickly, a stronger current is produced for a shorter time." From the static electric charge he could see lines of force reaching—though weakening as the square of the distance—to wherever in the universe they found the opposite charge. Space was "stressed" everywhere by such lines and fields of force.

The action of the electromagnetic force, Faraday saw plainly, was not to be found in the conductor or the magnet, where classical mechanics would locate it, but in the space between those material bodies. The generation and collapse of the electrical and magnetic fields attending such experiments must therefore, he concluded, generate waves in their extended fields or lines of force. He saw them rippling from the source of the disturbance indefinitely out into space—as light travels from the stars. He was sure that these waves would prove to be a kind of light.

For years, Faraday sought to establish an experimental connection between light and electromagnetism. In 1845, he succeeded: shining light through a slab of glass suspended in a strong magnetic field, he found it to be polarized. He showed further that the plane of polarization could be rotated by changing the orientation of the magnetic field. "Empty space," the theater of physics, was now pervaded with electromagnetic force. "All I can say," Faraday wrote in his notebook in 1846, "is that I do not perceive in any part of space, whether ... vacant or filled with matter, anything but forces and the lines in which they are exerted." In "thoughts on ray vibrations," he openly conjectured the identity of the waves in his fields with visible light. "The view which I am so bold as to put forth considers radiation ... as a high [frequency] species of vibrations in the lines of force which are known to connect particles and also masses of matter together." Still more boldly, he went on: "It endeavours to dismiss the aether, but not the vibrations."

Another 60 years passed before physicists could kick away the aether scaffold and accept this view of the fields and forces themselves pervading space. Not even James Clerk Maxwell, celebrated as the Newton of electromagnetism, could dispense with the aether.

The Maxwell equations

Maxwell was born in the year Faraday demonstrated the induction of electricity by a magnetic field. He had the education that Faraday lacked and mathematical talent equal to the capture of Faraday's vivid perception. Maxwell was to succeed in abstracting in a pair of complementary partial differential equations all the wealth of empirical observation recorded in Faraday's laboratory notebooks and papers. But first he had to construct elaborate mechanical models of Faraday's fields of force on paper and in his mind's eye [see illustration, page 76].

Maxwell's equations reckon also with the extraordinary result of an experiment in 1856 at Göttingen by Wilhelm Weber and Friedrich Kohlrausch. They had sought to determine the "ratio of the electromagnetic [unit] to the electrostatic unit of electricity." The electrostatic unit had been established by Coulomb as that parcel of electricity which "feels" the classical force of 1 dyne when placed in a unit electric field. To establish the corresponding unit of electromagnetic force Weber and Kohlrausch measured the strength of electric current that would generate an electromagnetic force of 1 dyne.

The dyne—Cavendish's term for the value of the gravitational constant [see page 51]—is the force that accelerates 1 gram to a velocity of 1 centimeter per second in one second. The unit of energy expended to that end is an "erg."

Given that the electromagnetic force is exerted by the motion of electric charges, then, in Maxwell's words, "the number of electrostatic units of electricity in one electromagnetic unit varies ... directly as the unit of time." The number to be determined was, therefore, a velocity, and the "number representing this velocity will be the number of electrostatic units of electricity in one electromagnetic unit." Measuring the deflection of a compass needle, Weber and Kohlrausch had found a value of 31,074,000,000 electrostatic units. That gave the electromagnetic unit the subminuscule force of 1/31,074,000,000 of one electrostatic unit or, in round numbers, 3×10^{-10} dyne.

Maxwell saw at once that, taken as a velocity, the centimeter-per-second value in the Weber-Kohlrausch fraction of 1 dyne was close to the velocity of light (3×10^{10} centimeters/second = 300,000 kilometers/second). Confirming their result with his own ingenious experiment, he got a value closer to the velocity of light than recently established by Auguste Hippolyte Fizeau in France [see illustration, page 90]. Maxwell went on to perfect his mathematical statement of the principles of electromagnetic radiation in 1865.

"The conception of the propagation of transverse electromagnetic disturbances," Maxwell wrote, "is distinctly set forth by Faraday in his 'thoughts on ray vibrations.' The electromagnetic theory of light proposed by him is the same in substance as that which I have begun to develop in this paper, except that in 1846 there were no data to calculate the velocity of propagation.... This velocity is so nearly that of light that it seems we have strong reason to conclude that light itself (including radiant heat, and other radiations, if any) is an electromagnetic disturbance in the form of waves propagated through the electromagnetic field according to electromagnetic laws."

Confirmation of the existence of "other radiations" propagating in the electromagnetic field came within a decade of Maxwell's untimely death, at age 47, in 1879. At the technical institute in Karlsruhe, the elegance and significance of Maxwell's equations engaged the enterprise of young Heinrich Hertz. He saw that the spark that discharges an electrostatic potential jumps back and forth in the spark gap at high frequency. In the oscillating spark, Hertz envisioned the exchange of energy between the electric and magnetic fields described in Maxwell's equations that would surely generate Faraday's electromagnetic waves.

Hertz accordingly set up a powerful electrical condenser at one end of the physics lecture hall. At the other, he set up what anyone today would recognize as a dipole antenna. The brief pulse of electric current that might be induced in the antenna by the waves would, he hoped, be strong enough to jump a spark gap there. Hertz was rewarded with sparks "visible to an eye which has been well rested in the dark."

The electromagnetic spectrum

To show beyond doubt the identity of the waves with light, Hertz reflected them from a metal surface, focused them from a parabolic metal reflector and even refracted them with a prism made of pitch. Hertz's waves, about a meter in length, were short enough to approximate light waves in their behavior. This property of "microwaves" was to be exploited in the radar technology that directed the British antiaircraft batteries and night fighters in their successful contest with the Nazi Luftwaffe in the Second World War. The waves of FM radio are short enough to betray the same property in the scrambling, especially by buildings in a city, of their reception in an automobile.

In 1895, within a decade of the Hertz experiments, Guglielmo Marconi broadcast his first wireless messages, on much longer wavelengths. Hertz had died of cancer, at age 37, in 1893. The hertz, abbreviated Hz, has now joined the coulomb, ampere and volt in the nomenclature pantheon of electromagnetism to designate the frequency of waves across the entire electromagnetic spectrum.

The frequencies of the visible spectrum, first spread out by Newton's prism, span the octave between 4×10^{14} and 7×10^{14} hertz [see illustration, page 30]; that is, from just inside the ultraviolet, at 3.75×10^{-5} centimeter, to just short of the wavelengths of infrared or heat radiation, around 7.5×10^{-5} centimeter. On either side of the visible spectrum, the waves and inversely the frequencies of the electromagnetic spectrum reach uncounted orders of magnitude.

The Maxwell equations show something that Faraday himself did not suspect: electromagnetic waves, depending on their wavelength,

penetrate and are carried not only in empty space and the atmosphere
but in all the media—for example insulators—that do not conduct or
hold the electrical and magnetic forces acting separately. Proof of the
equations is found in the medical imagery of X rays and CAT scans, as
well as in microwave cooking and fiber-optic communication. In modern cities people are bathed in and washed through by radiation from
electrical wiring, radio, TV and now cellular telephones.

The field displaces the aether

Maxwell recognized implications in Faraday's field and its lines of
force beyond the electromagnetic spectrum. He saw that Faraday had
proposed a radical departure from classical mechanics: "Faraday, in his
mind's eye, saw lines of force traversing all space where the mathematicians saw centres of force acting according to distance. Faraday saw a
medium where they saw nothing but distance. Faraday sought the seat
of the phenomena in real actions going on in the medium; they [Maxwell
had in mind Weber] were satisfied that they had found it in the power of
action at a distance impressed on electrical fluids."   Maxwell was
impressed and relieved to realize, "When I had translated what I considered to be Faraday's ideas into mathematical form, I found that in general the results of the two methods coincided." Faraday, Maxwell suspected, saw in his field and its lines of force not only an amendment of the
prevailing conceptual scheme of physical reality but an alternative to it.

> He never considers bodies as existing with nothing between them but
> their distance and acting on one another according to some function
> of distance. He conceives of all space as a field of force, the lines of
> force being in general curved and those due to any body extending
> from it on all sides, their directions being modified by the presence of
> other bodies. He even speaks of the lines of force belonging to a body
> as in some sense part of itself, so that its action on distant bodies can-
> not be said to act where it is not.… I think he would rather have said
> that the field of space is full of lines of force, whose arrangement
> depends on that of bodies in the field, and that the mechanical and
> electrical action of each body is determined by the lines which abut it.

From Faraday's conjectures and Maxwell's equations, inquiry into the nature of the physical world entered upon the crisis that was resolved in the new physics of the 20th century. Senior physicists, Lord Kelvin for example, sought to reconcile the electromagnetic field, so beautifully described in the Maxwell equations, with the classical mechanical models with which Maxwell had wrestled in formulating those equations. The younger physicists, in the words of Albert Einstein, were finding that "the achievements of the field theory had already become too striking and important for it to be exchanged for a mechanical dogma." It had to be accepted that "space has the physical property of transmitting electromagnetic waves and not ... bother too much about the meaning of this statement."

Ultraviolet catastrophe

On Einstein's own testimony, it was crisis attending progress on the more strictly classical question of heat that stirred him to action in 1905. The crisis arose, quite unexpectedly, from the convergence of thermodynamics with electromagnetism in the consideration of the light radiated by an incandescent solid body.

At temperatures above absolute zero, all solid bodies emit some radiation. The motion of their constituent atoms—heat, as demonstrated by Count Rumford—sets electromagnetic waves radiating from them. Around 50 degrees C the radiation becomes detectable as "warmth" by a human body at 37 degrees C. Jostling of the electrical charges on the constituent particles of the atoms generates the radiation, in accordance with Faraday's intuitive perception, Maxwell's equations and the Hertz experimental demonstration of electromagnetic radiation outside the visible spectrum. As temperature climbs, pandemonium in the aggregation of atoms becomes evident in the visible wavelengths—the object glows first red-, then white-hot. The change in color signifies the radiation of progressively shorter wavelengths toward the blue end of the spectrum [see illustration, page 67]. Real solid bodies radiate more strongly on the wavelengths of the line spectra of the elements of which they are composed. They absorb radiation more strongly on the same wavelengths. This understanding received theoretical treatment at mid-century from

Balfour Stewart at Cambridge and Gustav Kirchof at Heidelberg. The theory postulated an ideal "black body." This body, in its ideal blackness, absorbs radiation without reflection indifferently on all wavelengths. In equilibrium with the radiation, at exactly the same temperature with it, the black body is an equally perfect emitter of radiant energy on all wavelengths.

Instrument-makers approximated this hypothetical body in a hollow cylindrical or spherical chamber, made of platinum and blackened inside and out. Heated up, it emits radiant energy through a tiny hole for study outside. The instrument finds use today in the metallurgical industries; the color of its radiation, tuned to match that of the light in a furnace, gives precise measurement of the temperature of the melt.

Elegantly managed measurement of the intensity of radiation at different temperatures showed that, as the temperature rises, the peak of intensity moves across the spectrum from the longer toward the shorter wavelengths. At around the 500 degrees C of a hot iron, the intensity peaks outside the visible spectrum in the far infrared. That peak still lies outside the visible in the near infrared at the 2,000 degrees C of white-hot iron. The blinding light of the Sun, at 6,000 degrees C, reaches peak intensity in the yellow wavelengths of visible light. Smoothing out such measurements yields the ideal black body and the law relating heat and electromagnetic radiation. The energy radiated (and absorbed) by a black body increases as the fourth power of the temperature, as is seen in the blinding light of the Sun as compared to the light radiated by white-hot iron. Real bodies, depending on their composition, depart by larger or smaller margins from that rule.

The curves plotting the intensity of radiation across the spectrum for real bodies and for black bodies resembled those Maxwell plotted for the average velocities of atoms and molecules at different temperatures. This was a surprise. It had been generally accepted that the principle of equipartition of energy held for radiation as well as for matter. When curves were plotted for equal allocation of energy to each wavelength, however, they took off for infinity.

James Jeans at Cambridge University recognized that this unsatisfactory result must follow from overstretching of classical theory. By definition, 10 times as many more wavelets crowd in each order of magnitude of frequency upward from the red to the blue end of the spectrum. Equipartitioning of the energy across the spectrum would necessarily allocate more of the energy, therefore, to the higher frequency wavelets. By that reasoning, the ideal black body would radiate its energy outside the visible in the ultraviolet and even more numerous shorter wavelengths beyond. Jeans called this reduction of the classical theory to absurdity the "ultraviolet catastrophe" [see illustration, page 67].

In 1899, Max Planck proposed a correction to the principle of equipartition as applied to radiation. He showed that the energy of radiation bears a constant ratio to its frequency. The yellow light of sodium has a frequency of 5.1×10^{14} hertz. It carries a measured energy of 3.4×10^{-12} erg [erg: see page 79]. A single wavelet at that frequency has a duration or period of 1.9×10^{-15} second. Multiplying the energy by the duration of the wavelet yields 6.6×10^{-27} erg. This parcel of energy—designated h—is Planck's constant. Multiplying the frequency of the sodium light by h yields back its energy—3.4×10^{-12} erg. Multiplying by h the high-frequency violet light just inside the visible spectrum, at 8×10^{14} hertz, gives it an energy of 5.0×10^{-12} erg—nearly twice the energy of yellow sodium light. Whether violet or yellow, each wavelet carries the energy of the Planck constant, 6.6×10^{-27} erg.

Here was the explanation of the ultraviolet catastrophe. The law of conservation of energy requires that an atom absorb at least as much energy as it might radiate. The curve of the intensity of their radiation approximates the curve of the distribution of their velocities in accordance with Maxwell-Boltzmann statistical mechanics.

Planck called his constant the "quantum of action." In the word "action" he called on the early vocabulary of classical mechanics. Poetically, the term evoked the exertion of force by one body on another, the transfer of motion from one body to another. Planck's constant gives radiant energy its irreducible atom. Multiplied by the frequency of the radiation, it measures the value of the unit or quantum in which energy is conveyed at that frequency. Planck lived to see his quantum of action become the pivotal constant of modern physics, the breakthrough to radical revision of understanding of the nature of physical reality. In

1899, he was satisfied that his explanation had brought electrodynamics into agreement with classical mechanics.

The quantum: a particle of energy

Albert Einstein, as he recalled the moment many years later, did not share Planck's satisfaction "that the ... energy of radiation can be transferred only in such quanta." The idea was "in contradiction to the laws of mechanics and electrodynamics.... All my attempts ... to adapt the theoretical foundation of physics to this proposition failed, but completely.... The longer and more despairingly I tried, the more I came to the conviction that only the discovery of a universal formal principle could lead us to assured results."

At the time here in Einstein's recollection, from 1900 to 1905, he had received his diploma from the Federal Institute of Technology in Zürich and been refused a teaching assistantship there; he was making his way upward in the expert ranks of the Swiss patent office, and he was boldly employed in seeking mathematical answers to the deepest questions besetting the community of physics. Then in 1905, at age 26, Einstein published four papers. Two of them helped the community to agree at last that matter is composed of atoms—atoms that some physicists were on the verge of taking apart. The end of the repercussions of each of the other two papers is not in sight.

In the first, published in April, Einstein propounded his conception of Planck's quantum. Radiation, he boldly declared, must "possess a kind of molecular structure." In the second paper he showed, furthermore, that a quantum of radiation has momentum—an effective mass that is a function of its electromagnetic energy. Something like the Newtonian particle of light, a particle of matter-energy, conducts the transactions, he was saying, between radiation and matter.

Known but mysterious effects supported this radical idea. Einstein cited the "photoelectric effect" that had been observed but not understood. It is familiar now in the photovoltaic effect that powers a handheld calculator. Maxwell had pointed out the qualitatively different consequences of shining light of different wavelengths on different materials, such as the darkening of photographic film and draining of electric charge from metals. Hertz had been able to enhance the sparks in his receiving diodes by shining light on the metal at the spark gap.

Einstein's quantum of radiant energy triggers the photoelectric effect when it carries energy of a frequency at or above the threshold frequency of the energy that binds an electron in the atoms of the material at hand. Radiation of higher intensity at lower (or higher) frequency has no effect on the material. Yet, weak radiation at the resonant frequency excites a gush of electrons, measured in electric current from it, and the weakest radiation at that frequency has its best detector in the photoelectric effect in the same material. As James Jeans explained it, "You can't kill two birds [two different materials] with one stone; you can't even kill one bird [the wrong material] with two stones!"

Considering that each wavelet carries the same 6.6×10^{-27}–erg quantum of action, it might seem that it would take all of the waves arriving in one second to deliver in full the energy carried at each frequency. James Jeans wondered how a quantum could survive "massacre" in the two-wheeled turbine with which Fizeau had measured the speed of light. The imagination must learn to see radiation traveling as waves and acting as quanta. It may help to consider that the wavelet of blue light has a shorter duration than the wavelet of yellow light that packs the same quantum of 6.6×10^{-27} erg but delivers less energy. It can be seen that the blue wavelet delivers the quantum in a shorter time—with greater shock, what French artillerists call *brisance.*

People have lived with the almost equally contradictory notion of light as waves and as rays. In effect, the "ray" has proved to be a photon. Wave-particle duality is a reality to be lived with, much as it may escape comprehension or visualization [see illustration, page 113].

Special Relativity

In his third 1905 paper, Einstein undertook to state "a universal formal principle." *On the Electrodynamics of Moving Bodies* is better known as the paper on Special Relativity. Satisfying his criterion of "great simplicity," Einstein reset the foundations of physics on two "postulates." He gathered the mechanics of Galileo and Newton and the electrodynamics of Faraday and Maxwell in one internally consistent system. He brought space and time, matter and energy together as inter-dependent variables in a single continuum.

By the end of the second paragraph in that paper, Einstein has declared the two postulates to be sustained by the argument that follows, and has already disposed of absolute space. He observes first that, as Faraday had shown, motion of the magnet relative to the conductor and of the conductor relative to the magnet, indifferently, generate electric current and the electromagnetic field. "Examples of this sort, together with the unsuccessful attempts to discover any motion of the Earth rel-atively to the aether suggest that the phenomena of electrodynamics as well as of mechanics possess no properties corresponding to the idea of absolute rest. They suggest rather that ... the same laws of electrody-namics and optics will be valid for all frames of reference for which the equations of mechanics hold good."

Einstein's first postulate sets the idea of absolute space, in which Newton had anchored classical mechanics, aside. The laws of nature are invariant in the infinity of frames of reference in motion relative to one another that now succeed the single absolute. Accordingly, Einstein concluded, the aether will be seen to be "superfluous." Years later, reflecting on the malaprop uses of "relativity" in countless other venues, Einstein wished he had called it the "principle of invariance."

Relativity is "special" in this paper because it addresses constant motion only. This is the kind of motion Galileo found indistinguishable from rest. Echoing Newton's third law, the frames of reference in Special Relativity are called "inertial." Motion under acceleration, as by the gravitational force, had to wait for General Relativity in 1916.

For the second postulate of Special Relativity, Einstein offers no preliminary argument. He simply declares "that light is always propa-gated in empty space with a definite velocity, c, which is independent of the state of motion of the emitting body." This means that light emit-ted from a body will arrive at 3×10^{10} centimeters per second, regard-less of whether the emitting body is approaching or receding. In sum, the velocity of light is not relative. Absolute space and absolute time have yielded to the constant velocity of light.

Yet, experience shows that the velocity of light is no absolute.

Light travels in different media at different velocities, even if not enough different to amend the round number of 3×10^{10} centimeters per second. What is more, as Fizeau showed by another ingenious experiment in 1851, the velocity of light changes with the velocity of the medium in which it is traveling, as in a stream of water [see illustration, page 90]. Of great and wider significance, developed by Einstein in General Relativity, frequency and wavelength of electromagnetic radiation vary with variation in the gravitational field in which the radiation is traveling.

Relativity and the constancy of the velocity of light were ideas in wide circulation in the community of physics at the end of the 19th century. Einstein claimed no patent on either. Well schooled in the writings of Ernst Mach, he was ready to work with new ideas. Mach, memorialized in the "mach number" for the speed of sound and its multiples, was a founder of the Vienna school of "logical positivism." He was the most doctrinaire of those empiricists: nothing in the mind but what enters by the senses! He held in scorn the aether and the absolute space that the aether was contrived to save. Einstein was primed to embrace the consequences of their overthrow.

For his senior contemporaries, the overthrow came with the experiment of A. A. Michelson and E. W. Morley, first performed in 1881 when Einstein was three years old. Repeated over succeeding decades without success, its object was to detect the motion of the Earth against the "aether wind." It was to be detected much as a downhill skier senses "the wind." With the exquisite interferometer designed by Michelson, a beam of light was split, and the two daughter beams sent on different tracks, of identical length, precisely "tuned" to bring the beams together in resonance at the target [see illustration, page 91]. On one track, the beam was directed east, in the direction of the turning Earth and so "against the wind"; on the other, north or south, at right angle to the motion of the Earth. The aether, if it were there, would reduce the velocity of the beam pointed eastward. The experiment produced a null result. At the target where the two beams converged, no fringes (such as were first observed by Young [see page 56]) could be found to evidence any difference between them in velocity.

The consternation at this result in the community of physics has

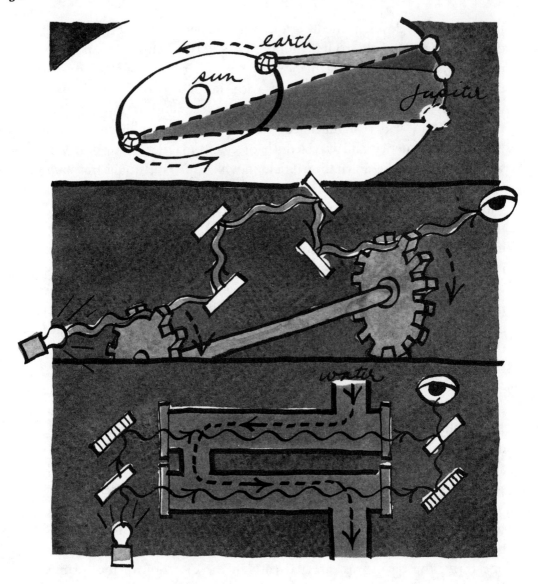

Velocity of light *was first calculated at 200,000 kilometers per second by Roemer in 1673 from differences in the time between eclipses of a satellite of Jupiter when Earth was on the near side and on the far side of the Sun [see page 54]. It was more accurately calculated at 300,000 kilometers per second by Fizeau from the angle at which the setting of teeth on two wheels stopped passage of light at known speed of rotation. He also found that motion of the transmitting medium adds to or subtracts from the velocity of light [see page 89].*

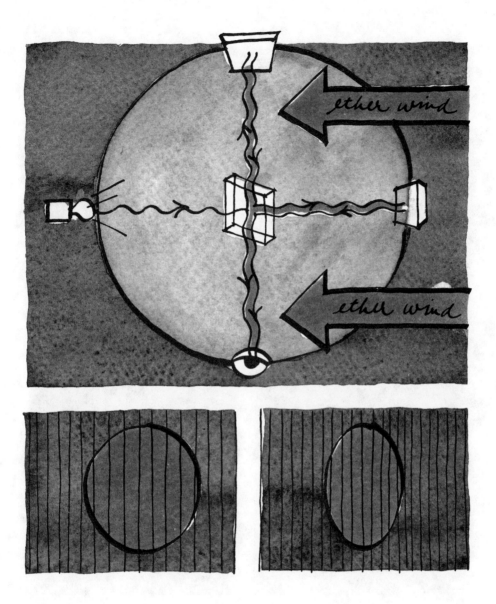

Attempt to measure aether "wind" *brought the most accurate measurement of the velocity of light [see page 89]. Michelson's interferometer split a beam of light onto two courses of same length to target instrument. On the course pointed in direction of Earth's rotation, the aether was to have slowed the velocity of light. Fitzgerald explained the null result as the consequence of contraction of dimensions aligned in the direction of Earth's motion. Lorentz explained it as contraction of spacetime in direction of motion [see page 94].*

Effective mass of the photon, _first postulated in Einstein's Special Relativity, explains the photoelectric effect: radiation of the right frequency and so the right energy, even at low intensity, kicks an electron free from an atom [see page 86]. The massless photon thus carries the "mass × velocity" mechanical force of momentum. Experimental proof, in the Compton effect, shows an X-ray photon in a billiard-ball rebound at lower frequency from a collision with an electron, which it kicks out of its orbit [see page 130]. In the symmetry of the interaction_

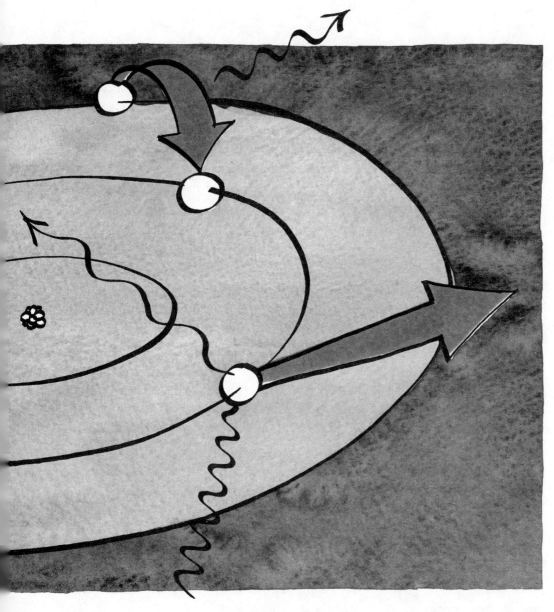

of light and matter, an atom emits light when an electron falls from a higher- to a lower-energy orbit. The mechanical force of light, "photophoresis," works on the cosmic scale. In interstellar space, starlight pushes dust particles together in clumps [top left]. In the shadow of clumps, the accumulation of dust masses proceeds to gravitational collapse in stars [see page 209]. Photons of sunlight erode the substance of a comet [inset] out into its tail, pointing always away from the Sun. This is a visible demonstration of the identity of matter-energy.

its measure in the radical proposal that came from the Irish physicist G. F. Fitzgerald. He proposed that the slowing of the beam against the aether wind could not be detected because the eastward track on which it was pointed is shortened by the motion of the Earth. By the "Fitzgerald contraction," all measuring rods would thus be shortened when pointed in the direction of their motion. The shortening of the eastward-pointing track in the Michelson-Morley experiment had concealed the decrease in the velocity of light imputed to resistance by the aether. In the prevailing conviction in the aether hypothesis, there were those who welcomed this proposition.

Fitzgerald-Lorentz contraction

The Dutch mathematical physicist H. A. Lorentz, quite independently, conceived the same explanation of the Michelson-Morley null result. He was not motivated, however, by the wish to save the aether hypothesis. On the contrary, he was ready to accept the velocity of light as a constant. He calculated that the contraction would be proportional to the square of the velocity of the body in question divided by the square of the velocity of light. The Lorentz-Fitzgerald contraction, accordingly, has a limit: at 3×10^{10} centimeters per second, the length of the meter stick must shrink to zero.

In 1904, the French mathematician Henri Poincaré called for a relativistic mechanics that would accept the velocity of light as "an unpassable limit." Hindsight shows that, if Einstein had not done it, his contemporaries would have succeeded in combining universal relativity of motion with constancy of the velocity of light in the creation of the new physics. They would not have done it with Einstein's *fraîcheur*— the term his biographer Abraham Pais found to characterize his unabashed address to the business at hand.

His two postulates, Einstein declares, "will suffice for the attainment of a simple and consistent theory of the electrodynamics of moving bodies." From them, as Euclid from his axioms, he draws theorems. Nothing in this artless procedure suggests the momentous implications for the physical world. Those surprises manifest themselves, each in its logical place, as the argument proceeds.

Having disposed of absolute space, Einstein addresses absolute time. In the positivist rigor of Mach his treatment of motion with respect to a frame of reference calls for a set of rigid x, y and z coordinates in the three dimensions of space, a rigid measuring rod, a clock and a clear understanding of what is meant by time. "If, for instance, I say 'That train arrives here at 7 o'clock,' I mean something like this: 'The pointing of the small hand of my watch to 7 and the arrival of the train are simultaneous events.'"

While simultaneity can be established—subject to unavoidable "inexactitude"—between events within a frame of reference, simultaneity between events in different frames of reference in constant motion relative to one another cannot be established. Owing to its finite speed, light takes time to convey information from one frame of reference another. Observed from either one of two frames of reference, lengths must in consequence be seen to be shorter and clocks to run slower in the other frame of reference. Within either frame of reference, because the measuring rod must contract along with what it measures (and lacking a Universal Clock against which to check the clock installed in either frame of reference), no such evidence of its own motion can be observed—as Michelson's disappointment plainly showed.

The Fitzgerald contraction thus emerges as a first derivation from Special Relativity. It somewhat complicates the transposition of measurements from one frame of reference to another. Against the background of absolute space and time, Galilean relativity has a simple procedure for such "transformations." In 1904, in order to fit such contraction—and the velocity of light that limits it—into the procedure, Lorentz had devised a whole new set of transformation equations. The "Lorentz transformations" are central to Special Relativity, and Einstein always cited them by their author's name.

Transformation of energy to mass

An important correction to Newton's law of inertia follows from the Lorentz transformations. The inertia of a body moving with respect to the observer increases subject to the same inverse ratio of the square of the velocity of the body to the square of the velocity of light that governs the contraction of the measuring rod. By that ratio, a force applied to increase the velocity of the body yields a declining increase in velocity. The energy of the impelling force goes increasingly, therefore, to increase the mass of the body (in the case of electrons brought to peak velocity in high-energy accelerators, up to 40,000 times). Energy of motion thus transforms to energy of mass. At the unattainable velocity of light, the energy-mass of a material particle goes to infinity. On the way to the "unpassable limit," the equation $m = v^2/c^2$ shows mass and energy to be interchangeable manifestations of one another, seesawing in interdependence with velocity. The two variables, space and time, in which velocity is expressed fuse in a physical variable: "spacetime."

From the frame of reference of the observer, it is the observed frame of reference that keeps the wrong time, cannot measure lengths correctly and gains mass. As they observe each other, they may well ask, Which frame of reference is right? The constant motion of the two frames of reference makes it impossible to bring their observations of one another into comparison. It would be necessary to turn one around and speed it up to rendezvous with the other; acceleration of the faster-moving frame of reference would take it out of constant motion and so out of its Special Relativity to the other.

If the observer has made careful measurements, then, as far as that observer can tell, the observing frame of reference is right. But, what is true of the frame of reference observed must also be true of the frame of reference observing. Its clock, its measuring rod, its mass must all vary with motion. The "experience" of such a thought experiment was enough for the alert physical intuition of Albert Einstein. Others have since been able to demonstrate tangible evidence of the interchangeability of matter and energy.

To confront his thought experiment with physical reality, Einstein considered the "peculiar consequence" that clocks run slower in faster-moving frames of reference. He managed the forbidden comparison of two clocks in constant motion in different frames of reference by accepting an approximate answer. "Let us assume," he says, "that the constant component of motion of a body moving at constant velocity on a continuously curved trajectory approximates constant motion within the terms of the principle. Comparison would then show that a clock at the equator must go more slowly, by a very small amount, than a precisely similar clock situated at one of the poles."

Einstein's thought experiment has been elegantly confirmed by comparing the timekeeping of clocks flown around the world with that of clocks on the ground. Clocks on satellites have confirmed as well the corresponding prediction of General Relativity.

Transformation of mass to energy

Having established the conversion of energy to mass in the "Kinematical Part" of his paper, Einstein shows that the same rules must obtain in the "Electrodynamical Part." The Lorentz transformations of the Maxwell-Hertz equations affirm what Ørsted, Ampère and Faraday had shown: "that electric and magnetic forces do not exist independently of the state of motion of the system of coordinates." The magnitudes of the forces are the function of relative velocities. Given that the velocity of light is constant, an observer in motion relative to a distant source of light perceives a shift in the frequency of the light—by the Doppler effect, familiar in the perception of approaching and receding sound— toward higher frequencies if approaching and toward lower frequencies if retreating. The degree of shift, either way, is a measure of relative velocity. That means, with Planck's constant in force, that the observer encounters quanta of higher energies if approaching and of lower energies if retreating. The amplitude of the energy is thus a function of relative motion: "to an observer approaching a source of light with the velocity c, this source of light must appear of infinite intensity."

Einstein considers next the transformation of electromagnetic energy to mechanical energy. He asserts that light shone on a perfectly reflecting mirror must exert pressure on it. He shows the pressure to be exerted by the inertia of the light quanta that corresponds to their energy and proceeds to calculate its force. As a function of their energy—that is, the Planck constant h multiplied by the frequency of the light—the inertia of the photons is equal to the square of that energy divided by the square of the velocity of light.

Photophoresis—the exertion of pressure by light—has a familiar demonstration in the tail of a comet. It is that pressure, reinforced by the solar wind of high-energy particles, that points the tail of the comet away from the Sun [see illustration, page 92].

In September 1905, Einstein published an emphatic after-thought: "Does the Inertia of a Body Depend on Its Energy Content?" There he shows: "If a body gives off the energy E in the form of radiation, its mass diminishes by E/c^2 ... so that we are led to the more general conclusion that: The mass of a body is a measure of its energy content; if the energy changes by E, the mass changes in the same sense by $E/9 \times 10^{20}$ [9×10^{20} centimeters second^{-2} = c^2] the energy being measured in ergs and the mass in grams." The unit electromagnetic force of 3×10^{-10} dyne, established by Weber, Kohlrausch and Maxwell, thus finds its complement in the gram of matter that binds 9×10^{20} ergs.

9×10^{20} ergs to the gram

Einstein had now repealed the law of conservation of mass. The cosmic balance sheet is balanced, however, by the supersymmetry of mass-energy conservation. To confront the consequence of Special Relativity on this side of the ledger, he proposed an experiment. "It is not impossible," he suggests, "that with bodies whose energy-content is variable to a high degree (e.g., with radium salts) the theory may be successfully put to the test."

Later large-scale experiments have left no doubt about the matter. The energy of 9×10^{20} ergs is approximately that released in the explosion of 17,000 tons of TNT. This was the force of the explosion at Alamogordo, New Mexico, on 16 July 1945, which forever impressed on the public mind that matter may transform to energy. The force of that explosion indicates that 1 gram of the 8 kilograms of plutonium in the weapon transformed to energy.

The devastation of two Japanese cities registered no more than a fraction of the total release of energy in the sphere of those explosions. No intelligible image conveys what is contained in the expression 9×10^{20} ergs. Recalling that an erg is the unit of energy that accelerates 1 gram to a velocity of 1 centimeter per second in one second and converting grams to more ponderable tons and centimeters to more visible kilometers, one may attempt to picture the acceleration of 9×10^8 tons to a velocity of 10 kilometers per second in one second. It helps, perhaps, to picture 9×10^8 tons as approximately five times the weight of the population of the United States.

Fields and ergs

It is less well understood that the transformation of mass to energy is universal. In the striking of a match, in the flame of a fire, in the light from a lamp, in the radiation from an antenna, in every chemical reaction, including those engaged in the reading of this page, the same transformation is proceeding—at a rate, of course, below detection.

Of all the 19th-century physicists, Michael Faraday would have been least surprised by the cosmos of Special Relativity. His fields and lines of force have the substance of the 3×10^{-10} erg of the unit electromagnetic energy. This thin spatial substance has its irreducible atom in the quantum of action, the 6.77×10^{-27} erg carried by each photon of electromagnetic radiation. It is continuous with the bodies of matter, in which matter-energy condenses to 9×10^{20} ergs per gram.

Faraday's lines of force can thus be seen, in Maxwell's words, "belonging to a body as in some sense part of itself, so that its action on distant bodies cannot be said 'to act where it is not.'" Action—electromagnetic radiation—as Faraday anticipated, moves in this space at the velocity of light. He would concur further in Maxwell's assessment of his views to the effect that "the mechanical and electrical action of each body is determined by the lines which abut it." Confirmation of those views awaited the work of Albert Einstein who, in longer historical perspective, succeeds Maxwell as Faraday's Newton.

Special Relativity did not win instant assent in the community of physics or overnight celebrity for its author. Recognition came, however, fast enough. According to Abraham Pais, it was Max Planck who first queried a point in the paper, not long after its publication. Planck was also the first to bring Special Relativity into discussion at a formal colloquium and the first to supervise a doctoral dissertation on it. In 1906, in no recognition of his extracurricular career, Einstein was promoted to the rank of "technical expert, second class," at the patent office.

The mathematician Hermann Minkowski, one of Einstein's teachers at Zürich and admired by his pupil, was the first to give formal recognition to the new conception of physical reality advanced by Special Relativity. Space and time stood no longer as absolutes outside of physics; they were now as much engaged as matter and energy in the events of the physical world and subject to equations that describe it. At the 80th Assembly of German Natural Scientists and Physicians at Cologne in September 1908, Minkowski laid out the new geometry of spacetime [see illustration, page 191], declaring:

> The views of space and time which I wish to lay before you have sprung from the soil of experimental physics, and therein lies their strength. They are radical. Henceforth space by itself, and time by itself, are doomed to fade away into mere shadows, and only a kind of union of the two will preserve an independent reality.... The validity without exception of [the two postulates of Special Relativity] is the true nucleus of an electromagnetic image of the world, which, discovered by Lorentz, and further revealed by Einstein, now lies open to the full light of day.

The percolation of Special Relativity into the consciousness of the community—the course of the revolution in physics—is thereafter charted by the succession and rising distinction of Einstein's academic appointments. In 1908 Einstein received the first, a marginal one, as *Privatdozent* at the University of Bern. He had meanwhile been publishing on a range of topics, including the first paper to bring quantum theory into the physics of the solid state. In 1909, the University of Geneva bestowed his first honorary degree. The University of Prague called him to a professorship in 1911. A year later, he was happy to return to a professorship at the Federal Institute of Technology in Zürich, his alma mater. There he was able to resume his important collaboration with the mathematician Marcel Grossman on difficulties he was encountering on the way to General Relativity. All at once, in 1913, Einstein was invited to a professorship without teaching obligations at the University of Berlin, a research appointment in the Prussian Academy of Sciences and the directorship of a Kaiser Wilhelm Institute.

By that time Special Relativity and the quantum had conquered the field in Europe. For a "paradigm shift" of such immensity, seven or eight years is not a long time. Given objective subject matter and common assent to reason, the community of science can arrive at consensus with alacrity. Einstein held his Berlin appointments until he emigrated to the United States in 1932, on the eve of Adolf Hitler's ascent to supreme power in Germany.

General Relativity

The publication of General Relativity in 1916, in the darkest days of the First World War, set off a different chain of events. Through the channels of communication in the international community of science, copies of that number of the proceedings of the Prussian Academy of Science found their way to subscribers on the far side of the Western Front. At Cambridge, Arthur Stanley Eddington began counting days to the total solar eclipse, scheduled for 29 May 1919 across the middle latitudes of the planet that would test a proposition from General Relativity. Eddington and Andrew Crommelin, of the Royal Observatory at Greenwich, then set up solar cameras on the island of Principe off the coast of Africa and across the Atlantic at Sobral in northern Brazil.

At a ceremonial joint meeting of the Royal Society of London and the Royal Astronomical Society in early November, the British astronomers were able to confirm the bending of starlight by the gravitational field of the Sun: it displaced the apparent position of a star by the nearly two seconds of arc predicted by General Relativity [see illustration, page 19]. This was twice the .75 arc-second displacement predicted by Newton. J. J. Thomson, president of the Royal Society, epitomized this historic occasion: "The deflection of light by matter, suggested by Newton [see page 54] ... would itself be a result of first-rate scientific importance; it is of still greater importance when its magnitude supports the law of gravity put forward by Einstein."

Einstein had to live the rest of his life as the most famous scientist in the world. He met the unsought obligations of oracle with humor and candor, seizing appropriate opportunity to speak for his causes and convictions. Once, in 1940, he agreed to the invocation of his oracular authority to influence a decision of state. For colleagues acting on the knowledge that Hitler's scientists understood what $E = mc^2$ meant, Einstein signed the letter they drafted for his signature that urged Franklin D. Roosevelt to start up what became the Manhattan Project.

Einstein's quest for the "formal principle" did not end with the triumph of 1905. For all its grasp of mechanical and electrical forces, Special Relativity did not comprehend the gravitational force and its apparent action at a distance. His work on electrodynamics had taken young Einstein into the mainstream. In his undertaking to comprehend electromagnetism and gravity in a single formal principle, he was alone.

To begin with, few contemporaries had active interest in the classical force of gravity. In his isolation from the academic community, Einstein was ignorant of the one significant contemporary inquiry into that force. Roland von Eötvös, at Göttingen, had found good reason for Newton's embarrassment in the dual, contradictory nature of mass. In the third law of motion, mass is the measure of a body's inertia or resistance to force; the same mass is the measure of a body's gravitational, attractive force. In an adaptation of the Cavendish experiment, Eötvös balanced heavy balls of two different metals at opposite ends of a torsion balance. These gravitational masses, indistinguishable in the Earth's gravitational field, proved to be equally indistinguishable as inertial masses aboard the rotating Earth on its orbit around the Sun. The balance gave no hint of torsion whether the Sun was "above" or "under" the Eötvös laboratory. To a high certainty, the two kinds of mass were equal. Knowledge of the Eötvös experiment might have accelerated Einstein's arrival at General Relativity. He would have had earlier confidence in his suspicion that the two forces, gravity and inertia, were not only equal and equivalent but one and the same.

"... happiest thought"

The patent office still preempted his days when, in 1907, Einstein had "the happiest thought of my life!" It occurred, he recalled, "in the following form ... for an observer falling freely from the roof of a house there exists—*at least in his immediate surroundings*—*no gravitational field*" (Einstein's emphasis). The gravitational field could, accordingly, have no more than a relative existence, like that of the magnetic field set up by the motion of electric charges. Weightless astronauts confirm the experience that Einstein imputed to his hapless observer; they sense a "switching off" of the Earth's gravitational field in the weightlessness attending their accelerating fall toward Earth's surface that just balances the constant motion imparted at takeoff and keeps them securely in orbit.

The force experienced in a gravitational field, Einstein concluded, must be a "pseudo-force," not unlike the centrifugal sensed by an observer in a frame of reference undergoing acceleration. The sensation of the downward pull of gravity is commonly replicated by upward acceleration of the observer in an elevator. The way to understanding of the gravitational force was by study of accelerating motion in spacetime.

The gravity of energy

That way opened with Einstein's consideration of "the gravity of energy." This necessarily pseudo-property stems from the inertia of radiant energy disclosed in Special Relativity. It followed that "the principle of the constancy of light in its customary version holds only for spaces with constant gravitational potential." To make sense of what followed from this radical perception, Einstein had to learn a new geometry.

The gravity of energy is observed in the gravitational redshift: the spectral lines radiated by elements in light from the Sun are shifted ever so slightly toward the lower-frequency, red end of the spectrum as compared to spectra of Earth-bound elements. Einstein saw that the shift could be explained by slowing of the frequency of the radiation at the point of its origin, in the gravitational field of the Sun. By similar increase in inertia, clocks are observed to run slower in frames of reference under constant or accelerated motion. The "clock," in the case of the Sun or other stars, is the frequency of radiation itself; slower frequencies of a given spectral line are observed in stars than in the same line on Earth. The stretching out of the oscillations in time reflects the drag, to that infinitesimal extent, of the solar mass on the inertia of light at its emission. Einstein found, accordingly, that "the principle of the constancy of light in its customary version holds only for spaces with constant gravitational potential."

In this coupling of electromagnetic and gravitational fields, Einstein could see that four-dimensional spacetime required no ordinary geometry. He was more physicist than mathematician. Late in life, distressed by quantum theory erected in baroque mathematical edifices, he protested, "it was not clear to me as a student that the approach to a more profound knowledge of physics is tied up with the most intricate mathematical methods." He turned for help to his fellow student and friend Marcel Grossman, now a professor of mathematics at Zürich.

Non-Euclidian geometry

Grossman introduced him to a four-dimensional geometry that was there on the shelf, ready-made to serve his purpose. It was the work, in the 1840s, of Georg F. B. Riemann, a student of Gauss. Spaces of four and more dimensions cannot be pictured in the mind's eye or on paper in two or even three dimensions; the reader may be assured that not even the most practiced geometer can visualize them. Such spaces are faithfully described, however, by the algebraic equations from their corresponding Cartesian, or analytic, geometries. As plane geometry must yield to spherical geometry in the mapping of sufficiently large regions of the Earth's surface, so Riemann framed his four-dimensional geometry to visualize the arrangement of objects in the distant cosmos. He understood that light came in from there at finite velocity. He did not, however, specify time as his fourth dimension.

The Cavendish experiment had demonstrated that matter is the source of gravitational fields, just as the electric charge was later shown to be the source of electromagnetic fields. The gravitational force was measured by that experiment at 6.67×10^{-8} dyne per gram. It does not exert the pull of the quantum-force fields, however, nor does it repel. Einstein's introduction of gravitation into Riemann geometry showed that force warping the structure of four-dimensional spacetime. The degree of warp, the tightness of the local curvature of the gravitational field, varies by a constant dyne per gram with the mass and density of matter present. Observed on a sufficiently large scale, the gravitational force is now seen to accelerate and decelerate the motion of masses on the four-dimensional gradients of spacetime in the gravitational fields surrounding those masses. By their motion masses send waves rippling in the fields at the velocity of light across the universe—exorcizing the action at a distance that so distressed Newton.

Star-track and parabola

Starlight betrays the curvature of spacetime. It travels the spacetime geodesic—at the velocity of light, the track of least distance and time—between the star and Earth. In the vicinity of the Sun the curvature of space is observed in the bending of the path of starlight that displaces the apparent position of a star. The starlight geodesic corresponds in plane geometry to a straight line, the shortest distance between two points. On the surface of a sphere, the geodesic is the line struck in the surface by a plane through the center of the sphere—the navigator's "great circle" [see illustration, page 19]. Starlight—deflected this way and that in the gravitational labyrinth of the cosmos—traces the ultimate extension of Galileo's parabolic trajectory. Recognition that this geodesic is the straight line in the Riemann four-dimensional (impossible-to-visualize) geometry twists the curvature of the trajectories of starlight into the gravitational fields of spacetime.

Like the surface of a sphere, but whatever its shape, spacetime is unbounded but finite. It was a work of Gauss himself that showed Einstein how to go about determining the ultimate curvature of the cosmos. The "intrinsic" curvature of a surface in three (or more) dimensions can be derived, Gauss had shown, from its geometry analyzed in two (or more) dimensions. Thus, the finding that the angles of a triangle add up to more than two right angles is a sure indication of positive, perhaps spherical, curvature of the surface in three dimensions. The intrinsic curvature of the spacetime cosmos, Einstein saw, is a function of its density. That remained to be determined by observation.

In his equations Einstein found another, distressing, uncertainty. They implied an expanding or contracting universe. To stabilize the universe, Einstein introduced a fudge factor into the equations. His "cosmological term" is the eye of the storm presently disturbing cosmology.

The deflection of starlight, confirmed by Eddington in 1919, provided a sufficiently romantic occasion for the celebration of General

Relativity and the apotheosis of Albert Einstein. In the mechanics of the solar system, however, Einstein had already found a surer test of his theory. From the time when observation of the solar system had made such exactitude possible, it had been known that Mercury's perihelion (the point at which its orbit comes closest to the Sun) shifts, or precesses, about 574 seconds of arc in a century. Calculation from classical theory had shown by 1900 that the precession should amount to 531 seconds of arc per century. The 43-second difference between the observed and calculated precession of Mercury coincided almost exactly with Einstein's calculation of Mercury's motion on the steepest (because closest to the Sun) gradient of curvature in the solar gravitational field.

General Relativity, embracing its vast territory, is not easily subject to laboratory experiment. The astronomical observations that sustain it satisfy by small margins the Einstein dictum: "Experience alone can decide the truth!" Throughout the century, physicists and astronomers have challenged General Relativity with experiments designed to yield more exact answers.

In the 1960s Robert Pound and Glen Rebka at Harvard University demonstrated the gravitational redshift in the modest gravitational field of the Earth. They showed that a radioactive isotope of iron emitted its narrow, almost "monochromatic," band of gamma rays at lower frequency at the bottom of a 70-foot tower than at the top. Their measurements came within 5 percent of the value predicted by the theory. At the Owens Valley Radio Observatory and the Goldstone Space Tracking Station in 1970, both under the corporate aegis of the California Institute of Technology, two groups of astronomers measured the shift in the apparent position of a quasar (a starlike object with a very large redshift) observed in the sky near the Sun. With the quasar radiating at radio wavelengths, there was no need to wait for an eclipse to darken the Sun, and there was time to achieve precision. The measurements came within 5 percent of Einstein's prediction, a considerable improvement on the 1919 stellar observations.

Astronomers at the University of Massachusetts found a virtual laboratory for the test of General Relativity in a binary pulsar (a pulsar is a pulsating radio source associated with a rapidly rotating neutron star).

With the 1,000-foot radio telescope at Arecibo, Puerto Rico, they kept the pulsar under observation on its orbit around its silent companion for a half decade. They found, in analogy with the shift of the perihelion of Mercury, a shift in the periastron of that orbit around the binary center of gravity and in precise accordance with the theory. This measurement could be made with precision, because the revolving member makes 1,100 trips around its companion in one Earth year. Numerous measurements of the gravitational redshift in the light from the two bodies strongly affirmed what solar observation and the Pound-Rebke experiment had shown in confirmation of the theory. Over the cosmic instant of four years they were able to establish and measure as well a decrease in the period of revolution. This squared with another proposition from the theory, which shows that mass is lost to waves of gravitational energy, as it is to electromagnetic.

Elusive formal principle

General Relativity had served meanwhile as the general theory that organized the metamorphosis of astronomy to observational cosmology, which began with the showing by Edwin Powell Hubble in 1924 that the Andromeda Nebula hangs in space outside the Milky Way, a galactic universe thronging with as many billions of stars as the Galaxy itself. Soon galaxies were counted in billions. A new cosmological redshift added to the gravitational showed this abruptly expanded universe to be expanding. For all of this and new knowledge of the cosmos that began to come in on electromagnetic frequencies outside the visible, General Relativity was ready.

Special Relativity had unified mechanics and electromagnetism. It had collapsed space and time in spacetime, a physical variable of the electromagnetic field, and materialized the action of the electromagnetic wave in the quantum. General Relativity fused gravity and inertia. It gave spacetime the structure of the gravitational fields exerted by the presence of mass. This field theory recognized, however, no quantum of action. No bridge crossed the two conceptual universes. Understanding required that they be united or their separate existence explained. Deep in this void was the antithetical duality of particle and field. For Einstein this gap in understanding was the important question. The mining of the frontier opened up by his own work on the quantum of action that engrossed his contemporaries did not command his interest.

Most troubling to Einstein was the acceptance, in quantum theory, of probability as sufficient in the accounting of events in the physical world. He remained committed to the classical mission of seeking the laws that determine those events, firm in the conviction that they are so determined. After 1916, Einstein had no part in the penetration by quantum theory of the nearest and farthest reaches of nature open to experience. Physics had, after all, gone far on the incomplete premises of the Newtonian order. "Momentary success," he wrote, "carries more power of conviction for most people than reflections on principle."

Einstein spent the rest of his working life attempting to complete the statement of the universal formal principle that he broached with such confidence in 1905. In his Generalized Theory of Gravitation, published in 1949, his 70th year, he hoped he had found a way at last to bring the electromagnetic field of Special Relativity into unity with the gravitational field of General Relativity.

The theory, Einstein conceded in the account he wrote for SCIENTIFIC AMERICAN, was "heuristic." It did not yet yield "conclusions that can be confronted with experience." It would invite, he hoped, the development of "new mathematical methods."

Einstein's successors have now developed the confidence to address his purpose. They hope to secure the unity of the forces of nature, not by the mathematical methods he had in mind, but in quantum theory. That requires, ultimately, demonstration of the quantum of the gravitational force.

3

Light and Matter

*There is a limit to the fineness of our powers of
observation and the smallness of the accompanying disturbance
—a limit which is inherent in the nature of things.*
P. A. M. Dirac

Max Planck established the quantum of the interaction of
light and matter. In the quantum, Albert Einstein found
the equivalence of energy and matter; in the invariant
velocity of light, he found space and time fused in the physical reality of
spacetime. On the foundations of physics thus reset, their successors
proceeded swiftly to comprehension, at mid-century, of the physical
order underlying diversity in the nearer world familiar to our senses.

Quantum electrodynamics accounts in full—its acronym is
QED—for the role of the electromagnetic force. It explains, therefore,
the structure of the atom external to its nucleus and arrays the 92 varieties of naturally occurring atoms in their periodic table. Therewith it
accounts for the self-organization of the atoms in chemical compounds.
QED, above all, accounts for the interaction of light and matter.

Sunlight captured by photosynthesis in the chemical bonds of
elementary organic molecules sustains life on Earth. Another interaction
of light and matter selectively reflects to the eye each of the colors of the
visible spectrum from the world around. At the retina, still another
interaction excites vision. Life is quantum transactions among the large
and small molecules of the living cell. In QED, physicists have arrived
at a final theory of the narrow cosmic circumstance of human existence.

During the second half of the century, physicists took the conceptual apparatus of QED into inquiry outside its own very considerable territory. Their confidence that they will find order and simplicity underlying the complexity of nature was at times daunted. First they encountered two hitherto unknown forces of nature. Then the "atom-smashers" that were to dissect out the ultimate particle produced instead eruptions of brief-lived particles. In their decay, physicists encountered for the first time an asymmetry in the structure of the natural order.

Resolution of the disordered experimental findings in the theory of quantum chromodynamics has redirected the mission of physics. From the quest for the irreducible particle of matter inquiry has turned to the question of how matter came to be. Converging here with cosmology, it has reached deep into the past, close to the very beginning of the observed universe. There it has brought in sight a ghostly symmetry. Symmetries are central to perception and understanding. One familiar symmetry balances equations on their two sides. Symmetries are perceived everywhere in nature, in radial symmetry as in jellyfish and in the bilateral symmetry of the vertebrates. They are sought in the underlying order of the physical world and found in the conservation laws. The new ghostly symmetry is the ultimate. So symmetrical as to have no discernible feature, it becomes manifest only after it is broken. The primordial kernel of the universe had only high density and high temperature. In the next instant, it appears, a succession of symmetry breakings—phase transitions, as water to ice or to steam—let go the matter-energy, particle-wave dualities known now to human experience.

The construction of the QED conceptual apparatus brought a succession of the brightest and toughest minds into collaboration across generations through the first half of the 20th century. Its equations meet tests by experiment within the smallest margins of error in human experience. Technologies drawn from the theory itself give these experiments their unprecedented exactitude. In the solid-state electronics of the computer chip and the fiber-optic cable alone this supreme intellectual achievement of the 20th century has set technological, economic and cultural change in unprecedented acceleration.

At Columbia University in the 1930s, I. I. Rabi established one of the first centers for cultivation of the new physics in the United States. After half a century, he confessed:

> I feel my generation and the current generation have not devoted the time and profound effort to make the extraordinary phenomena of relativity and quantum mechanics accessible to the intelligent, educated person. I am sure it can be done because that's the way I understand it. This failure to make the subject accessible to the general educated public has, to my mind, resulted in driving science, particularly physics, out of the secondary schools. Unless a great effort, a really great effort, is expended in this direction, the outlook for the future is bleak.

Too few people have known the deep unifying connection among the diversity of things, processes and experience that is held in understanding by QED. To share this experience in some way is a reason to be alive at the beginning of the third millennium.

Never at rest

Quantum physics takes inquiry into physical reality at the magnitude of the atom (10^{-10} meter) and of atomic events (10^{-21} second). In the busy interior of an atom, at 10^{-30} cubic meter, particles of 10^{-36} cubic meter can hardly be imagined standing in one place longer than 10^{-20} second. In the cubic-meter world of common experience, the scaled-up time would be a billion years. Down there, the particles are forever—in the lecture refrain of R. P. Feynman at the California Institute of Technology—"jiggling, jiggling." So long as they do not jiggle in unison the laws of classical physics remain in force up here. Of the corresponding apparent disorder, it is the mission of quantum physics to make sense. To that end, it has established that quantum events observe the essential conservation laws of classical physics.

To this netherworld David Bohm, an American physicist expatriated (in the "atomic secrets" tumult of the 1950s) to Birkbeck College, London, opened a wide door: "at the quantum level of accuracy, an object does not have any 'intrinsic' properties (for instance wave or particle) belonging to itself alone; instead, it shares all its properties mutually and indivisibly with the systems with which it interacts. Moreover, because a given object, such as an electron, interacts at different times with different systems that bring out different potentialities, it undergoes continual transformation between various forms (for instance wave or particle form) in which it can manifest itself."

By "for instance," Bohm perhaps allowed that events at the quantum scale in spacetime might involve forms other than the wave and the particle perceptible to human beings. He did not allow that this distant world can be regarded as disjoined from the near and more familiar one. On the contrary, Bohm insisted: "It is only in terms of well-defined classical events that quantum-mechanical potentialities can be realized. Moreover, this interdependence is reciprocal, for it is only in terms of a quantum theory of its component molecules that the large-scale behavior of a system can be understood. Thus, large-scale and small-scale properties are both needed to describe complementary aspects of a more fundamental indivisible unit, namely, the system as a whole."

That has to be the case. The quantum world would otherwise be out of reach of experience. The instruments that excite quantum events make them perceptible in sought-for mechanical, electric or electromagnetic effects. Surfacing, as they do, in such effects, quantum events are to be perceived at all times everywhere.

High in the sky, solar photons that make it through the life-protecting ozone layer excite oscillation in the electrons on outer orbits around molecules of nitrogen and oxygen. An oscillating electric charge, as Faraday showed, radiates electromagnetic waves. Away from the Sun, a clear blue sky can be seen to be luminous; it is reemitting sunlight. At the ground, the white light of the Sun is largely absorbed and reradiated as heat. From the absorption of the peak energy of sunlight, in the yellow-red wavelengths, by molecules of chlorophyll, nearly all life on Earth takes its energy. The leaf reemits the unabsorbed sunlight most strongly in the green wavelengths. Atoms and molecules in leaves and flowers, paint and dyestuffs, gold and silver thus selectively reradiate the frequencies of sunlight perceived by human beings as the colors of the spectrum. "Reflection" is reemission. There are no passive intermediaries in the cascades of events in the quantum world.

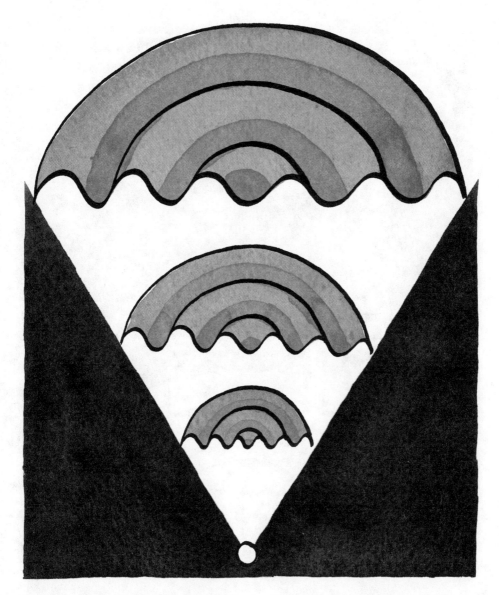

Wave-particle duality is pictured here as the enlargement of the wave structure of a putative particle taken as the point from which the structure is enlarged. Observed in experiments with light waves and matter particles, the duality, so offending to common sense, requires recognition of the particle or wave as an event rather than a "thing." In this graphic metaphor, the particle might have a diameter of 10^{-1} millimeter, the wavelength of a photon at 10^{12} Hz or a diameter of 10^{-24} millimeter, that of a low-voltage electron.

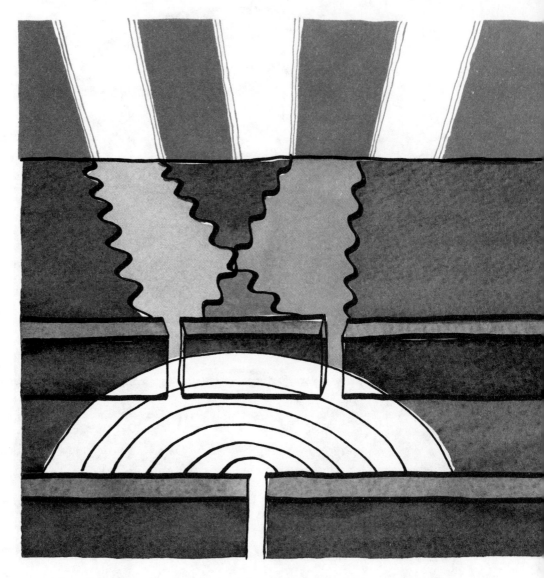

Wave nature of light *was demonstrated in 1817 in this experiment by Young [see page 56]. Into a closed chamber, he shone light through a slit. Inside the chamber, the light flooded through two slits. On the far wall of the chamber the light made a pattern of bright and dark vertical fringes. Young showed that when waves from the two slits, traveling different distances, arrive in phase at a given point, they reinforce to make the bright fringes; when they arrive out of phase, they cancel each other, making dark fringes.*

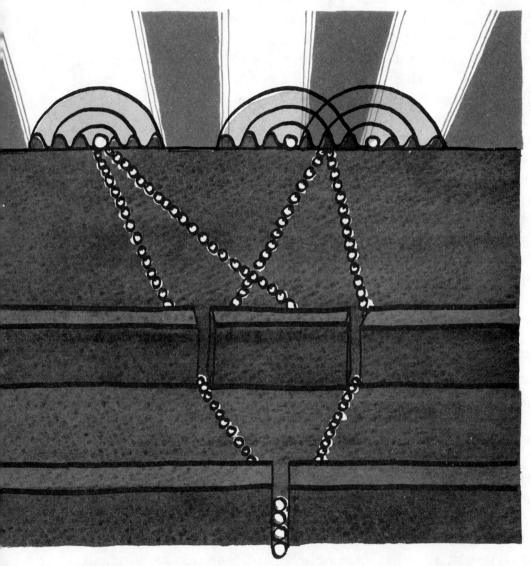

Particulate nature of light *was demonstrated in 1908 in this experiment by Taylor [see page 116]. Into a closed chamber he allowed light from the dimmest source he could manage to leak one photon at a time, perhaps hours apart. On the far wall, over days of exposure, photographic film registered the classic Young fringes. Each photon, Taylor and confirming experiments by others showed, traveled through the two slits (its wave nature) and then, traveling different distances to the film, must have registered in or out of phase with itself.*

Newton found it easier to think of light as corpuscles of some sufficiently small mass and possessing momentum [see page 54]. Early in the 19th century, Thomas Young troubled classical physics by demonstrating the wave nature of light [see page 56]. In 1908, Geoffrey I. Taylor at Cambridge University performed an experiment that caught the photon as both particle and wave. He exposed photographic film through two slits to a light source so dim that, calculation showed, photons would arrive one at a time, as much as an hour apart. After weeks of such exposure, the film exhibited the Thomas Young wave-interference pattern [see illustration, preceding page]. Any doubt that photons are simultaneously wave and particle was dispelled by an experiment by Robert Pfleegor and Leonard Mandel at the University of Rochester 50 years later. Sensitive counters showed that photons of monochromatic laser light were arriving one by one. It could be shown that the intensity of the interference pattern corresponded to the number of photons arriving, one by one, at a given point. The wave property had to be associated, therefore, with each photon, not the ensemble.

The quantum behavior of some atoms can be brought into view by restraint on their jiggling. Low temperature imposes such restraint, as can be seen in the phase-transition of all elements and some compounds (H_2O, for example) from the gas to the liquid to the solid state in step with the reduction of temperature. Helium remains a gas down to the very low temperature of 4.2 degrees K. At Cambridge University in the 1930s, Pyotr M. Kapitza showed that helium, cooled down another 2.2 degrees toward absolute zero, makes a unique transition to superfluidity. Moving now in otherwise improbable unison, helium atoms climb the wall of their container as a thin film. They fill a test tube to the level of the bath of helium in which the test tube is placed. They conduct heat as a steep wave, or pulse, with the speed of sound. In a universe that cold, there would be no observers to record such violation of the laws of classical physics. Those laws obtain in the narrow temperature regime that permits the processes of life on Earth [see illustration, page 7]. The surprises to be found elsewhere in the physical order, it must be understood, come naturally.

Foremost in public attention, as quantum theory made its conquest of the physical world, has been inquiry into the "strong" force that binds the nucleus of the atom—its "strength," by now, well popularized. This third force of nature, first measured in 1931, acts inside the 10^{-14}–meter diameter of the nucleus. There it holds the nuclear particles together against their mutual repulsion by the electromagnetic force, approaching maximum at that close range. Also encountered in the 1930s was another short-range force, the so-called weak force. Its function may be described as opposite to the strong. The weak force engages in the emission of particles from the nucleus.

It is perhaps remarkable that two forces of nature could have gone unrecognized until so recently. But their discovery could not have happened sooner. People had to master first the mechanical and electromagnetic forces. Only then could they fashion instruments to reach into the dimensions at which the short-range forces act.

Over the past 25 years, theoretical physicists have brought the plurality of forces and the multiplicity of particles into some order. They have fitted the two short-range forces into the conceptual scheme of QED. Each of the new forces now has its quantum theory. The quanta of their action—cousins to the photon, the quantum of the electromagnetic force—have been demonstrated by experiment.

The weak force is already united in theory, validated by experiment, with the electromagnetic force. The union of the "electroweak" force with the strong force is proposed in the "grand unified," called GUT, theories. Experiment to test this proposal calls, however, for energies 10 or more orders of magnitude beyond the reach of presently demonstrated accelerator technology.

No irreducible atom

The quest for the irreducible ultimate atom has settled for the finding that there is no such thing. Ordinary matter reduces not to one but to four different particles, two each of two kinds, quark and lepton, that have so far proved irreducible. If the finding of four particles fails the classical ideal of one, the four are enough. All the diversity of nature in common experience arises from two quarks and the two leptons in their interaction through the four forces.

These particles, it has to be admitted, have their antiparticles. But antiparticles are not part of the structure of ordinary matter. They make their appearance only rarely in nature. Manufactured in particle accelerators, they serve in study of the extraordinary matter from which, it has been learned, the particles of ordinary matter descend. If the quest ever reduces to one kind of particle—and the fusion of quarks and leptons is forecast in GUT theory—it will surely have an antiparticle.

As will be seen, the quarks and leptons of ordinary matter, together with two more generations of increasingly massive and energetic quarks and leptons, account for the multiplicity of particles that confounded physics in the 1950s. For micro-instants, at their microscopic targets, accelerator beams were creating conditions found in the interior of stars and in cataclysmic events in the lives of stars and galaxies. As the power of accelerators increased, the energy of events at their targets were generating conditions that prevailed in an earlier, smaller, denser, hotter universe.

The approach of particle physics, via the interior of the atom, to earlier times in the history of the universe brought its work into convergence with cosmology. The theory of General Relativity had predicted an unstable universe. The first fixes on galaxies external to the Milky Way showed the universe to be expanding. Backward runs of the tape of universal history then showed that the younger universe must have been not only smaller but hotter, pervaded with the energies attained momentarily at the targets in the physicist's accelerators. The highest energies attained thus far reach, it is calculated, into the first second of the history of the observed universe. Deeper in the first second, particle physicists expect to find the conditions projected in GUT theory that unite the three quantized forces in one primordial force.

In those same first instants of universal time, over the course of the Big Bang, cosmologists seek events that could account for the observed large-scale structure of the universe. With particle physicists they have begun the collaboration that addresses the ultimate unification: how to bring the particle physicists' three quantum forces into union with the cosmic force of gravitation. That union is predicted on the energy scale two orders of magnitude beyond that at which the trinity of quantum forces fuse in one. Failing the ultimate union, so far beyond the reach of experiment, the "standard model" of particle physics and General Relativity separately comprehend all that experiment and observation can say about the physical world.

Not what but whence

No longer: What is the ultimate particle? The question is: Whence the particle? Matter and energy being one, the question more precisely is: Whence the rest mass of the particles of matter?

Rest mass is distinguished from the mass that particles acquire from energy invested in their motion, as in a particle accelerator. Energy invested to accelerate the motion of particles goes to increase their mass because no amount of energy can make them exceed the velocity of light. The photon, the quantum of electromagnetism, has no rest mass; in its flight at the ultimate velocity, it is never at rest. It acquires effective mass, however, from the energy that generates it, in accordance with the Special Relativity equation, $m = E/c^2$—the inverse of the familiar $E = mc^2$. A particle of matter at rest or in constant motion possesses the classical Newtonian potential energy of its location in a gravitational field; it has rest mass. That is what makes it a particle of matter. Special Relativity has shown that the rest mass possesses another potential energy: that of its transmutation to radiant energy.

By 1990, the growing power of theory and of the accelerators to test it had brought the origin of the rest mass into sight and nearly in reach. It was to be found in a breaking of the ghostly symmetry from which the universe took its design. The last symmetry-breaking at creation—the first in reach of experiment—was to have set the weak force free from the electromagnetic. A so-called Higgs particle, the quantum of the then-prevailing cosmic energy, would have drawn its rest mass from that event. The test of this proposition appeared to be in reach of the enormous energy that was to be generated by the Superconducting SuperCollider—4×10^{13} electron volts, 40 trillion electron volts, 40 TeV—then under construction at Waxahachie, Texas.

Midway in its construction, in 1994, the SSC was cancelled by the 103rd Congress. The case for public support of physics by the U.S.

government was never taken to the electorate. Awed by the might of the nuclear weapon, the military establishment had more than enough surplus in huge Cold War weapons-development budgets to finance physics in the universities and in new national laboratories. The Department of Energy, custodian of the paramilitary atom, continues to supply their principal support. As Erwin Schrödinger observed in 1953, such expenditure by the "various state ministries of defense" was to have hastened the "realization of the plan for the annihilation of mankind which is so close to all our hearts." The electorate should long before have heard from the community of physics the testimony Robert R. Wilson—architect of the country's presently most powerful accelerator at Batavia, Illinois—to the U.S. Congress: "It has nothing to do ... with defending our country except to make it worth defending."

Discovery of the electron

The atom did not command the attention of physicists until the discovery of the electron late in the 19th century. To that discovery, Michael Faraday's work on electrochemistry supplied the decisive clue [see page 70]. In 1881, the physicist Hermann von Helmholtz observed: "Now the most startling result of Faraday's laws is perhaps this: if we accept the hypothesis that the elementary substances are composed of atoms, we cannot avoid concluding that electricity also, positive as well as negative, is divided into elementary portions which behave like atoms of electricity."

Others were already persuaded to this idea. A vacuum pump designed by the German glassblower Hermann Geissler made it possible to draw down the pressure inside glass vessels to small fractions of an atmosphere. With two electrodes setting up high-voltage potentials across such vessels, William Crookes caused the rarefied gases inside to glow. The color of the glow depended upon the kind of gas. ("Neon lights" today excite economic activity all around the world.)

Setting the voltage sufficiently high and the gas pressure sufficiently low, Crookes saw the diffuse glow gather in a narrow ray that emanated from the negative electrode. Deflecting this "cathode ray" from its course in a magnetic field, he discovered the ray to be negative-

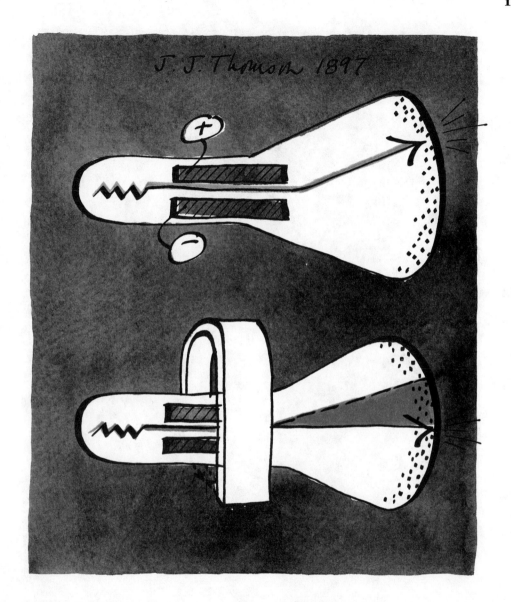

Charge-mass ratio of the electron *was established by J. J. Thomson in 1897 [see page 123]. Crookes had shown that a glowing beam would emanate from a high-voltage source in a vacuum tube [see page opposite]. An electric field deflected the beam toward its positive pole [top], evidence that the beam was composed of negative particles. Measuring the force of a magnetic field that countered the deflection, Thomson showed the ratio to be 1.77×10^{-8} of the mass Faraday had shown for positively charged hydrogen ions [see page 71].*

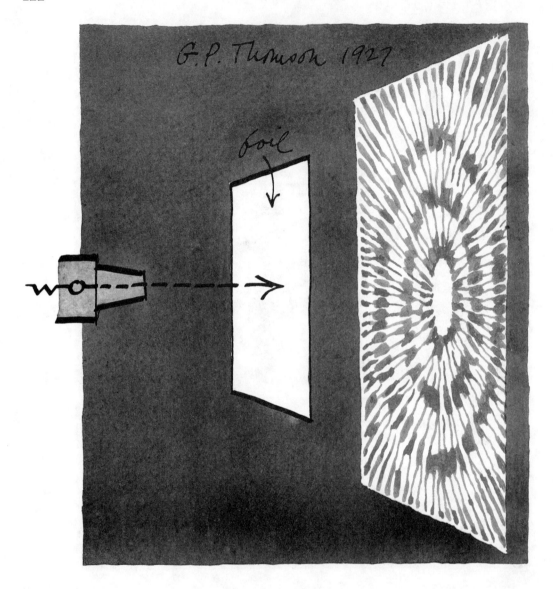

Wave nature of the electron *was established by this experiment in 1927 by G. P. Thomson, son of J. J. Thomson, who demonstrated the particulate nature of the electron [see illustration, preceding page]. An electron beam penetrating a metal foil registered on photographic film a pattern of interference fringes like those Young and Taylor found for light [see illustration pages, 114–115]. Davisson and Germer demonstrated the wave nature and correspondence of electron wavelength to voltage by diffracting electrons from a crystal lattice [see page 42].*

ly charged. In France, Jean Perrin found that a metal plate placed in the ray cast a shadow in the greenish glow and acquired a negative charge. On what an increasing number of observers surmised to be negatively charged particles traveling in the cathode ray, the Irish physicist George Johnstone Stoney had, in 1874, bestowed the name "electron."

In 1897, in the Cavendish laboratory at Cambridge University, J. J. Thomson established the existence of the electron. He installed a second pair of electrodes in a cathode ray tube at a right angle to the passage of the ray [see illustration, page 121]. The electric field deflected the ray in the direction of its positive plate. He measured then the electric force required to overcome deflection of the ray by a magnetic field introduced in opposition to it. Thereby, Thomson secured the ratio of the charge to the mass of what were now clearly particles in the ray. He found the ratio to be 1.77×10^8 coulombs per gram.

At Leyden, Pieter Zeeman placed a sodium flame between the poles of a strong electromagnet. With a diffraction grating, supplied perhaps by H. A. Rowland, he observed that the bright, double yellow line of sodium was broadened by the force of the field. Variation of the strength of the field caused the lines, in the case of some elements, to split. H. A. Lorentz joined Zeeman in calculating the ratio of the strength of the field to the change, wavelength by wavelength, in width of the lines. That yielded another measurement of the charge-to-mass ratio of the electron. It agreed nicely with Thomson's results. The experiment demonstrated conclusively as well that electromagnetic radiation originates in the motion of the negatively charged electrons.

The charge-to-mass ratio did not yield separate values for the mass and the charge of the electron. The ratio could be compared, however, with that of hydrogen, which Michael Faraday established in the 1830s. He had shown that the passage of 96,500 coulombs would accumulate a gram of positively charged hydrogen at the negative pole of the chamber [see illustration, page 60]. It could be seen that collection of a gram of electrons required nearly 2,000 times as much electric current. The charge-to-mass ratio of the electron (1.77×10^8) was thus nearly 2,000 times larger than that of hydrogen (9.65×10^4). [$1.77/9.65 = 1.834 \times 10^3$.] The electron had to be 2,000 times lighter, therefore, than the hydrogen atom, the lightest and smallest of the atoms. Now more precisely set at 1/1,836 of the proton nucleus of the hydrogen atom, this tiny rest mass establishes the electron as a particle of matter as well as a particle of electricity.

The atom was thus shown to consist of component particles. The opposite electric charges of the tiny electron and of the principal mass of the atom bound them together in neutral charge. In the cathode ray tube the electric potential dragged electrons into the ray from the metal in the cathode. The next question was where to fit the electron in the structure of the atom.

X rays and spontaneous radioactivity

Two unlooked-for discoveries opened the way to evidence on this question. Wilhelm Konrad Roentgen in Munich found that fluorescent materials light up in a cathode ray tube, especially when he placed a metal plate in the ray. Attempts to block the radiation with opaque materials showed it to be highly penetrating. Failed attempts to deflect it in a magnetic field showed it to be true radiation, an extension of the electromagnetic spectrum beyond the ultraviolet. Roentgen had discovered X rays. In 1895, he demonstrated their potential usefulness with the first X-ray photograph—of the bones of Frau Roentgen's right hand.

Roentgen's discovery prompted Henri Becquerel in Paris in 1896 to find out whether the fluorescent materials he was investigating might emit X rays along with visible light. On photographic film shielded from sunlight by heavy black paper he placed pieces of a uranium-containing mineral, looking to the sunlight to excite fluorescence. He found rewarding spots on film. On a sheet of film overlooked in a desk drawer with a piece of the mineral resting on it, he discovered an even more rewarding fuzzy black image of the whole piece of mineral.

Becquerel's finding was later confirmed by Pierre and Marie Curie's discovery of two other radioactive elements: polonium and then radium. Each turned out to have a characteristic rate of emission. Some average percentage of the atoms present in a sample were emitting at all times. For this spontaneous radiation there was no ready explanation. And there was no imaginable way to explain why a particular atom in a sample should emit radiation at a given time. For such indeterminate as well as spontaneous events classical physics had no precedent.

Ernest Rutherford, who had come from New Zealand to apprentice with J. J. Thomson, trained his formidable attention on this radiation. He found at least three qualitatively different kinds of radiation emanating from uranium and thorium. What he called "alpha" radiation he determined to be particulate—heavy, positively charged particles. His "beta" rays turned out to be Thomson's electrons. The "gamma" rays he identified as true radiation, of higher frequency and energy than Roentgen's X rays [right]. With his McGill colleague Frederick Soddy in 1903, Rutherford then showed that upon emission of the particulate alpha or beta radiation, "the resulting atom has physical and chemical properties entirely different from the parent atom." The parent elements had transmuted to chemically different elements. Rutherford and Soddy had shown how the alchemists' ambition to turn dross into gold might be realized. They had affirmed also the conjecture of William Prout in 1815 [see page 56] that the elements in all their variety must be composed of different numbers of the same ultimate atom, most likely hydrogen.

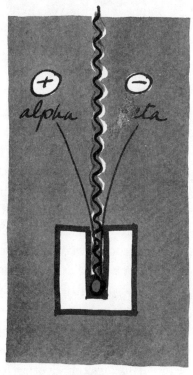

Not long after, at Manchester University in England, Rutherford identified his alpha particles as ions of helium, with the mass of four hydrogen atoms. In these particles he recognized a means for close-up inspection of the positively charged principal mass of an atom.

Gifted with the visual imagination of a Michael Faraday, Rutherford was at ease in the subterranean dimensions at which he worked. To a skeptic of the materiality of the electron, he declared, "Why, I can see the little buggers as plain as I can see that spoon in front of me!" Mutual repulsion of the positive charges of the target atom and the alpha particle, Rutherford saw, would scatter the alpha particles from the target.

By measuring the angles at which they scattered, he could get some idea of the size of their target.

Scattering experiments remain the principal tool of particle physics. The particle accelerators, scintillation counters and computers that do this work today simply scaleup—but by orders of magnitude—and automatize the experiments first devised by Rutherford. In 1911, he contrived to get a beam of alpha particles to issue from a hole drilled in a block of lead in the bottom of which he embedded a few grains of a radioactive element. He aimed the beam at a target of gold. To detect the rebounding alpha particles, he placed a fluorescent screen, like a filter, on the lens of a low-power microscope mounted to rotate at measured angles around the target [see illustration, page 146].

The model of the atom advanced by Thomson—with the negative charge bound to the surface and the positive charge diffused throughout the volume of the atom—would have yielded a pattern of low-angle scattering. Rutherford was astonished to find some alpha particles rebounding at high angles, some on nearly reverse course. "It was as though you had fired a 15-inch shell at a piece of tissue paper and it came back and hit you!"

Rutherford's calculation showed that "the positive charge of the atom is concentrated into a minute center or nucleus and the compensating negative charge is distributed over a sphere of radius comparable with that of the atom." He calculated the diameter of this "nucleus" at 10^{-12} meter. Rutherford had found Helmholtz's conjectured positively charged particle of electricity. Not only did electricity come in particles; it now appeared that the atom was made of particles of electricity.

The diameter of the atom had been established at 10^{-8} meter by experiment proposed in one of Einstein's first two 1905 papers. Later measurement has set it at 10^{-10}. At that dimension compared to 10^{-12}, solid geometry shows the nucleus occupying a tiny fraction of the volume of the atom ($10^{-10} - 10^{-12} = 10^{-2}$; $10^{-2} \times 10^{-2} \times 10^{-2} = 10^{-6}$). The electrons take up an even more negligible fraction of it. Inside the atom Rutherford had found a vacuum, a vast void. Scaled up to the size of the Sun, the nucleus would find its nearest satellite electron orbiting out beyond Pluto. The mass of the atom had to be concentrated in its tiny nucleus. Its density could be calculated at an unimaginable 10^{14} times the density of water. A drop of the nuclear stuff would outweigh all the ships on the oceans.

It was difficult, of course, to see substances of everyday experience composed of solar-system atoms. Arthur Stanley Eddington, as an early popularizer of the new physics, cited his writing table: "mostly emptiness ... in that emptiness ... numerous electric charges rushing about with great speed ... their combined bulk ... less than a billionth of the bulk of the table itself ... when I lay the paper on it the little electric particles in their headlong speed keep on hitting the underside, so that the paper is maintained in shuttlecock fashion at a nearly steady level."

Rutherford's solar-system atom had deeper problems. The electrostatic force, Coulomb had shown, weakens as the square of the distance. Conversely, it approaches its maximum at the infinitesimal distance between the electron and the nucleus. According to classical mechanics, the momentum—and hence velocity—of the electron on orbit around the nucleus must give it sufficient momentum to resist that relatively enormous force. An object in orbital motion undergoes acceleration. According to classical electrodynamics, an accelerating electric charge must generate radiation. To radiation, then, the electron must constantly lose energy. The electron must spiral into the nucleus, therefore, at an ever faster rate. James Jeans, whose ultraviolet catastrophe had challenged Max Planck [see illustration, page 67], calculated that a universe made of such atoms would have a lifetime of 10^{-20} second.

Electrons on quantum orbits

Niels Bohr, an apprentice from Denmark in Rutherford's laboratory, wondered whether Planck's answer to Jeans's first catastrophe might have relevance to this second one. The atom was a perpetual motion machine forbidden by classical mechanics. Its electrons did not spiral into the nucleus but moved on stable orbits. Bohr turned to the question of which orbits, among the infinite possible, are allowed.

Einstein had shown that a quantum of radiation carrying enough energy—equal to the product of its frequency multiplied by Planck's constant h—will kick an electron loose from a susceptible atom [see page

85]. In its susceptibility, the atom had to be somehow resonant to just that frequency of radiation and the energy of its quantum. Bohr boldly proposed that the stability of electron orbits must also be governed by the quantum of action: $h = 6.6 \times 10^{-27}$ erg. The mass of the electron multiplied by its velocity and multiplied again by the distance it travels in one orbit constitutes an "action" in classical mechanics. It is the action that offsets, for that moment, the electromagnetic attraction of the positive charge of the nucleus. Bohr postulated that such electronic action must equal h or some whole-number multiple of h. A quantum of radiation emitted or absorbed by an atom would then carry the difference in energy between two allowed electron orbits.

The frequencies and so the energies of radiation emitted by hydrogen had been well established by spectroscopy from its emission-line spectrum. Taking those energies to equal the gaps between allowed orbits, Bohr calculated the energies of the orbits. He found that the energy from orbit to orbit does indeed increase by a whole-number multiple of h. It was apparent that, upon emitting a quantum of radiation of a given frequency, the hydrogen electron jumps from a higher allowed orbit to a lower one. Conversely, upon absorbing a quantum of radiation, the electron must jump from a lower to a higher orbit or, with enough energy, free from the thrall of the nucleus in accordance with the Einstein photoelectric effect.

The electron does not travel from one orbit to the other; it disappears at one and appears at the other, with no guarantee that it is the same electron. The discontinuity of the jump must, of course, seem strange to a world that knows continuous motion. This may be taken as testimony to the indivisibility of the quantum of radiation. That indivisibility may be taken as expressing the irreducibility of the 6.77×10^{-27} erg that is the atom of action. Planck's quantum, with Bohr's help, thus ensured to the universe a life expectancy of 10^{20} seconds.

To Bohr's atom, Arnold Sommerfeld in Munich made an important correction in 1914. The velocity of electrons on their orbit appeared high enough to require considerations from Special Relativity: increase in their mass had to be reckoned with. In the hydrogen spectrum, and in the spectra of other elements, many lines exhibit "fine structure." First observed by Pieter Zeeman in 1896, such infinitesimal splitting of one or another line can be induced by strong magnetic fields. Sommerfeld decided that this must be due to change in the relativistic energy and mass of the electrons—evidence of change in their velocity. Electrons giving such signals had to be traveling on elliptical orbits, slowing down at farthest retreat from the nucleus and speeding up on closest approach. Calculation from Sommerfeld's postulate landed squarely on the observed values. Plainly, electrons obey Kepler's third law; like the planets on their orbits around the Sun, they comply with the law of conservation of angular momentum [see page 48].

While the solar-system atom remains fixed in public imagination, inquiry into order at quantum dimensions in spacetime soon made substantial correction of the Rutherford-Bohr-Sommerfeld model. The Bohr atom had, in the first place, but a single electron. The model could not fit into its structure the two, three and many more electrons bound to atoms heavier than hydrogen. At a deeper level, it had to be shown "why" the Bohr postulates work. The compliance of stable orbits to the constant h required manifestation in something like the electric charge that Faraday had supplied to the chemists' purely conceptual valence.

The need for metaphor

In the two decades that followed the First World War, the perfection of quantum theory brought it more surely in command of its netherworld and in need of more elaborate metaphor from common experience. Thus, the Dutch physicists Samuel Goudsmit and George Uhlenbeck endowed the electron with quantum "spin" necessary to account for its magnetic moment. Another quantum number, this spin made it possible for two electrons to revolve about the nucleus at the same orbital distance. That had been forbidden by the Wolfgang Pauli "exclusion" principle published just the year before. Pauli showed that two electrons cannot occupy the same energy level—metaphorical orbit—much as classical physics says that no two bodies can occupy the same space. With the spin of one electron pointing up and the other down, however, Goudsmit and Uhlenbeck made it possible for two (but not more) electrons to occupy the same orbit. With this amendment the atomic model could accommodate as many electrons as were needed, on elliptical as well as circular orbits, to neutralize the positive charges of the heaviest elements [see illustration, pages 134–135].

Arthur Holly Compton, at Washington University, St. Louis, in 1923 produced evidence to demonstrate the momentum, the inertia, of the quantum. This was Einstein's seminal contribution to quantum theory: the photon has the effective mass of its electromagnetic energy and so packs the mechanical force of momentum as well. At the "soft" X-ray frequency of 6×10^{18} Hz, a quantum carries momentum of 10^{-9} dyne and, so, a relativistic mass of 10^{-30} gram. That is an appreciable—10^{-3}—fraction of the 10^{-27}–gram rest mass of the electron.

The Compton effect is a super-photoelectric effect. Compton directed X rays at metal foils and found that they kicked electrons loose at remarkably high velocities. More important, he found that the X rays were reflected from the target at lower energy, signified by longer wavelength. The energy of the reflected—reemitted—radiation equaled the excess over that imparted by the X ray to the electron [see illustration, page 93]. The Compton effect thus invokes the elastic collision of classical physics. From such collision, two billiard balls each carry away their share of the conserved collision energy. Here was conclusive confirmation of the particle-like nature of the quantum, a proposition resisted by many physicists, including at first Planck and Bohr.

The wave nature of the electron

A historic doctoral dissertation in 1924 by Louis de Broglie explained "why" electrons travel on quantized orbits. Citing Special Relativity, he argued that, if a quantum of radiation has mass—as Compton had shown—then an electron must have properties of radiant energy, such as wavelength. The higher the velocity of an electron, the greater its energy; the shorter its wavelength, the higher its frequency.

Accordingly, de Broglie proposed, the circumference of a stable orbit must be just equal to the wavelength (or a whole multiple of it) of the electron occupying that orbit [see illustration, page 133]. In classical mechanics, such a wave would be a standing wave and so self-regenerating; that would help, at least metaphorically, to account for the stability of the orbit. As Bohr had established, the product of the mass and velocity of the electron and the circumference of a stable orbit (i.e., de Broglie's electron wavelength) must equal the constant h or a whole multiple of it. Wavelengths shorter or longer than the circumference of the allowed orbit would be self-interfering, not self-regenerating.

Confirmation of the wave nature of the electron had to wait for the experiment performed by C. J. Davisson and L. H. Germer at Bell Telephone Laboratories in 1927. They directed beams of electrons at crystals. The crystal lattices, serving as diffraction gratings, sorted electrons out according to their frequency. George Thomson, son of J. J. Thomson, conducted much the same experiment with the same results in the same year [see illustration, page 122].

The uncertainty principle

Meanwhile, Werner Heisenberg sought to dispense with the mechanical metaphor and electron orbits in need of explanation by de Broglie. This resolute positivist, at age 23, produced a formal mathematical description of the atom based exclusively upon spectroscopic evidence of the energy states of atoms. His abstract, apictorial equations successfully accounted for all the energy states recorded by spectroscopy. Engaging at once the interest of his seniors at Göttingen, Max Born and Pascual Jordan, and of Wolfgang Pauli, the "matrix mechanics" of Heisenberg soon comprehended in a general theory all that was known about the atom and was open to embrace more.

Heisenberg relaxed his positivism to explain his "uncertainty principle" to the lay public in this thought experiment: A photon illuminates an electron for inspection in a microscope. The higher its frequency and the shorter its wavelength, the more precisely can the microscope establish the location of the electron. Colliding with the electron and thereby establishing its location, the energetic photon imparts momentum to the electron, making measurement of its momentum uncertain. A photon of longer wavelength and lower energy gets a better fix on momentum but a blurred measurement of location.

The uncertainty cannot be charged to the design of the experiment. Absent Maxwell's demon, people are constrained to exchange

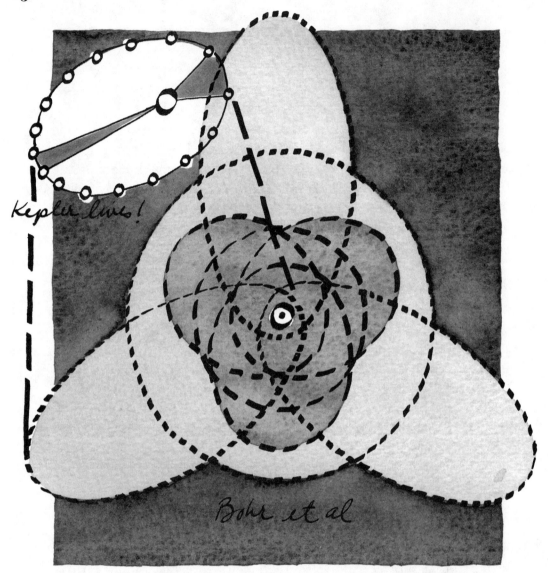

Kepler lives!

Bohr et al

Atom as solar system *was the metaphorical image invoked to visualize the tiny nucleus, only 10^{-12} meter, inside the 10^{-8} meter diameter of the atom associated with metaphorical electron orbits. Bohr showed that electrons are restricted to allowed energy levels or "orbits" that are fixed by whole multiples of the Planck energy quantum [see page 127]. On elliptical orbits, Sommerfeld showed, electrons obey the mechanical law of conservation of angular momentum and, like Kepler's planets, sweep out equal areas in equal times [see pages 48 and 128].*

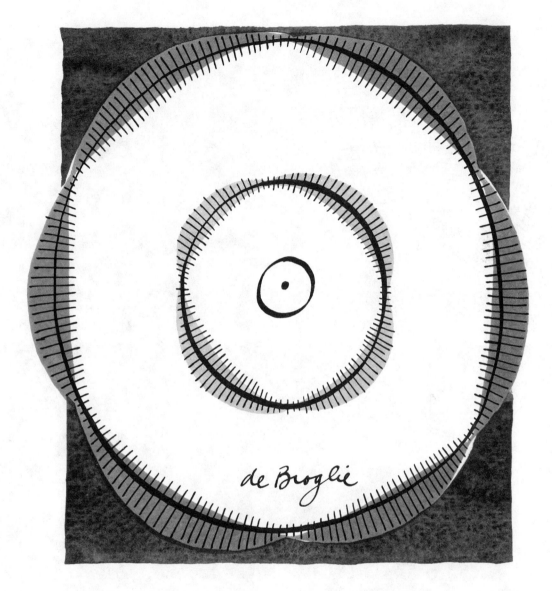

Circumference of electron orbits is always, de Broglie showed, a whole multiple of the electron wavelength [see page 130]. The energy carrying an electron around one orbit is always in turn the same whole-number multiple of the Planck quantum. This is the "action," the energy expenditure, that offsets for that moment the electrostatic force exerted by the positive charge of the nucleus. Two wavelengths in the inner orbit here indicate the expenditure of two quanta; on the higher-energy outer orbit the expenditure is four quanta.

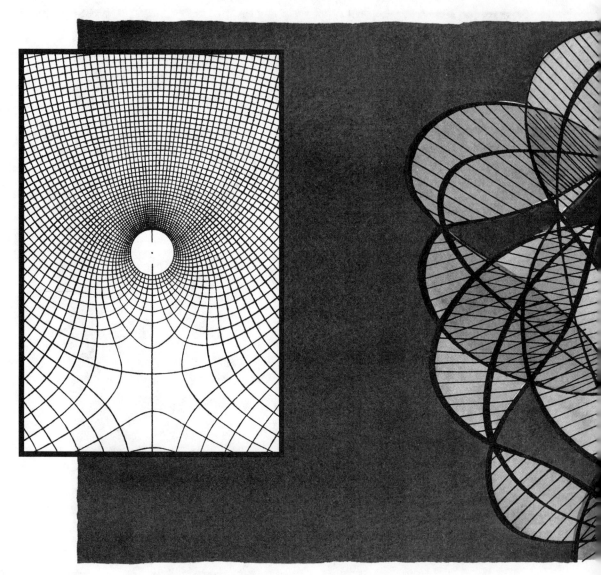

Electron-wave-cloud atom *pictures the work of Heisenberg, Schrödinger and Dirac. The austere, apictorial equations of Heisenberg, describing the quantum mechanics of the atomic structure, asked for no such metaphor [see page 131]. Schrödinger's wave equations gather the Heisenberg uncertainty principle and the de Broglie wavelength rules that govern the circumference of electron orbits to enclose the atomic nucleus in a more or less symmetrical electron-wave cloud [see page 136]. One electron serves the purpose: traveling at 10^{-2} the velocity of*

light around its 3×10^{-10} orbit, it makes 7×10^{15} revolutions in one second and, effectively everywhere at once, wraps the nucleus in its waves. To this insubstantial cloud, Dirac supplies the substance of the Faraday-Maxwell-Einstein field [see inset and page 99]. That manifests itself in the ergs of electromagnetic energy continuously traded in virtual photons between the electron and the nucleus that keep this perpetual motion machine going [see page 137]. In spontaneous emission virtual photons become actual.

energy with the object of any experiment at the quantum scale in space and time. The exchange must perturb the observation, and h is the limit to "the smallness of the disturbance" [see Dirac, page 109].

Inspired by de Broglie and repelled by matrix mechanics, the Austrian physicist Erwin Schrödinger produced in 1926 a complete wave-equation model of the atom. It sets de Broglie waves vibrating on planes in all three dimensions around the nucleus and it wraps the nucleus around in a cloud of electron waves [see illustration, preceding page]. The Schrödinger equations thus smear out the location of the electron in wave packets of an indeterminate breadth of frequencies in accordance with Heisenberg's uncertainty principle.

Between Heisenberg and Schrödinger it can be seen that indeterminacy is rooted in the structure of the physical world. The first uncertainty is whether the electron is to manifest itself as a packet of waves in a field of electromagnetic force or as a charged particle. The resolution of this uncertainty depends upon which identity is sought. With some instrumentation the electron interacts as wave, in which case its momentum can be measured. With other instrumentation, it interacts as a particle, the position of which can be measured.

Bogus epistemology

Uncertainty in quantum theory has excited renewed philosophical dispute about the reality of external reality. If the experiment makes the choice of wave or particle, does not the experiment fix the nature of the reality observed? Some dogmatic positivists argue that it does. Still others find the boundary blurred between the event and its observation. Does not the instrumentation itself engage in the event? The scientist as well? And the observer of the scientist?

Such argument within the scientific community has encouraged "postmodern" talk about the ground of knowledge. Does not quantum uncertainty place poetic insight on a par with physics? Does not the engagement of the observer in the observed evoke the holistic vision of nature from the ancient civilizations of the Orient?

Physicists accept the results of their experiments to show that, independent of observation, the electron, in interaction with the background quantum underworld, functions at times as particle, at times as wave. In the canon of strict positivism, of course, their faith has no standing. Nothing can be said about an event that is not substantiated by observation of that same event. The observation that shows the electron as particle has nothing to say about the state of the electron in the instant before or the instant after that event. Unlike observations of the moon, experiments at quantum dimensions cannot show that they have ever seen the same electron twice. It is possible, moreover, to take two different measurements of the moon at the same time. Physicists who work at it long enough, however, are as comfortable with the floating identity of the electron—whether particle or wave—as astronomers with the tangibility of the moon. Their lay observers may comfortably regard the situation with the same *sangfroid*.

David Bohm commended a yet cooler attitude to his students. He observed that knowledge even of the world in reach of the unaided senses is charged with uncertainty. The equations of the calculus applied to fixing the position of objects in motion yield Newtonian "infinitesimals" that never shrink to zero. Bohm's students were ready then to hear: "We cannot visualize simultaneously a particle having definite momentum and position. Quantum theory has shown that it is not necessary to try, because such particles do not exist."

Review of the seemingly antithetical Heisenberg and Schrödinger equations soon showed them to be mathematically congruent. They could be employed alternatively to address one question or another raised by the expansion and perfection of quantum theory.

An electrical engineer named P. A. M. Dirac, discouraged by job prospects in the depressed British economy, sought a fellowship in physics in 1926 at Cambridge University. His sponsors were amply rewarded. In equations as austerely barren of waves and particles and other pictorial metaphor as Heisenberg's, he unified quantum mechanics with Special Relativity in "quantum field theory." The Dirac equations improve upon the de Broglie standing wave to sustain the perpetual motion of the electron. With the vacuum-occupying electromagnetic field that is generated by its spin and orbital motion, the electron continuously exchanges quanta of energy—"virtual photons." It thus draws the energy necessary to sustain its motion from the field and balances the account by returning the photon to the field. The strength of the electromagnetic force and the mass of the electron set the period of exchange, in accordance with the uncertainty principle, at 10^{-21} second. Visible only in the equations, the virtual photons keep the law of conservation of energy in force. In spontaneous emission, however, virtual photons sometimes declare their presence as actual photons.

The Heisenberg-Schrödinger-Dirac model presents a new metaphorical picture of the atom. Its diameter of 10^{-10} meter gives it a circumference of 3×10^{-10} meter. The electron in Bohr's hydrogen atom travels, in its stable, lowest-energy level, at 2.2×10^6 meters per second, a 10^{-4} fraction of the velocity of light. At that speed it makes 7×10^{15} round-trips in a second. This is also its de Broglie frequency. Freed from a single orbit by the Schrödinger equations, the electron has time in one second to be effectively everywhere at once around the nucleus. In place of the solar system, there emerges a more or less symmetrical cloud of negative electricity anchored to an equal positive charge gathered in a point at its center. Rutherford's vacuum surges with energy that brings Dirac's virtual photons winking into and out of momentary existence [see illustration, pages 134–135].

Here, the valence of chemists gains physical substance. The electron is the valence. In a molecule, it spends most of the time equally and alternately with the two nuclei, which take turns in the ion state. In the time that it spends in the space between them, its negative charge balances the mutual repulsion of the positive charges of the nuclei across a space of about 10^{-10} meter or an atomic diameter.

The Dirac restatement accounted successfully for all the then-known anomalies in the line spectra of the elements. The magnetic moment of the electron found its origin in the relative motion of its electric charge as spelled out in the "Electrodynamical Part" of young Einstein's paper on Special Relativity. In the exchange of virtual photons the magnetic moment is displaced by a value that Dirac computed at 1 "**g**." In 1948, experiment set the value of **g** at 1.00118.

The vacuum discovered by Rutherford between the nucleus of the atom and its nearest electron becomes, in the Dirac equations, continuous with the vacuum of interstellar and intergalactic space and the vacuum in the interior of atoms in the most distant galaxy. For all its emptiness, the luminiferous vacuum of space is charged with the energy of gravitational and electromagnetic—and, as will be seen, Higgs—fields. The scene also of the continual eruption and evanescence of virtual particles in the fields, the vacuum of quantum physics does as heavy duty as the aether of classical physics and has real substance.

Antimatter

To the incredulity and distress of the community, the Dirac equations also posited the existence of an anti-electron bearing a positive charge. This was the first particle to make its first appearance in an equation. For all of the others there had been experimental evidence that asked for mathematical connection to the preexisting order. Here was a mathematical invention asking for materialization in an experiment.

Dirac would seek to reduce alarm among his colleagues by speaking of the particle as a "hole" vacated by the electron; as such, it served the mathematics perfectly well. The existence of the "positron" was confirmed by the U.S. physicist Carl Anderson in 1930. He was observing the showers of particles from the collision of cosmic rays—high-energy particles that plunge into the Earth's atmosphere from elsewhere in the universe—with atoms in the atmosphere. The magnetic field in his detector deflected electrons in one direction and matching positrons in the opposite direction.

Dirac had discovered more than a particle. The discovery of antimatter effectively doubled the number of particles physics has to reckon with. It also maintains the conservation of matter-energy. When particle collides with antiparticle, they mutually annihilate in the blaze of an energetic photon. By the reverse process, a sufficiently energetic photon will give rise to a particle-antiparticle pair.

The quantum theory was now all but complete as a final theory of the provinces of the physical world governed by electromagnetism. A few eager workers, the Russian physicist George Gamow at the Cavendish laboratory and Edward U. Condon and Ronald Gurney in the United States, were already finding it useful in attack on questions about the nucleus. Recruiting an increasing number of young enthusiasts, the community pressed on to enlarge and apply the theory.

At periodic international conferences of all engaged, there was time to hear the misgivings of the elders, principal among them Albert Einstein. If physics would settle for measurements of position and momentum taken at different times and probably not of the same particle, did it not fail its undertaking to secure complete description of the physical world? A theory that accepted such scanting of the mission was framing the wrong experiments.

Niels Bohr, sensitive to Einstein's concerns, undertook to respond. Experiment dictated the theory, Bohr asserted. The separate determinations of position and momentum report "complementary" aspects of nature. Together, the two findings give the complete description of the physical world.

At the conference in 1930, Einstein proposed a thought experiment to demonstrate that physics is not constrained to complementarity. An ideal clock [left] would open an ideal shutter in an ideal box hung on an ideal spring balance to admit a single quantum of radiation. The clock and the balance (measuring the energy-mass of the quantum) would thereupon have made simultaneous measurements of both time and energy. From Special Relativity itself, Bohr drew his refutation: in falling, the ideal clock would take on ideal mass, making its time-keeping less than ideal [see page 96].

Einstein was to intervene once more, from exile in the United States, with an argument in support of another thought experiment proposed by two young admirers there. Summarily stated, two particles of opposite spin have departed from an event. Would not the determination of the spin of one, Einstein and his friends asked, tell you the spin of the other? In response it was said that this would show merely that the collapse of the wave function of one particle by the experiment had simultaneously collapsed the wave function of the other. Einstein dismissed this argument as "spooky action at a distance."

The Einstein-Podolsky-Rosen thought experiment has now been conducted more than once as an actual experiment. The separated particles show the predicted opposite spin. In no concession to Einstein, the result is put down to unexplained "entangled states."

Quantum electrodynamics arrived at its present state of perfection around 1948. Perfection came of a crisis brought on by the mastery of the microwave spectrum in the development of radar fire-control and navigation instruments during the Second World War. At Columbia University, the pupils of I. I. Rabi had brought wavelengths of 3 centimeters and less into the command of radio technology. In 1945, Willis Lamb used this instrumentation to measure the quantum that would induce a change in the energy-state of hydrogen for which Dirac had estimated a wavelength of 2.74. Lamb found a value of 3.3 centimeters. Between theory and the experiment, the discrepancy was 20 percent; it involved, moreover, the mass of the electron itself. The community turned at once to reexamining the foundations of the theory.

The cure was soon found not in the foundations but in the mathematical procedures for relating experiment to theory. Freeman Dyson, R. P. Feynman, and Julian Schwinger in the United States and Sin Itoro Tomonaga in Japan showed that the values for charge and mass of the electron entered in the equations could be "renormalized." Essentially by substitution of the last experimental finding for the naively calculated infinity, renormalization resets the theoretical terms for the next round of experiment and recalculation.

The Dirac "**g** factor" in the magnetic moment of the electron testifies to the success of the strategy. Lamb's colleague Polycarp Kush, using much the same microwave instrumentation, found the first experimental value of 1.00118 in 1948. The latest calculation stands at 1.01159652190. The latest experimental value is 1.01159652193. With some uncertainty in the last two digits of both numbers, agreement between theory and experiment now reaches at least eight places to the right of the decimal point.

Feynman called renormalization a "dippy process" and wondered whether the renormalized theory is mathematically legitimate. Nonetheless, as he readily attested, it works. Henry W. Kendall and Wolfgang K. H. Panofsky, who took QED into the structure of the proton during the 1960s, declared:

> The laws of electricity and magnetism as they are now embodied in the equations of quantum electrodynamics represent the one and only one area in physics where a single quantitative description has proved valid over the entire range of experiments for which it has been tested, from cosmic dimensions down to 10^{-15} m.

Quantum theory is cited often as the premier "paradigm shift," the acrobatic turn of thought that jettisons the old for the new. In fact, the constant h had its origin in measurement that contradicted classical theory about the color of radiation from an incandescent solid. In the photoelectric effect, the quantum demonstrated its material reality. Quantum field theory keeps its actors under the governance by the classical conservation—of mass, energy and charge—and respects its symmetries. It thereby relates events in the newly understood quantum world with the more familiar events in the nearby and brings the two worlds into synchrony with one another at their respective limits.

Work on the nucleus of the atom had meanwhile proceeded on a course of methodical experiment in search for a theory. From the isolation of the nucleus in 1911, Rutherford and the Cavendish laboratory continued to lead the way to the resolution of its structure. Before the First World War, his junior colleague, H. G. J. Moseley, probing the inner electron shells of the heavier elements had set the ground for present understanding that the nuclei of atoms above hydrogen hold two kinds of particles, positively charged and neutral; that the number of positively charged particles establishes the chemical identity of an atom, its "atomic number"; that the number of these and of the neutral particles establishes its "atomic weight." Chemically identical "isotopes" of a given element are distinguishable by weight, fixed by difference in the number of neutral particles in their nuclei. It remained to establish the identity of that neutral particle. Moseley's work established the configuration of the electron shells of the elements across the Mendeleyev table, including that of elements then not yet isolated.

In 1915, at age 27, Moseley was killed in action in the assault on the "soft underbelly of Europe" at Gallipoli. That goes to explain the asperity with which, it is said, Ernest Rutherford responded to a request from the First Lord of the Admiralty. Young Winston Churchill wanted the Cavendish laboratory to set aside its work and take on the problems of antisubmarine warfare. Rutherford told him: "It's more important than your damned war!" The laboratory then did help with the location of submarine targets for depth charges.

Nitrogen transformed to oxygen

Rutherford's next objective was to measure the force that binds protons and neutrons in the nucleus. He looked first for higher-energy alpha particles that could overcome the electromagnetic force of repulsion. He found them emitted by the isotope radium C′. For his target he chose nitrogen. Its nucleus, with only seven protons, would present a relatively low barrier to the onrushing radium C′ particles.

In a chamber filled with nitrogen, Rutherford saw scintillations at some distance from the alpha-particle emitter. He correctly deduced that some particle chipped either from the alpha particles or the nitrogen atoms was signaling its departure from the collisions. By magnetic deflection, Rutherford established that the fleeing particle was a proton, the first ever jarred loose from an atomic nucleus by human agency.

Later experiment established that the other product of the reaction was an isotope of oxygen, $_8O^{17}$. With one proton and two neutrons of the alpha particle incorporated in the target atom, nitrogen had transmuted to oxygen, specifically, the weight-17 isotope of oxygen. The departing proton balanced the particle equation: $_7N^{14} + {}_2He^4 = {}_8O^{17} + {}_1H^1$. Of this alchemy, Eddington was moved to say: "What is possible in the Cavendish laboratory may not be too difficult in the Sun."

Not long after, observation of such reactions at the Cavendish laboratory was facilitated by the cloud chamber, the ingenious invention of C. T. R. Wilson. In saturated vapor inside the cloud chamber, particles sign their passage in condensation trails. It was in a cloud chamber that Carl Anderson recognized the track of the positron. Much easier to see than scintillations, these tracks convey much more information about the reaction and can be photographed. Such a photograph of an alpha particle–nitrogen collision made by P. M. S. Blackett in 1925 fully confirmed the 1919 findings [below].

The proton now presented itself as a suitable projectile for attack on the nucleus. Its single positive charge would suffer smaller repulsion by the positive charge of the target nucleus than the alpha particle's two charges. Calculation showed that protons would gather formidable

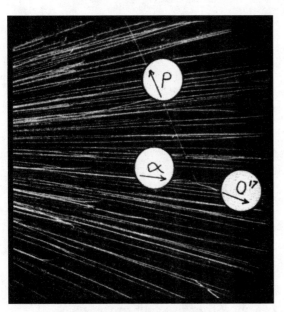

striking power upon acceleration through an electric potential of 1 million volts. To John Cockroft and E. T. S. Walton, Rutherford assigned the task of building the first particle accelerator.

By 1931 they got their "voltage multiplier" up to 0.6×10^6 volts. At this power, they could accelerate protons to energies on the order of 10^5 electron volts. An electron volt, "eV"— the energy imparted to an electron in passage across an electric-potential gap of 1 volt—is a tiny amount of energy. It takes about 10^{23} eV to light a 100-watt bulb for one minute. That energy is spread, however, among millions of atoms in the incandescent filament and its surroundings. To pump that much energy into a single subatomic particle in an accelerator beam exceeds the reach of present technology by 11 orders of magnitude.

Cockroft and Walton aimed protons at targets of lithium, the third element up from hydrogen, and boron, the fifth element up. In the cloud chamber, lithium yielded a two-prong track at the point of collision, signifying its breakup into two alpha particles. The three-prong track from boron evidenced three alpha particles ($_3Li^7 + _1H^1 = 2\ _2He^4$; $_5B^{11} + _1H^1 = 3\ _2He^4$).

The breakup of lithium and boron provided the first experimental proof of the $E = mc^2$ equation from Special Relativity. The length of the particle tracks gave a measure of the combined kinetic energy of the two alpha particles: 17.3 million electron volts (MeV). The mass of the lithium atom plus the proton is a little more than 7 + 1 atomic mass units; precisely, 8.0263 such units. The mass of two helium atoms is 8.0077 atomic mass units. The difference of 0.0186 atomic mass unit is equal, by $m = E/c^2$, to 17.2 MeV. The mass lost in the reaction was just equal to the energy of the reaction.

That energy was, moreover, the measure of the strength of the strong force that holds the particles of the atom together. The magnitude of the energy that binds the particles in the nucleus is of the order of millions of electron volts. That compares to everyday experience with the 1 to 10 electron volts of chemical binding secured by the electromagnetic force; that is, the electromagnetic force is only $.5 \times 10^{-2}$ (.05) the strength of the strong force.

The success of the Cavendish laboratory activated the construction of "atom smashers" in all the industrialized countries. The electrostatic generator built by Robert J. van de Graaf for the Massachusetts Institute of Technology in 1936 reached four times that power. To public wonderment, lightning bolts leaped between its towering 15-foot spherical condenser terminals and between them and the structure of the aircraft hangar that housed it.

The cyclotron invented by Ernest O. Lawrence at the University of California, Berkeley, opened the way to really high energies. The cyclotron accelerates the projectile particle by pulsing the input of electric power to the acceleration of the particles. In the words of Robert R. Wilson, it works by "giving them a sequence of small pushes instead of one big push." On this principle, even before the Second World War, amplified the support of physics, particles were accelerated to millions of electron volts (MeV) [see illustration, page 147].

James Chadwick, in 1932, established the existence of the neutral

Nucleus of the atom *has been the target of assault at ever-higher energy ever since Rutherford aimed alpha particles at it in 1913 [see page 126]. From angles at which alpha particles rebounded from their target, he calculated the diameter of the nucleus at 10^{-12} meter, a 10^{-4} fraction of the then-estimated diameter of the atom. The first particle accelerator, the "voltage amplifier," raced protons across single voltage drop to energy of .6 MeV [see page 144]. With this instrument, Cockroft and Walton secured the first experimental proof of $E=mc^2$.*

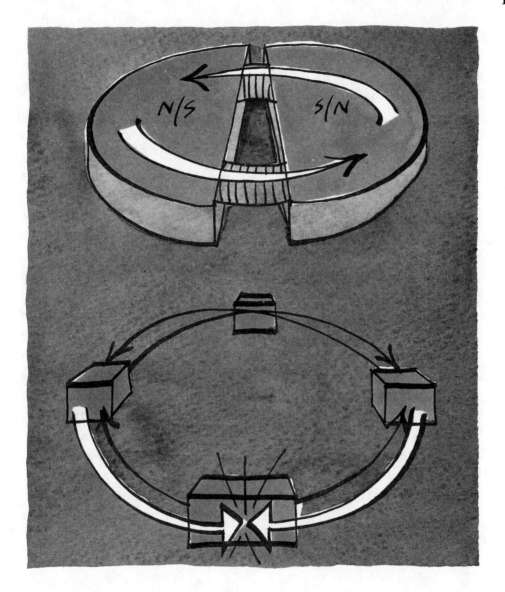

Particle accelerators *have reached the trillion electron volt range on the principle, first demonstrated by the cyclotron [top], of giving particles many strong pushes—in the cyclotron by alternation of magnetic fields as the particles crossed from one "D" magnet to the other [see page 145]. Acceleration of particles in opposite directions into collision with one another, first demonstrated by Budker at the Akademgorodok in Novosibirsk [see page 153], doubles the effective power of accelerators, but at the cost of the number of productive collisions.*

nuclear particle, which had already been given the name "neutron." He found the particle in emissions from the breakdown of certain radioactive elements. Outside the nucleus of the atom, the neutron proved to be unstable. It has a half-life of 10.3 minutes. In that time, half of a population of neutrons will have undergone the beta emission first observed by Rutherford in the disintegration of radioactive elements [see page 125]. The emitted electron carries off the negative charge, and the former neutron settles down as a stable proton. The proton has a half-life of not less than 10^{32} years at latest measurement. The transformation of a neutron to a proton in the nucleus of an atom undergoing beta decay attracts an additional electron to that atom's outer shell and gives the atom a new chemical identity.

Beta emission presented a problem that was resolved by Wolfgang Pauli in 1930. Einstein, in his afterthought on Special Relativity, had proposed that periodic weighing of a sample of a radioactive element might show whether mass is lost to energy and so provide a test of whether $m = E/c^2$ [see page 98]. The measured mass loss by beta emission affirmed the equation, but it exceeded the mass and the energy of the most energetic of the emitted electrons. Pauli proposed that the excess mass was carried off in the energy of an unobserved particle (now called a neutrino), carrying no charge and little or no rest mass, traveling with sufficient velocity—namely, energy—to balance the books.

The particle went undetected until the 1950s. Lacking charge entirely and mass very nearly, the neutrino interacts almost not at all with other particles. The rest of Pauli's explanation of beta emission worked so well that physicists were content to let the hypothetical neutrino go on balancing the books for more than 25 years before its existence was experimentally established.

Discovery of the weak force

The troublesome beta emission led Enrico Fermi in Rome, early in the 1930s, to the discovery of the fourth, "weak," force. He took beta emission to be evidence of a quantum-mechanical process under way at all times within the tiny radius of the nucleus. In accordance with Dirac's equations, the proton and neutron continually trade identity there by the exchange of virtual electrons (or positrons) and the new Pauli particles. At times the virtual electrons and neutrinos can gain energy to become actual in spontaneous beta emission.

For the new force, Fermi set up a complete quantum-mechanical model. In the interaction of the charged particles of the atom, the force would require two quanta, positive and negative. Their effective range, acting inside the nucleus, would be on the order of the 10^{-15}–meter diameter of the nuclear particles. To act at such short range, the uncertainty principle requires that a quantum have mass, a mass inverse to the range at which it acts and, therefore, very large. (Conversely, the photon, with its infinite range, has zero rest mass.) Fermi left the mass of the weak-force quanta unspecified. The virtual exchange of the massive quanta would necessarily proceed at low frequency, which he estimated at 10^{-9} second, compared to 10^{-21} second for the electromagnetic force. When found, a half century later, the quanta proved massive indeed, far heavier than the proton.

Fermi's success in this extension of the quantum theory prompted Hideki Yukawa at the University of Tokyo to propose in 1935 a quantum theory for the strong force. Again, exchange of virtual quanta between the nucleons and the field of the strong force would satisfy the law of conservation of energy. Yukawa's calculations, taking account of the strength of the force, the 10^{-15}–meter range at which it acts and so the period of the virtual exchange between the nucleon and field—10^{-23} second—showed that its quantum must have a rest mass close to 300 electron masses. To bind positively charged protons and neutral neutrons in the nucleus, Yukawa's "meson"—midsized particle—would require a neutral as well as positive and negative electric charge.

At this early stage, quantum theory was yielding insights into old questions about the natural world. For the first time, there was an explanation for starlight. Eddington and other astrophysicists showed that stars are masses of hydrogen and helium, with average temperatures of 1.5×10^6 to 3.0×10^6 degrees K in their gravitationally compressed interiors. At such temperatures atoms are stripped of their electrons. They constitute a high-density "plasma," a gas in which their positive electrical charges complicate high-velocity Newtonian mechanical interaction.

In the interior of sufficiently massive stars, Hans Bethe at Cornell University in 1938 found pressures and temperatures sufficient to fuse nuclei of hydrogen, in a few steps, into nuclei of helium—protons into alpha particles. Such fusion reactions proceed differently for the three isotopes of hydrogen in different combinations with each other. The mass loss, from the ingoing hydrogen nuclei, he showed, must yield energies of 1.4 MeV to 20 MeV per helium nucleus produced.

Atomic fission

From such questions at the end of the 1930s, the exactly opposite nuclear reaction—fission as opposed to fusion—diverted the attention of the community. The availability of neutrons, following Chadwick's isolation of the particle in 1932, set this line of inquiry in motion.

The neutral neutron encounters no barrier at the positive charge of the nucleus. In 1932, Enrico Fermi had systematically exposed samples of the table elements to low-energy, or "slow," neutrons from the unstable element thorium. When his survey reached uranium, the heaviest element in the table, neutron exposure produced an ambiguous result. In the debris of the reaction, Fermi thought he detected traces of a short-lived "transuranic" element, heavier than uranium. With more immediate questions about other elements on his agenda, he left the uranium experiment to that surmise for the time being.

In the nuclei of the heavier elements, the nuclear binding force per nuclear particle declines steeply. Natural uranium occurs in three isotopes: $_{92}U^{238}$ (99.3 percent), $_{92}U^{235}$ (0.7 percent), $_{92}U^{234}$ (0.01 percent). Thus, the binding force must hold 92 protons together in the nucleus against their mutual electrostatic repulsion, plus 146, 143, 142 neutrons, respectively. That is why the heavier elements tend to undergo spontaneous, radioactive breakdown. The least stable of the uranium isotopes is U^{235}, prone to disruption by slow neutrons.

Early in 1939, Niels Bohr, in the United States for scientific meetings, received an urgent message from the physicist Lise Meitner. She wrote from Stockholm, where she had taken refuge from her native Germany, now the Third Reich. Her Berlin colleagues Otto Hahn and Fritz Strassman had repeated Fermi's uranium experiments and had told her of an entirely unprecedented result. With her nephew, the physicist Otto Robert Frisch, then in England, she had confirmed the result: upon absorbing a slow neutron, the nucleus of U^{235} "fissions" into two daughter elements: barium, of atomic number 56, and krypton, of atomic number 36 (56 + 36 = 92, the atomic number of uranium). It could be calculated that the mass loss in this reaction is 0.2 atomic mass unit—equivalent to 200 MeV—per fission. Of further import, it was noted, this reaction yields two new neutrons.

Fermi, a refugee in the United States from fascism in Italy, and his new colleagues at Columbia University confirmed these findings. With John A. Wheeler, Bohr published a paper on the phenomenon of nuclear fission in the autumn of that year. Soon after, at the urging of the Hungarian physicist Leo Szilard, publication on the subject was suspended in the scientific press until after the end of the Second World War.

Szilard had found his way to the United States, having fled Berlin—"on the first train," he said—on the morning after the Reichstag fire in 1933. In London in 1936, he reflected on the sensitivity of the unstable elements to intrusion by neutrons. Stepping off a sidewalk one day, as he recalled, he was seized with the idea that a nuclear reaction set off by a single neutron might release two new neutrons. Those two would set four neutrons loose on the next round; the four would multiply to 16 neutrons.... In the chain reaction [above], Szilard recognized a mechanism by which the enormous energy stored in the rest mass of the atomic nucleus might be released in titanic explosion. He filed a patent on his idea in the British Patent Office and assigned it in secret to the Admiralty. It was Szilard who drafted the letter for Einstein's signature that helped to persuade Franklin D. Roosevelt to set in motion the production of the nuclear weapon.

At the war's end, the physicists demobilized with the eyes of the world on their next undertaking. Their mysterious enterprise had produced the absolute weapon and now promised an inexhaustible supply of energy for free. Especially in the United States, physicists had the national, in particular the military, budget at their command. The U.S. physicists who had apprenticed in Europe, with many of their teachers in exile from Europe working with them, made U.S. universities the centers of the next great undertakings in quantum physics.

The scale of their enterprise went immediately into exponential expansion. The 184-inch cyclotron built by Lawrence at Berkeley before the war was the last such instrument financed by philanthropy (the Rockefeller Foundation) in the United States. The diameters of its first successors after the war were measured in feet, then in hundreds of feet, and the last, built underground at Batavia, Illinois, has a diameter of 1.25 miles and so a circumference of nearly 4 miles.

The lesser curvature of a longer circumference reduces the loss of energy to radiation as the particles race around the circular track, the loss being the cost of deflecting the particles from motion in a straight line. The lost energy radiates tangentially from the deflected beam on a wide band of frequencies and is called "synchrotron radiation," after the type of accelerator in which it first came to note. By reducing loss to synchrotron radiation, larger diameter brings delivery of a higher percentage of an accelerator's energy to its target. LINAC, the linear accelerator at Stanford, California, speeding electrons down its 3-kilometer straightaway, averts such loss of energy. Its peak energy is limited, however, to the velocity reached through that distance. In a circular accelerator, particles may make many more than one trip around the circumference, traveling farther and receiving many more R. R. Wilson "pushes."

No longer could such machines be housed in university laboratories. The big accelerators are themselves national laboratories managed by university consortia. Only the federal treasury could support the order-of-magnitude increase in the energy output of accelerators that came every six years for four decades after 1945. Output escalated from

5 MeV to 500 MeV, through two, five, ten and then hundreds of GeV (G for giga or billion; i.e., 10^9 eV) to the TeV (T for Tera or trillion; i.e., 10^{12} eV) of the accelerator at Batavia.

Application of quantum theory to study of matter at low temperatures brought big increases in the yield of MeVs and GeVs from the electricity that powered these machines. Superconductivity multiplies the strength of the magnetic fields that steer the particle beam securely on course at much higher momentum around the circumference of the same high-energy accelerator.

A true quantum jump in yield came with the colliding of accelerator beams. Energy directed at a fixed target is lost in the momentum of the onward travel of the products of the collision. The effective yield at the target declines steeply, furthermore, with increase in power: a 10 GeV beam gets 2.9 GeV; a 100 GeV beam, 10 GeV; at 1 TeV, a mere 35 GeV. In the colliding beam, the energy is potentially doubled. This stratagem was first made to work by Gersch I. Budker at the Akademgorodok created by Soviet physicists in Siberia near Novosibirsk in the 1950s. Colliding beams in the Tevatron at Batavia and its rival at CERN have reached energies close to 2 TeV. Nature exacts, however, its inevitable cost: the number of desired collisions falls far below those in the fixed target. That is offset, in turn, by putting some of the energy into "luminosity," the number of particles clustered in the pulsing beam; this stratagem secures increase in the number of collisions, but at somewhat lower energy. Where the terms of the experiment permit, the collision of particles with antiparticles adds to the power of the accelerator the energy of their mutual annihilation.

Rutherford's fluorescent screen and low-power telescope and Wilson's cloud chamber have long since yielded to other sensors and counters of the hundreds of millions of events set off by an experiment. For the crucial experiment at the CERN accelerator that demonstrated the quanta of the weak force, a three-story instrument was wheeled on railroad tracks into the zero point. It housed cubic yards of sensors that counted and tracked charged particles in flight and tons of material to capture the masses of the charged and uncharged particles. Computers caught the torrent of data, equivalent for each experiment to visible fractions of the contents of the Library of Congress, to find and report the looked-for events. The looked-for events are found among thousands of events enacted over months and years of operation.

The USSR was the only other nation that could sustain on its own a comparable escalation of energy and size of industrial plant required to carry particle physics forward. In Europe, a consortium of nations is now completing construction of a larger TeV accelerator at CERN, near Geneva, aiming at 15 TeV at the zero point.

The multiplicity of particles

In 1947, confirmation of Yukawa's 1935 prediction of the quantum of the strong force illuminated the postwar return of the physicists to their proper mission. The British physicist C. F. Powell enlisted cosmic rays in place of an accelerator beam in this achievement. On an expedition high in the Bolivian Andes, he captured particles of 270 electron masses that also otherwise fit Yukawa's specifications. Experiments in accelerators soon confirmed the strong interaction of this meson with the nucleons. As the predicted quantum of the nuclear strong force, this meson, or pion, would seem to round out, rather than complicate, the list of fundamental particles and quanta.

The neutral pion does not disturb this picture, decaying in 10^{-14} second to two gamma rays. In the 2.6×10^{-8} second of their existence, however, both the positive and the negative pion decay to a neutrino and an entirely new particle called the μ-meson, or muon. A giant lepton cousin of the electron, the muon has a negative charge (its antiparticle, a positive charge) and a mass of 210 electrons. It decays in the next 2.1×10^{-6} second to an electron or positron and two neutrinos.

Of the muon Rabi is quoted as asking: "Who ordered that?" Through the 1950s, the number of new particles—similarly unstable and brief-lived—increased into the hundreds. As the postwar accelerators came on-line, the particle beams were no longer merely scattering from their targets. They were breaking up the target nuclei and contributing their own energy to the creation of new particles in the scatter. The more energy the accelerator beams poured into their targets, the greater the mass of the unstable particles that scattered from them. Within larger or smaller fractions of a second, these particles decayed by various routes to familiar stable particles such as protons and electrons, to equation-balancing neutrinos and to photons of electromagnetic energy. Classified by the forces with which they interact, by mass, lifetime and decay products, the baryons (as protons, neutrons and heavier particles came to be called) and mesons ran through the Greek alphabet invoked to name them. By 1985, there were 400 of them.

A conservation law broken

In 1957, the community had to sustain an abrupt and perhaps more severe test of its morale: the "overthrow of parity." This constituted nothing less than the violation of a conservation law. T. D. Lee at Columbia and C. N. Yang at Princeton had undertaken a collaborative study of the weak force, suspecting it might violate one of the basic symmetries of nature, specifically that of parity. Parity may be likened to handedness, right or left. There is no up or down, no north or south, in the spacetime continuum. The universe should be equally indifferent to left and right. Lee and Yang proposed, however, that electrons issuing from the neutron in beta emission break the "parity" symmetry, that beta emission shows directional preference.

Chien Shiung Wu of Columbia took on the test of this proposition. For her elegant experiment, she enlisted the low-temperature laboratory of the National Bureau of Standards. In the superpowerful field of a superconducting magnet she placed a sample of a radioactive isotope of cobalt. Thereby, a big percentage of the spinning cobalt nuclei were caused to point north. Thus pointed, they were spinning either to left or right. The object then was to discover whether those nuclei in the regimented sample that happened to decay during the test would eject electrons *preferentially* north or south or *randomly* north and south. Lee and Yang's suspicions were confirmed; the electrons ejected in one of the two directions in a flux sufficient to rule out randomness. The neutrons in those nuclei were handed—either right- or left-handed, but not indifferently so [see illustration, next page].

Leon Lederman, a colleague of Lee and Wu at Columbia, had made the university's 400 MeV accelerator serve as "a pion factory" in

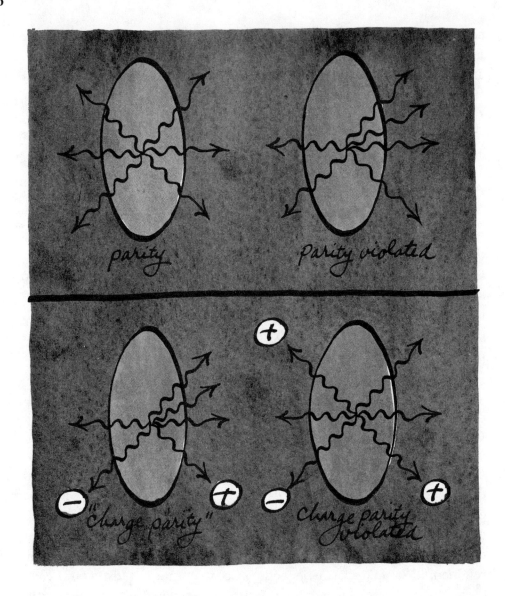

Overthrow of parity was an unanticipated violation of the symmetries looked for in nature [see page 155]. As cosmos has no up or down, physical processes should prefer neither right nor left [top left]. In decay, certain particles fired particles in preferred directions [top right]. Symmetry was to be restored by association of the anomalous decay with the particle's charge. Experiment then found violation of this charge-parity supersymmetry; explanation is sought in the slight excess of matter over antimatter at the Big Bang [see page 233].

pursuit of other questions. It struck him to ask whether pion-to-muon decay might also violate parity. In that case, the offspring muons would all be spinning either left or right. If the muons, in their decay thereafter, emitted their electrons preferentially in one direction, that would violate parity again. The double violation would constitute the decisive overthrow of parity. That is just the way the experiment came out. The weak force does not conserve parity.

In a subsequent experiment, Lederman and his Columbia colleagues Melvin Schwartz and Jack Steinberger turned the more powerful accelerator at the Brookhaven national laboratory into a muon factory. Through 40 feet of scrap armor plate set up to absorb the parent pions, the muons and their daughter electrons and other particles, they filtered out two kinds of neutrino. These two neutrinos were the electron's known antineutrino and a new muon neutrino.

The weak force has been shown to have a strength of only 10^{-14} that of the strong force. It is the only one of the four forces involved in the interaction of all the known particles [see illustration, pages 166–167]. Now it was found to be engaged in the violation of parity.

Any violation of symmetry of nature is an occasion for urgent reflection. It implies violation of the natural order. Wu spoke rightly when, invited to address a conference on high-energy physics, she began, "I am here on the strength of the weak interaction."

Einstein had shown that matter is not conserved; what is conserved is the supersymmetry of matter-energy. Theoretical physicists looked for such a supersymmetry in charge-parity (CP) conservation. They pointed to evidence that the emission of positrons, the electron's antiparticles, violates parity in the handedness opposite to the emission of electrons in the usual beta decay. Charge and parity might be conserved in a CP supersymmetry. Val Fitch and James Cronin of Princeton conducted an experiment, however, that showed the weak force does not invariably, by 100 percent, conserve CP symmetry.

This violation of charge symmetry was welcomed by astrophysicists. They had been asking how it was that the universe, otherwise governed by conservation laws and symmetries, should depend for its very existence upon the asymmetry of matter and antimatter. Only in the high temperature and high density of energy at accelerator targets do particles of antimatter make momentary appearance and vanish in mutual annihilation with particles of matter. The Fitch-Cronin experiment established that the excess of matter over antimatter had to be a slight one. In the first moments of universal history, calculation shows, particles of matter exceeded those of antimatter by a 10^{-8} fraction of their number. From their mutual annihilation in the instant of their appearance, there remained that 10^{-8} excess of particles of matter. That slight excess, Lederman has observed, "accounts for all the matter in the presently observed universe, including us."

The plurality of forces

The weak force was now implicated in the origin of matter; that is, of the rest mass of the particles. But the weak force was at least one force too many. Four could not be the irreducible number of forces holding together the unity of the natural order.

The unification-in-theory of the weak and the electromagnetic force was accomplished in simultaneous and independent publications in 1967 by Steven Weinberg at Harvard and the Pakistani physicist Abdus Salam at Trieste, Italy. (Salam had established in Trieste a freestanding graduate institute to intercept promising young physicists from the developing countries and return them home equipped to do independent work and to teach their country's next generation.) The Weinberg-Salam work drew on studies by Julian Schwinger, who had attempted to establish the mass of the charged quanta of the weak force, and his graduate student Sheldon Glashow. In 1961, Glashow had proposed that Fermi's charged quanta must have an uncharged sibling for which he estimated a mass. Particles subject to the weak force would thus, like those subject to the strong, interact through three quanta.

The prospective unification owed to Peter Higgs at the University of Edinburgh the radical proposal that a field of static energy—neither attracting nor repelling—pervades the cosmic vacuum. From that reservoir would come the rest mass of particles.

The physicists of the 20th century were on the verge of an achievement comparable to the unification of electricity and magnetism in the electromagnetic force by the physicists of the 19th century. The "electroweak" theory would reduce the plurality of forces by one. That promise went to renew confidence that the mission of physics was attainable, that an ultimate unity and simplicity of principle will be found to underlie the natural order.

The experimental physicists were already at work. The electroweak theory predicted an interaction involving "weak neutral currents." Its energy was within reach of the then–most powerful accelerators—in the 100 GeV range—at CERN and Batavia. Evidence had to be of the most indirect kind, because the experiment called for directing a beam of featureless neutrinos at a target, which thereupon would issue another generation of featureless neutrinos. Under the leadership of Carlo Rubbia, who held concurrent appointments at Batavia, CERN and Harvard, the teams at the two accelerators secured evidence of the weak neutral current with sufficient assurance in 1973–1974. That in turn established the existence of the neutral quantum of the weak force first proposed by Glashow. Weinberg, now at the University of Texas, called it the "Z" particle—for "zero" and, he hoped, "last" particle—in his statement of the complete theory.

The unification of the electromagnetic and weak forces was thereby affirmed. It was affirmed again in 1983–1984 when Rubbia and his collaborators induced the appearance of the positive and negative "W" (for weak) and the neutral Z particles in the accelerators at CERN and Batavia, with their energies meanwhile stepped up another order of magnitude. These quanta proved unexpectedly massive; no longer were the masses expressed in MeV, but in GeV. The Ws have a mass of 80 GeV and the Z, an even larger 91 GeV. The most massive particles then encountered, they showed that the weak force increases rapidly in strength as the energy of the experimental interaction increases. Such increase in strength would bring it into convergence with the strength of the electromagnetic force in another few orders of magnitude. Their convergence would bring the fusing as well of their leptons, the electron of the electromagnetic force and the muon of the weak force.

Success of the electroweak theory lends strong support to the GUT theories that seek unification of the electroweak with the strong force. The strength of electroweak force is seen in theory to increase to convergence with the strong force at the much higher energy of 10^{15} GeV. The successive and cumulative convergence of forces in the electroweak and grand unifying theories now brought the Higgs fields into active service. The Higgs field—in analogy with the positron, the first particle to make its first appearance in an equation—is the first field to originate in an equation. In dispassionate anticipation of confirmation by experiment, it has been said that a Higgs field "is added for the specific purpose of spontaneously breaking its symmetry."

In the new cosmology emerging from the collaboration of particle physicists with cosmologists, the Higgs field has assumed a central presence. It is the energy at the temperature and density of energy in the target of the accelerator and at the corresponding moment in the genesis of the universe. It is the energy of the vacuum whence all else came. In the first instant of time, it had the featureless symmetry of the undifferentiated energy at incomprehensible density and temperature that was all the universe was. In successive symmetry breakings or phase transitions, at discrete temperatures and densities, it gave rise to the universe. The last of them, the breaking of the electroweak force that liberated the weak and electromagnetic forces, is now coming in reach of experiment. Well out of reach of experiment is the phase transition proposed in the GUT theories that separated the electroweak from the strong force.

If there is a Higgs field in reach of experiment, there has to be a Higgs quantum. The masses of the W and Z particles predicted a still more massive Higgs particle. No less than a quantum jump in accelerator power, two orders of magnitude from the 100 GeV to the 10 TeV range, was called for now. For such investment in accelerator power there were other demands. The Higgs field and particle, implicated in the origin of rest mass, was as central to investigation of the strong force as the weak. For cosmologists, the prospective insight into the creation and evolution of the universe held no less meaning. At Waxahachie, however, the 100-kilometer tunnel stores irrigation water. Hope for early simulation of the primordial Higgs field turns the installation of new gear in the 27-kilometer tunnel at the CERN establishment and the amplification of the power of the Batavia accelerator now under way.

Investigation of the strong force had been proceeding meanwhile without consensus on a guiding theory. It had soon been conceded that the accelerators, with their increasing power, were not splitting the atom into ever more fundamental particles but, on the contrary, were manufacturing composite particles in increasing number and weight.

The theoreticians were presented with the humble task of taxonomy. The first order of business was to distinguish and classify these brief-lived particles. They divided into two groups: middle-weight mesons, cousins of Yukawa's pion, and the heavy "baryons," cousins of the proton and neutron. Through the late 1950s, much time and talent went into breaking the clan down into families, characterized by mass, charge, spin, lifetime, decay routes and end products.

The Eightfold Way to the quarks

In 1961, employing the higher mathematics of group theory, Murray Gell-Mann at the California Institute of Technology and Yuval Ne'eman—physicist, Israeli army colonel and president of the University of Tel Aviv—independently discovered that the families of baryons could be assembled into eight superfamilies. Gell-Mann called this superfamily system the "Eightfold Way." The allusion to the teachings of Buddha excited momentary interest in physics among the restless undergraduates of the day. Not until 1964 did Gell-Mann find physical significance in the number eight. He did so simultaneously with George Zweig, a Caltech colleague away at the time at CERN. They recognized that the eight groups could be accounted for by just three different entities in combinations of up to three members. Gell-Mann called these entities "quarks," this time in literary allusion to James Joyce's *Finnegan's Wake.*

Gell-Mann felt constrained at first to regard quarks as bookkeeping devices. Their acceptance as prospectively fundamental particles entailed the embrace of electromagnetic heresy. To endow the proton with its positive charge—or the neutron with its neutrality—the three quarks had to carry fractional electric charges, never observed in nature.

The three quarks were called "up" and "down," for no reason, and "strange" in reference to the "strangeness" of the prolonged decay time of some baryons in which, it was postulated, the strange quark would appear. In the theory, it was the 2/3 positive charge of each of two up quarks combined with the negative 1/3 charge of a down quark that would endow a proton with its 3/3 unitary positive charge. One up and two down quarks, correspondingly, would make the neutron neutral. In other combinations, the three quarks and their antiquarks sufficed to account for all the baryons catalogued. They accounted, as well, for the mesons. As to heresy, R. P. Feynman observed that Franklin could just as well have set the unit of electric charge at three.

Here the quanta of the weak force play their role. Coupling with the quarks, the W quanta change their charge. In the decay of a neutron to a proton [see page 148], a W quantum transforms a down quark to an up quark. The new proton has thereby acquired its 3/3 integral positive charge, and the electron and neutrino depart from the scene.

With the strange quark, the down and the up could be teamed in combinations to match the catalogue of stable and transient particles on record and transient mesons and baryons not yet on record. The discovery in 1964 of the strangest possible particle, the omega-minus, Ω^-, made up of three strange quarks, offered a welcome first confirmation of the Gell-Mann–Ne'eman–Zweig scheme.

Acceptance of quarks as real particles required reconciliation with the more affronting condition that no one has or apparently ever will see a quark. The effort to pry quarks loose from one another pressed hard on the increasing power of accelerators. As their energy approached the top of the GeV scale, however, the accelerators poured an increasing percentage of it into the creation of more massive particles in which the quarks continued to hide.

Yielding to the impossibility of driving the quark into the open, physicists turned to establishing its existence *in situ*. The experiment that did so in 1967 was a novel variation on the scatteration and geometry by which Rutherford measured the diameter of the nucleus of the atom in 1911. Henry Kendall and Jerome Friedman from MIT and Richard Taylor employed the 20-GeV electron beam of the Stanford Linear Accelerator (SLAC) as their probe. Feynman, their consulting theoretician, called the quarry particle inside their proton targets a "parton," a more plainspoken name than quark.

The experiment was arranged so that the electron would "feel" the partons, if any, inside the proton. Accelerated to velocity high enough to elude the electromagnetic attraction of the parton-concealing proton, the projectile electron would loose a photon at the partons and depart at lower energy and at a different angle from the angle of incidence. Recording the energy and angle at which tens of thousands of electrons rebounded from such hits, the experiment not only established the presence of three parton-quarks inside the protons but showed them to be pointlike—electron-like—and so irreducible, at least at the energy employed to detect them. The experiment also found them in a perpetual dance with one another.

It is the force that binds the quarks that gets them dancing. The strong force that binds the nucleons is recognized now as the residual of the force that binds the parton-quarks inside them. It had been observed that the strong force falls to zero, perhaps even reverses to a repulsive force, just inside the radius of its maximum strength at 10^{-15} meter. That behavior has its explanation now at the primary site of the strong-force interaction, inside the nucleon. The SLAC electron probe found the strong force falling in strength when the dancing of the partons brings them closer than the threshold 10^{-15} meter. The force ascends abruptly to maximum at 10^{-15} meter; energy then poured in to drive the quarks apart goes to the creation of new particles.

Quantum chromodynamics

Perhaps because the range of the strong force is thus effectively infinite, like the photon of the electromagnetic force, its quanta have no rest mass. It is quanta, not quantum, because the interquark strong force is mediated by eight quanta, to which the resourceful but non-classical philologist Gell-Mann gave the name "gluon." The plurality of gluons serves to bind the quarks, with their fractional charges, together. To allow for the presence of two up-quarks in a proton despite the Pauli exclusion principle, Gell-Mann gave the strong-force charges different "colors." The Gell-Mann metaphor calls in the primary colors—red, green and blue—that Newton found in light. Each color signifies one of three kinds of strong-force "charge" and its opposite. In the proton, red and blue up-quarks combine with a green down-quark to make white; that is, no color. So, a red and an anti-red quark make no color in a meson. "No color" is the satisfied integral charge of the strong force and makes all of the quark compounds white.

The gluons give the quarks their color. In the Gell-Mann scheme, each gluon carries a color and an anticolor. Their emission and absorption changes the color of a quark; thus, the absorption of a red–anti-green gluon turns a red quark anti-green. While nine such combinations are possible, eight conduct the interactions in what Gell-Mann called "quantum *chromo*dynamics."

The extravagance of metaphor marshaled by Gell-Mann and his colleagues speaks for the interchangeability of identity and relationship among particles and forces. As for "particles" and "forces," the David Bohm caution is in order: "At the quantum level of accuracy, an object does not have intrinsic properties … instead, it shares all its properties mutually and indivisibly with the system with which it interacts."

The color charge distinguishes the gluons from the quanta of other forces. The gluons are mutually attracted. The enormous energy they carry is evident in the violence of their interaction. At accelerator targets, they decay to "jets" of baryons and mesons flying off in three directions. Here are objects sharing properties mutually and indivisibly with the system in which they interact. Quantum chromodynamics dissects out the quarks and the gluon quanta of the strong force by the properties in accordance with which the experiment finds them interacting.

At this point, the search for unity and simplicity in the foundations of the physical world had established two generations, so to speak, of quarks and associated leptons. The up- and down-quarks in association with the electron and its neutrino supplied the constituents of the nuclei of the atoms, stable and unstable, at everyday energies. The muon and its neutrino and the strange quark were of the next generation. They made their appearance in the first gush of unstable particles from the accelerators. This generation was lacking, however, a quark partner for the strange quark, as the down-quark in the first generation had its up.

The GUT theory advanced by Glashow and Howard Georgi at Harvard for the grand unification of the electroweak and the strong force called for a fourth quark, to which Glashow gave the name "charm," to consort with the strange quark and complete the second generation of particles. Indeed, the scheme left open the possibility of a third and more generations to come.

Experimental physicists set out on the hunt for the fourth quark. In November 1974, experiments at the powerful Brookhaven accelerator conducted by Samuel C. C. Ting and his Columbia associates and at SLAC by a Stanford group led by Burton Richter simultaneously produced the heaviest particle then on record. At SLAC, colliding beams of electrons and positrons attained the necessary 3-GeV energy; the hugely energetic photon from their annihilation rematerialized in (among other particles) a particle its discoverers designated ψ, "psi." Ting called it "J." Glashow was happy to call it "charmonium," on the calculation that it was composed of his proposed charm and its antiquark. In 1976, a particle that combined a "naked" charm with a down-quark established the existence of the fourth quark beyond doubt. Therewith, another aspect of this family of particles came sharply to notice. At 1.8 GeV, it fit the trend of increase in quark masses by an order of magnitude at each step up from the .005 GeV mass of the bottom quark. With four quarks to match the four leptons, the symmetry was restored in the existence of two matching families [see illustration, page 167].

The restoration was momentary. It occurred to Martin Perl at Stanford to see whether, at the new high level of energy available in the colliding electron-positron beams, the SLAC accelerator might be producing a massive particle that decayed by the weak rather than the strong interaction. With an "all-purpose" detector designed by colleagues at Berkeley, he found evidence in 1975 for what he had been calling the tau (τ) particle, "t" in the Latin alphabet, for third electron. Confirming evidence soon came from a new powerful accelerator at Hamburg, Germany. Though the tau neutrino was yet to be observed, the tau lepton and a possible third generation of quarks had to be reckoned with. Needless to add, the tau proved even more massive than the muon. With 1,860 eV mass-equivalence, the tau—the electron of this generation of particles—is heavier than the proton.

In 1977, the discovery of the fifth quark rewarded the investment

Particles	symbol	mass	charge	lifetime	decay products	electromagnetic	strong	weak
Electron	e	1.0	−1	∞	stable			
Neutrino	λ	0.0	0	?	stable?			
Proton	P	1836.1	+1	10^{32} YRS	stable			
Neutron	N	1838.6	0	10.3 MIN	e λ P			

Forces	strength	range	reaction time	quantum	rest mass
Electro-magnetic	.05	∞	10^{-21} SEC	photon	0
Strong	1	10^{-15}	10^{-23} SEC	gluon	???
Weak	$10^{-2?}$	$10^{-?}$	10^{-9} SEC	W0+−	80−90 GeV
Gravity	10^{-39}	∞	???	graviton?	0

Four particles and four forces *and their interaction underlie physical reality in common experience. Particle masses are expressed in the cross-referential determination by J. J. Thomson that the charge-to-mass ratio of the electron is nearly 2,000 times that determined for the hydrogen ion—the proton—by Michael Faraday. The strengths of the forces in their action between particles are expressed in similar cross-referential terms, with that of the strong force set at 1. The tiny force of gravity shown is that between two particles.*

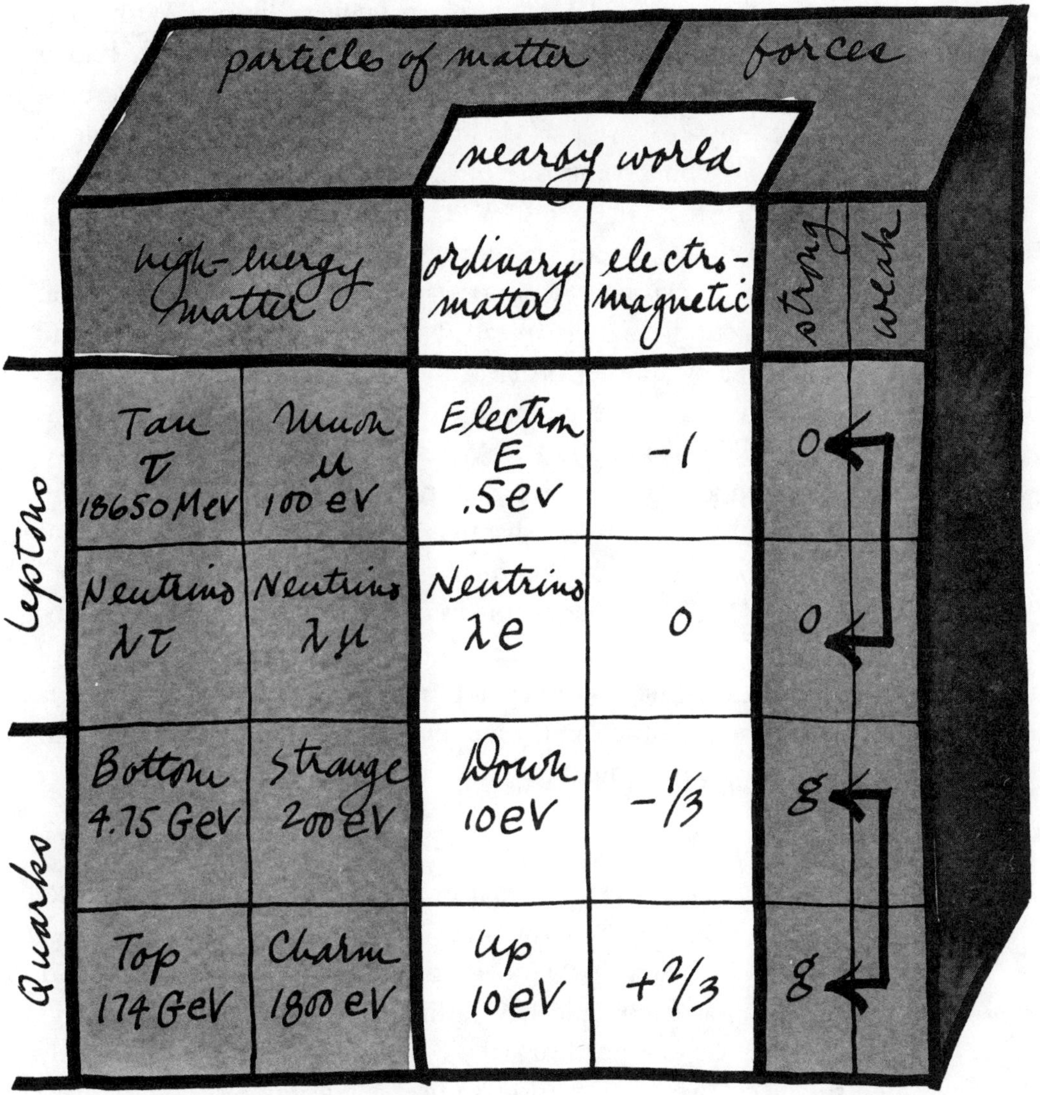

	particles of matter			forces	
	high-energy matter		ordinary matter (nearby world)	electro-magnetic	strong	weak
leptons	Tau τ 18650 MeV	Muon μ 100 eV	Electron E .5 eV	−1	0	
	Neutrino ντ	Neutrino νμ	Neutrino νe	0	0	
Quarks	Bottom 4.75 GeV	Strange 200 eV	Down 10 eV	−1/3	8	
	Top 174 GeV	Charm 1800 eV	Up 10 eV	+2/3	8	

Matter and energy are charted here in the interactions that underlie physical reality in common experience and in the state of the observable universe in the first second of its genesis. The ultimate (in present understanding) particles are two each of two kinds, lepton and quark; their masses are expressed in units of energy with which they are interchangeable. Absent is gravity for which its quantum is postulated in theory, not now in reach of experiment such as that to which all of the understanding displayed here has been subjected.

in the world's most powerful accelerator at Batavia, Illinois. With energy of 400 GeV available at the target, two orders of magnitude higher than at SLAC, Leon Lederman and his associates established the existence of a new most massive particle at 9.5 GeV. Called the "upsilon" (υ), it was shown to be a composite of the new b, for "beauty" or "bottom" quark, with a mass of 4.75 GeV, and its antiquark.

To match the beauty or bottom quark, there had to be a "truth" or "top" quark. Its existence was not established until 1994. By that time, superconducting magnets installed in the Batavia accelerator had increased its power by another order of magnitude, to 1 TeV. By colliding-beam strategy, that could be doubled to 2 TeV. The top quark was found at a mass-equivalent of 174 GeV [see illustration, page 167]. Two orders of magnitude more massive than the bottom quark and five and six orders of magnitude more massive than the up- and down-quarks that constitute ordinary matter, the top quark seems to answer one question. Its mass is so close to the lower estimates of the mass of the Higgs quantum that it leaves no room between them for another generation of mesons and baryons. Successful generation of the tau neutrino at Batavia in July 2000 completed the three generations of particles. There is confidence now that the next quarry is, indeed, the Higgs quantum.

The "God Particle"

That was to have been the first order of business at Waxahachie, Texas. The 40-TEV Superconducting SuperCollider was to look for what Lederman has called the "God Particle," at an energy of around 1 TeV. The Higgs particle would have affirmed the existence of a Higgs field, in this case at energy near 10^3 TeV. That would have doubly affirmed the electroweak unification and set a milestone on the road to verification of the GUT theories. The origin of rest mass and so of the everyday matter known to humankind would have had, accordingly, its theoretical explanation supported—or not—by experiment. The leptons and the quarks would today have more assurance of their common ancestry. Lederman's blasphemy, if any, was to the point.

The new accelerator at CERN and the upgrading at Batavia may yet, and even soon, bring in the Higgs particle. Success may encourage investment in the reach for another or even two orders of magnitude in accelerator power. The energy at which the electroweak and the strong force converge will lie still another 10 or 11 orders of magnitude beyond. Approach to the present limits of experiment and foreclosure on investment at nearer energies have confined the talent of the rising generation of physicists to theory.

Theories accordingly abound. Building upon or overleaping GUT theory, they reach for the ultimate unification embracing all four forces of nature. Steven Weinberg—observing that the electroweak and the strong force unify at "only" two orders of magnitude below the energy at which the three forces become one with gravity—has encouraged the overleaping strategy.

The "supersymmetry" family of theories introduces a superpartner for each of the rest-mass particles and antiparticles and the quanta. It fuses the families of the particles and quanta, moreover, into one sort of entity; it promises to bring gravitation into coherence with the three polar forces more naturally and it offers a variety of new particles to account for the missing "dark" matter that troubles the cosmologists. Some of its propositions might have come within reach of the SSC. Evidence for one of them will be looked for in the attempts to produce the Higgs particle at Batavia and at CERN. A "peculiar" form of supersymmetric matter figures in the inflation-prelude to the Big Bang propounded by Alan H. Guth [see page 237]. This amendment of the new genesis is strongly ratified by the isotropy of the 2.725 K cosmic background radiation observed, with departures of no more than 50 parts per million, around the vault of the sky above and below the plane of the Galaxy.

The "string" and "superstring" theories secure unification of the four forces by addition of six dimensions to the four dimensions of spacetime. Strings replace point-particles, getting rid of the infinities that beset points. The extra six dimensions curl up the strings inside the particles within a 10^{-33}–meter diameter in the four dimensions of spacetime. Vibration of the strings at critical modes, in the metaphorical operation of the equations, generates the forces and particles.

Such theorizing brings physicists into philosophical division. Some seniors have more sympathy than others for the enterprise of their successors. Steven Weinberg was welcomed not long ago to a colloquy at Harvard with a limerick by Howard Georgi, cheerfully declaring the positivist view:

> Steve Weinberg, returning from Texas,
> brings dimensions galore to perplex us.
>> But the extra ones all
>> are rolled up in a ball
> so tiny it never affects us.

More recently, with no prospect for approaching the energy required for GUT experiments, Glashow has conceded that theory may just as well proceed to the ultimate unification of the quantum forces with gravity at two orders of magnitude beyond. From the intricacies of string theory may yet come the design for a feasible experiment—what Einstein called "conclusions that can be confronted with experience."

Since the superquarks may be taken also as ancestors of the mere quarks, regenerated in the accelerators at energies prevailing in the first second of creation, particle physicists are seizing the occasion for collaboration with astrophysicists. They are particle physicists already engaged in cosmology. Together, they will look into the continuous blizzard of cosmic rays—high-energy particles—that plunge into Earth's atmosphere from their origin in cataclysmic events in faraway spacetime.

A tiny percentage of cosmic rays have been found to carry energies at the upper end of the accelerator range. They make their arrival known by collision with atoms in the outer atmosphere and resulting showers of daughter and granddaughter particles. Sufficiently large arrays of detectors on the ground ("sufficiently" = tens of thousands of detectors arrayed across half of Connecticut) may observe such events. If the capital for this investment can be gathered, then, after a decade or two of waiting and counting, the detectors might record plausible evidence of the arrival of a Higgs particle. If so, the partnership of particle physicists and cosmologists will then have reached into the first 10^{-20} of the first second of the Big Bang, whence the observed universe came.

4

Space and Time

We make bold to contemplate a Universe
in which all Newton's is but a speck.
D'Arcy Thompson

P article physicists, in their partnership with cosmologists, find gravity on their agenda. Throughout most of the past century, they were able to ignore this force of nature, the first in human experience. They were engrossed in study of electromagnetism, first recognized in lightning. Then, in the 20th century, they discovered the two short-range forces, the weak and the strong. They have successfully comprehended the three forces in the conceptual system—quantum field theory—that crowned their inquiry into the first of them. Inquiry into the role of these forces in the structure of Earth-bound matter has brought understanding of events transpiring on a titanic scale in the distant universe. Even out there, particle physicists could ignore gravity because it exerts between two charged particles a tiny 10^{-39} fraction of the force of electromagnetism. It is weaker even than the weak force, exerting only 10^{-25} the strength of that force between two particles.

Gravity nonetheless now compels their attention. In the particle-accelerator experiments that brought the three quantum forces into unity they discovered that they were reenacting the history of the universe from near its very start. Back there and then they encountered the cosmologists. They found the cosmologists working with an entirely different conceptual scheme. This is the theory of General Relativity. A field theory purely, it structures the four-dimensional spacetime of the universe in the gravitational fields of its contained masses. While the accelerating motion of a mass sets waves moving in its fields at the velocity of light, the theory recognizes no quantum of their action. General Relativity was nonetheless accounting for all that the cosmologists were learning about the structuring and the history of the universe in the thrall of all-pervading gravitation.

"So long as gravity remains an un-quantized force," says Paul Davies of the University of Adelaide, and everyone agrees, "there exists a devastating inconsistency at the heart of physics."

The force of gravity

In Galileo's "heaviness" is the human experience of force, of work and fatigue. Gravity organizes space on Earth in three dimensions; the falling of things sets the vertical coordinate from which people draw the horizontal. Twice a day, the tides testify to the gravitational bond between the Sun, the Earth and the Moon. It can be seen that gravity has shaped living forms, from flounders to eagles, at all times in the course of their biological evolution. Against this force the bipedal human stands erect.

Astronomers could not fail to keep gravity at the center of their attention. Now capacitated by their instruments as observational cosmologists, they see gravitation—the force in action—as the supreme organizer of the cosmos. Unlike the other three forces, gravity increases in step with the gathering of mass under its force. As astronomers learned only in the 20th century, gravity keeps 100 billion stars wheeling in the arms and thronging in the centers of spiral galaxies and densely massed in elliptical galaxies. It gathers these island universes by the hundreds and thousands in galactic clusters and superclusters. Within the 15- to 20-billion-light-year radius of the observed cosmos, gravitation contains an estimated 100 billion galaxies.

In the local Galaxy—the whole universe as recently as the 1930s—obscuring clouds of interstellar dust and gas, light-years in breadth, gather under the one-way attractive force of gravity. Here gravitation switches on the quantum forces. In denser globules within the cloud, increasing in force with the mass and density of the matter aggregating under its force, gravitation collapses the material into protostars. Heat and pressure within the infalling mass set the new star glowing, first in the infrared wavelengths of the electromagnetic spectrum.

Then, crushing atoms together at still higher density, gravitation ignites the thermonuclear alchemy of the strong and the weak force. Naked nuclei of hydrogen, the most abundant of the atoms in the universe, are fused to nuclei of helium, the next most abundant. The loss of mass to energy in the thermonuclear reactions sets up the counter-pressure of heat—the high-velocity motion of the naked atomic nuclei inside the star—sufficient to withstand the inward drag of the gravitational force. Starlight now radiates across both visible and invisible wavelengths of the electromagnetic spectrum.

An average star, of about the mass of the Sun, burns most of the ration of hydrogen it is fitted by its size to burn in 10 or so billion years. It exhausts the last of it in its briefer life as a Red Giant, inflated to as much as 250 times its original size. In the case of the Sun, that will be well beyond the Earth's orbit. Gravitation next collapses the Red Giant to the size of a small planet, ejecting much of its mass into interstellar space. To the inventory of elements in the universe the ejected mass returns almost all the hydrogen and contributes an infinitesimal quantity of helium and such other elements as this type of star cooks in its lifetime. Now a "White Dwarf," the star glows with the residual heat of its former glory and of its abrupt collapse. Resistance of the stellar mass to the further action of gravity, which has squeezed out all interatomic space, is now supplied by quantum mechanics. In accordance with the Pauli exclusion principle, electrons resist compression to the same energy level [see page 129]. A cubic centimeter holds 10 tons of this stellar mass. The White Dwarf cools down. Its light fades; it is a Black Dwarf.

Upon exhausting their fuel, bigger stars, up to three solar masses, collapse to the extreme density of their nuclear particles. They do so having first exploded as supernovae, blowing off into space the larger portion of their mass. For a few days they may outshine a whole galaxy. Under the enormous gravitational force in the interior of these stars during their lifetimes and the immense instantaneous pressure of their thermonuclear explosions, much of the rest of the elements in the periodic table reactions is synthesized. The huger remnant mass of a supernova collapses into a neutron star, an object smaller than a White Dwarf. The correspondingly huger gravitational force squeezes electrons into protons, transforming protons to neutrons. A neutron star has a diameter of 10 kilometers and a density 10^8 times that of a White Dwarf: a billion tons per cubic centimeter, the density of the atomic nucleus.

Such is the energy bound in matter, the steady and explosive generation of starlight over the history of the universe has burned no more than about 10^{-4} of its primordial mass. The ashes of the thermonuclear furnaces—all the elements heavier than hydrogen and helium—constitute around 10^{-3} of the mass of known present matter. A sample of this moiety is collapsed by gravity in the solid masses of the terrestrial planets of the solar system.

From stars in their afterlife as neutron stars or White Dwarfs gravity may extract more energy than they ever yielded in their eons of thermonuclear radiance. In conservation of angular momentum, gravitational collapse speeds up the period of the star's rotation from days down to minutes, seconds and even milliseconds. The star becomes a dynamo, the strength of its magnetic field correspondingly amplified. The field entrains the nearby interstellar atmosphere and the gases of the companion star, if there is one, in pyrotechnics on all frequencies that may outshine the star's former visibility. The radiation that discloses the presence of these objects is familiar to particle physicists as the "synchrotron radiation" that steals energy from particle beams under acceleration on circular orbits [see page 152].

The gravitational force, increasing without limit with the mass and density of the matter gathered in its thrall, as at the center of a galaxy, may ultimately overwhelm everything in reach. Above three solar masses, it is thought, the assemblage collapses into a black hole. Particles of matter-energy are crushed out of existence into a "singularity," a point of infinity, that conserves no more than the angular momentum of their descent into the hole and their net charge. From a black hole not even photons can escape. That is their blackness.

The black hole offers the best explanation of sources of radiant energy that outshine—not always on visible wavelengths—all others in the observed universe. Such are the quasars, short for quasistellar objects, now well identified as massive quasigalaxies. Quasars shine with synchrotron radiation flooding, it is thought, from the acceleration of thousands of solar masses to nearly the velocity of light inward to vanish in black holes at their centers.

Geometrodynamics

Gravity is, in sum, the supreme force in the cosmos. Gravity is not, however, the force so called in common human experience—as Einstein's hapless "observer" sensed in his fall from the roof and weightless astronauts can testify [see page 102]. The gravitational force is exerted, Henry Cavendish showed, by matter in proportion to the mass of matter present. It does not have the pull of the quantum-force fields, nor does it repel. The equations of General Relativity show gravity working in an entirely different way. Exerted by masses, gravity acts not on other masses, but on the spacetime in which they float. Yet, of course, gravity feels like a force acting on the body. The body similarly senses centrifugal "force." This is not a force in its own right, but a sensation experienced when the body's motion is deflected from a straight line.

Space and time, separate and absolute, are the theater of events in classical physics. The "unpassable limit" of the velocity of light collapses them, however, to four-dimensional spacetime. In Special Relativity, the energy imparted to the velocity—a variable measured in space and time—of a moving body increases the mass of the body without limit as the velocity approaches the limiting velocity of light [see page 95]. In General Relativity, the gravitational fields of stellar and galactic masses structure the dynamical medium spacetime.

Gravitation—happily called, by John A. Wheeler, the force of "geometrodynamics"—warps spacetime in proportion to the presence of mass. The geodesics traced by the acceleration and deceleration of masses on gradients in spacetime manifest the contours of the fields of gravitational force exerted by masses everywhere in motion in the universe. At the ultimate velocity, photons of radiant energy increase and decrease in wavelength and effective mass as they trace their geodesics. Taking these geodesic as straight lines, the well-known shortest distance between two points in plane geometry, four-dimensional geometry warps spacetime with their curvature.

The gravitational field of the Earth has a radius of curvature of about 10^{11} meters, about two-thirds of the distance to the Sun. In three-dimensional space the curvature of a line traced by such a long radius would be imperceptible even over long distances. Through four-dimensional spacetime in the Earth's gravitational field, the free fall of an object accelerates speedily in the time direction: in five seconds to 49 meters per second. But 49 meters per second divided by the velocity of light yields an imperceptible 1.6×10^{-7} meter in the space direction, its curvature still less perceptible. People find the effect of this curvature nonetheless easy to detect and learn to be wary of it.

The warping of spacetime around a center of mass goes to ever shorter radius of curvature with increase in mass and density. Density is of high consequence in setting the radius of curvature. While the Sun is 300,000 times more massive than the Earth, it is a gassy body; its radius of curvature extends only four times longer than that of the Earth. At the center of a black hole, where density approaches infinity, the radius of curvature approaches that of a geometric point. The acceleration of infalling masses there approaches the velocity of light. This evidence for the presence of a black hole has been observed in the rising frequency of radiation emitted by particles on their descent into it. More recently, it has been observed in the stretching of the wavelengths and then the extinction of light at that boundary whence not even photons return.

Geometrodynamics requires relaxation of the earthbound kinesthetic impulse that asks for muscular exertion to start motion. Nowhere in spacetime is there a body at rest or a point of rest. Masses at large in the universe are everywhere in free fall, as Wheeler puts it, in the gravitational field of their local spacetime. With no "up" or "down" or other coordinate in the cosmos, it cannot be said which way things are falling except with respect to each other. There is no need for the Newtonian "expression for force" [see pages 45 and 47] to get things in motion. On the cosmic scale, as well as on the scale of quantum events, motion is the state of nature.

Bodies on the surface of the Earth have been stayed in their free fall by that surface. That fall resumes upon its falling, jumping or being pushed off a surface prominence. All bodies on the Earth go with the planet on its free fall in the gravitational field of the Sun, with the Sun's free fall in the gravitational field of the Galaxy, with the Galaxy in free fall on the gradients of the presence of mass in the surrounding cosmos.

Newton's laws of motion faithfully account for motion to the limits of mass, velocity, and distance in common experience on Earth. Bodies can be, for convenience of the inhabitants, taken to be at rest or in constant motion with respect to the planetary frame of reference; they are accelerated from rest into motion by forces equal and opposite to their inertia. Newton's gravity likewise accounts, with minor correction, for the mechanics of the solar system. General Relativity comes in to make sense of the nearly incommensurable universe beyond.

In their respective realms Newton's and Einstein's physics show how the world works. The new physics asks what the world is made of and, now, how it came to be. Classical physics interdigitates, with the new physics on the vast as well as on the tiny side of its range of amplitudes; its conservation laws are conserved at both extremes.

General Relativity under test

In the 1970s, physicists engrossed with the discontinuous quantum were confronted with the power of the classical concept of the continuous field. Their inquiry into the nature of matter had taken them into cosmology. They subjected General Relativity to rigorous new test by experiment. Einstein's crowning invention has held its own down to ever smaller margins of error.

The Pound-Rebka experiment at Harvard University, for example, used the technology of quantum physics to confirm the gravitational redshift in the modest gravitational field of the Earth [see page 106]. At Princeton University, Robert Dicke and Carl H. Brans conducted an improved Eötvös experiment. Their results proved in closer conformity with General Relativity than with their own attempt to improve upon it. With reflection of laser light from "corner reflectors" placed for the purpose on the Moon, it became possible to measure the 400,000-kilometer distance to the Moon within an error of 30 centimeters—an error of 7 × 10^{-10}. The Earth and the Moon could thereupon be employed in a cosmic Eötvös experiment [see page 102]. The acceleration of the two masses toward the Sun proceeds at the same rate within the 30-centimeter margin of error.

Irwin Shapiro, at Massachusetts Institute of Technology, took advantage of the space program to demonstrate the time delay of light that attends its deflection in a gravitational field. He and his colleagues exchanged radio signals with a transponder aboard the Viking lander on Mars, comparing time of signal travel with the position of the planet in the sky closer to and farther from the Sun—after the fashion of the Roemer observations of the Jovian eclipses, which established the finite velocity of light [see page 54]. The measured variation in the time of the round-trip spread, in accordance with theory, by as much as .002

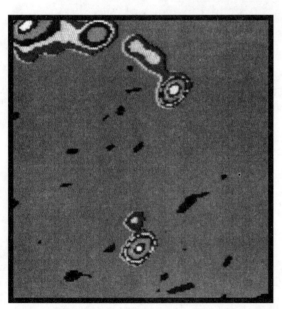

second. Measurement of the deflection by the Sun of grazing radiation from quasars now yields the deflection of electromagnetic radiation at all angles in the gravitational field of the Sun.

The discovery in 1979 of a "double quasar" [left] dramatically demonstrated the curvature of spacetime in the presence of gravitational mass. Two apparent quasars were shown to be one; its two images are refracted by the curvature of space by and around unseen masses in the line of sight from Earth. The detection of such gravitational lensing has become a technology. Multimirror telescopes are engaged in perfecting the measurement of the warping of spacetime within the Galaxy by the presence of unseen mass. This is the mass required, in addition to the luminous mass, to supply the gravitational field that organizes the Galaxy and gathers galaxies in clusters.

The 1970s brought secure evidence for perhaps the most crucial prediction from General Relativity. Masses that induce curvature in spacetime by their presence must, in accelerating motion, send gravitational waves through the curvatures at the speed of light. Such waves still await detection. An "average" gravitational wave should compress and stretch materials by one part in 10^{22}. A wave from a supernova might induce distortion of one part in 10^{18}. That would be 10^{-3} millimeter in a billion kilometers. So small an effect eluded years of ingenious experiment by Joseph Weber at the University of Maryland.

It was the six-year study in the 1970s of the binary pulsar using the 1,000-foot radio telescope at Arecibo, Puerto Rico, that produced the conclusive evidence for gravitational waves. Joseph Taylor and his University of Massachusetts colleagues established the loss of energy to gravitational waves in the shrinking of the tight, high-velocity orbit of a neutron star around a companion neutron star [see page 106].

Theory addressed to healing the "devastating contradiction at the heart of physics" postulates a "graviton," the quantum to go with the gravitational wave and bring the four forces of nature into unity. The graviton is embedded in quantum-field theories in which cosmologists now envision the universe. For its detection, no experiment has yet been devised; it makes no appreciable difference—except in theory.

General Relativity stands meanwhile on its own feet. "No inconsistency of principle has ever been found in Einstein's geometric theory of gravity." John A. Wheeler wrote in 1970, "No purported observational evidence against the theory has ever stood the test of time."

Cosmic past and future

There was a time, projected in theory from experiment and observation by particle physicists and cosmologists, when the entire cosmos was in the grip of gravity alone. Mass-energy—or whatever it was at that instant—was concentrated within the then nearly infinitely small radius of the curvature of spacetime. A cubic centimeter of the universe then contained 10^{100} grams of matter or, equivalently, 10^{120} ergs of energy. The temperature stood at an immeasurable 10^{32} degrees K. Gravity, if it could be called gravity in the absence of other forces to be distinguished from it, had perhaps overreached its strength. For the so-called Big Bang was already under way. Successive phase transitions—as from the solid to the liquid state—brought particles into interaction with quantum forces; protons and neutrons were soon settling out of the chaos of high-energy particles.

The primordial explosion continues to this day in the expansion of the universe. The supreme force of gravity may be yielding its grip on the whole. Until recent observation compelled consideration of that possibility, it was the density of the universe that preoccupied cosmologists. At the critical density, General Relativity showed spacetime to be flat; the expansion of the universe would continue, decelerating, forever. At density above critical, it showed the curvature of spacetime to be positive; expansion would reach a limit and the universe would then contract in a "Big Crunch." At density below the critical, the curvature would be negative and the universe would expand forever exponentially. The size, age and rate of expansion have argued strongly for flatness. If the density at the beginning were off the critical in either direction by the tiniest fraction, time would have rapidly magnified the departure, and the universe would long since have evaporated or collapsed.

The question of density has been wide open. The luminous mass comes to barely 1 percent of that required to account for the age of the universe and the rate of its expansion indicated by everything else that is observed. Now this question is overshadowed by evidence that gravity may be in contention with a hitherto unrecognized force of repulsion. Precise observation of the redshift of galaxies in the middle distance indicates that the rate of universal expansion has recently, on the cosmic time scale, come under acceleration. A "dark energy" is postulated. This would be an energy of the vacuum welling from the Higgs field, the source of the rest mass of particles [see page 160].

Even this turn of events had its foreshadowing in General Relativity. In 1917, Einstein was moved by "Cosmological Considerations in General Relativity" to introduce, in a paper with that title, a correction to his original equations. His "cosmological term," stabilizing the universe against collapse, did service as a force of repulsion.

The vision of the Big Bang gained persuasive evidence from the experiments of particle physics, as related in the preceding chapter. To affirmation of the vision observational cosmology has contributed two conclusive bodies of evidence: the cosmic redshift and the cosmic microwave background radiation.

Cosmic redshift

The redshift is established by observation of the spectra of the light from thousands of galaxies. Lines in these spectra are invariably shifted toward the red, long-wavelength end of the spectrum. Such lengthening in the waves of sound from a retreating source, as from the whistles of locomotives or nowadays the horns of 18-wheel tractor trailers, is familiar as the Doppler effect. The same effect is observed in the lengthening of wavelengths of light from cosmic sources. The explanation of the cosmic redshift begins with the further observation that it increases with the distance of the light-emitting galaxies from Earth. Light started from those galaxies longer and longer ago. At each increase in redshift, the universe was not only younger but smaller and denser. It is the expansion of the universe itself, the expansion of spacetime through which light travels at its constant velocity, that stretches the wavelengths in the spectra of the galaxies in proportion to their distance from their observers. The longer ago the light started from them, the greater the expansion of spacetime, the larger the redshift, the greater the apparent velocity of their retreat. Without precise measurement of distance to a particular galaxy, however, its redshift tells nothing about that galaxy's motion in space. From the fixed point of observation to which they are confined, moreover, observers must imagine their Galaxy presenting the same redshift to observers, if any, in a galaxy for which they have measured a redshift.

The second contribution from observational cosmology goes straight to confirm the abrupt expansion of the universe from its primordial atom, the Big Bang itself. From reflection on that event, first proposed by George Gamow in the 1930s, Ralph A. Alpher and Robert C. Hermann at Cornell University predicted that an echo must persist as background radiation pervading the cosmos. Accidental discovery of the radiation in 1965 confirmed the prediction. The spectrum of this signal from the genesis of the observed universe and the cosmic temperature it indicates fit closely to the calculation of those numbers from theory.

The cosmos now in understanding confronts human identity with still another Copernican revolution. Along with their dislocation to nowhere in particular in space, people have now learned that their existence is cast at no particular point in time. The species came along 10 or 15 billion years into the history of the universe. Cataclysmic events, dwarfing and paling any they have ever before imagined, have transpired in the local Galaxy and may still impend. Stars like the Sun have come and gone by the tens of millions. The Sun itself is fated, in about 5 billion years, to enter its Red Giant phase. The human species as well as the human individual must be reckoned as mortal. The cosmos will continue, evidently, on its heedless course long after its human observers are gone, as it did before they arrived.

The cosmos now understood allows, however, for a human future of longer duration than any people have ever planned for. There would seem to be time enough for any purpose. The attempt at civilization so recently begun has led, still more recently, to the organized acquisition of objective knowledge. In the nick of time, it may be, enough can be learned to keep the future open against habit and institutions that may interdict tomorrow.

Only in the 20th century did astronomy reach into the cosmos beyond the solar system and its immediate stellar neighborhood. As recently as 400 years ago, the sky was a dome of fixed stars in the perception and imagination of almost everyone. It occurred to Copernicus, when he placed the Earth on orbit around the Sun, that it might be possible to measure the distance to the stars. He saw that their apparent position might change when seen from one and then the other side of the Sun. He tried to detect the angle of parallax, the measure of that change [see illustration, page 185]. He failed but attributed the null result to the failure of his instruments at so tiny an angle. Tycho Brahe attempted the same measurement with his precision instruments. He took his null result to show that Copernicus had wrongly displaced the Earth from the center of the cosmos and amended the Ptolemaic system to accommodate the elliptical orbits on which his calculations placed the planets.

Some observers and thinkers did conceive of the possibility that the sky might go on and out forever and that stars range in space at long and different distances from the Earth. This was one of the heresies for which Giordano Bruno burned at the stake in 1600. Around 1660, Christian Huygens showed, from mathematical consideration of those point sources of light in the night sky, that the stars must be distant suns. The sky nonetheless remained closed for more than another century. As late as 1835, August Comte, protagonist of empiricism, declared, "the field of positive philosophy lies wholly within the limits of the solar system, the study of the universe being inaccessible in any positive sense."

Stellar parallax

In 1838, only three years later, objective measurement penetrated the vault of heaven. At the University of Koenigsberg, with a new instrument made by Joseph von Fraunhofer for accurate measurement of the angles at which it was pointed, Friedrich Wilhelm Bessel established the parallax of the star designated 61 in the constellation Cygnus. He determined its parallactic angle at around 0.3 second of arc. That set the distance to 61 Cygnus near 10 light-years.

Observation from opposite sides of the Earth's orbit subtends an angle of one second of arc at the distance of 3.26 light-years, a unit designated as 1 parsec. No star but the Sun is inside that radius from the Earth. The nearest is Alpha Centauri, at 1.32 parsec or 4.3 light-years. Bessel's observations of hundreds of stars showed a few to be closer and most of them more distant than 61 Cygni. These numbers explain why not even Tycho could resolve the parallactic angle of a star. Astronomers use the parsec (cosmologists, the megaparsec: 3.26 million light-years), rather than the light-year, as their unit of measurement. Thereby, perhaps, they keep forward in the mind's eye the geometry to which they must ultimately refer the very much longer and frailer yardsticks with which they reach into the cosmos.

By 1875 measurement of the parallax of 5,000 stars carried the known universe out to hundreds of light-years. The invention and refinement of photographic film then brought the discovery that the "fixed" stars of all prior human experience are in motion in all directions across the sky. Comparing photographs taken years apart, astronomers found thousands of stars in different locations in their star fields. The overwhelming number of stars remained "fixed," of course, too far away for detection and measurement of their motion. The "proper" motion—across the line of sight—of the unfixed star images on the photographic plates could be precisely measured. Measurement of the redshift of these stars disclosed the velocity of their motion away from the Earth. The two measurements combined yielded the distance to them. Geometry thus carried the observed edge of the universe out to 3,000 light-years by the end of the century.

Assured geometric measurement of distance made it possible to establish the "intrinsic" or "absolute" brightness of stars. With similarity of stellar spectra taken to establish similarity of intrinsic brightness, it became possible to estimate from their apparent brightness the distance to stars for which no proper motion can be detected. Falling off as the square of the distance, the apparent brightness of a star yields a measure of its distance upon comparison with the intrinsic brightness of a similar star of known distance. Astronomers now had a yardstick to take their measurements beyond the reach of geometry.

At the Harvard Observatory early in the 20th century, Henrietta Leavitt made a discovery that vastly extended the reach of this yardstick. The observatory's southern station in Peru had accumulated a large number of photographic plates of the Magellanic Clouds (Ferdinand Magellan's navigators had made note of these "nebulae," diffuse glows in the southern sky). Comparing plates exposed days or weeks apart, Leavitt discovered that certain stars in the clouds varied in apparent brightness from day to day and week to week. Comparing successive images of 25 of these stars, she found a linear relation between their apparent brightness and the period of their variation. Their presence in the Magellanic Clouds placed them all at the same distance from the Earth. Leavitt recognized that their comparative apparent brightness must correspond to their comparative intrinsic brightness.

(A few years later, another female astronomer employed by Harvard Observatory in the clerical duty of comparing stellar spectra made another major contribution to astronomy that sets the terms of

Measurement of parallax *of nearer stars anchors all of the cosmic yardsticks [see page 183]. Observation from opposite sides of the Sun subtends one second of arc, 3.26 light-years. At .01 second of arc, this yardstick brings now 300 light-years and 100,000 stars and globular clusters into reach of geometry. "Absolute" brightness of these objects at measured distances calibrates comparative brightness scales. Redshift scale [see page 189] is buttressed by newly established scale based on the flash-profile of Type 1a supernova [see page 215].*

cosmology. From the prominence of the emission lines of iron in the spectrum of the Sun and the commanding presence of this element in the Earth, especially in the magnetic-field-generating core, iron was held to be the most abundant of the elements. The Cecilia Payne Gaposchkin inventory of the spectra of countless stars established hydrogen in that status. After some controversy, her findings were confirmed, with hydrogen in overwhelming 93 percentage followed by helium at 7 percent of the atoms in the universe and all the rest of the elements, including iron, concealed in the rounding of the numbers.)

Variable stars turn up commonly in the globular clusters that adorn the local Galaxy. In these clusters, 100,000 to a million stars assemble in self-gravitating systems. At Mt. Wilson Observatory in California, Harlow Shapley deployed the 60-inch telescope, the first of the great California telescopes, to survey the known 100 globular clusters. He adapted Leavitt's period-luminosity relation to calibrate the distance to them. With the distance to the nearest clusters calculated by comparative brightness, he established an intrinsic brightness for the variable stars in these clusters. With this yardstick, Shapley measured the distance to the one-third of the clusters gathered in the great star cloud of Sagittarius on the southern horizon of the Milky Way. He calculated the region to be 17,000 parsecs, 55,000 light-years, from Earth.

With the publication of Shapley's finding in 1918, other large-scale observations fell into place. The concentration of stars in the Milky Way had persuaded some 19th-century astronomers that the universe might have the shape of a grindstone, seen edge-on from Earth. Shapley's observation put the center of this grindstone at a theretofore unimagined distance away. It also put the Sun and Earth in an unremarkable position near its apparent outer edge. In 1922, now director of the Harvard Observatory, Shapley reviewed Leavitt's photos of the Magellanic Clouds. It was clear that most of her variable stars were located in globular clusters in those nebulae. Shapley's measurement of the apparent brightness of the clusters showed the nebulae to be 75,000 light-years from Earth. Azimuthal measurement placed them outside the plane of the Galaxy. The Magellanic Clouds were, it was apparent, star systems outside of, smaller than, but otherwise comparable to that of the Galaxy.

With the new 100-inch Mt. Wilson telescope, Edwin Hubble used the Leavitt-Shapley yardstick to measure the distance to the globular clusters in a nebula visible as a diffuse patch of light in the constellation Andromeda. The first big-telescope pictures of this Andromeda nebula had already established it as the model spiral nebula [right], seen tilted somewhat from face-on, of the many such nebulae that were coming in through the big telescopes. The 100-inch telescope resolved its brighter individual stars. It more easily resolved the globular clusters concentrated, as in the Galaxy, around the center of the system—the star-crowded nucleus from which the spiral arms trail outward. Hubble's redshift measurement placed the nebula at the distance of a million light-years, two orders of magnitude more distant than the Magellanic Clouds.

At that long distance, the breadth of this nebula on the photographic plate had to be 50,000 light-years. Andromeda was, beyond question, an "island universe" fully comparable in size to the Galaxy that once had been thought to comprise the whole universe. Numerous other spiral and elliptical nebulae had by this time registered on the astronomical plates. They too had to be recognized as island universes, outside of and still more distant from the Galaxy.

Hubble called the island universes "nebulae," the term by which the first of them had come into astronomy. Shapley called them "galaxies." Today that term recalls his rivalry with Hubble for honors as opener of the door to the very much larger and grander universe.

If the Galaxy had to lose its standing as the whole universe, it was possible to imagine that, face-on, it might be seen to be as glorious as the great spiral in Andromeda. In the Milky Way, a spectacle that promises such splendor, our local island universe is seen edge-on. The Galaxy had now to be accepted as one of the perhaps countless galaxies that were waiting to be counted. In every patch of the polar sky outside the plane of the Milky Way grasped by the 100-inch telescope, the number of new galaxies registered was proving to be a function of the time exposure of photographic film. By high-precision pointing of the massive instruments at the same guide star, night after night, holding the star in place against the motion of the Earth, these exposures are lengthened far beyond the few best viewing hours in a night.

By configuration, these stellar systems sorted into three classes. There are the so-often-spectacular spiral galaxies. Astronomical plates captured these celestial pinwheels at all angles to the Earth, from edge-on to face-on. In their blazing central nuclei are concentrated 90 billion of their estimated average 100 billion stars. Lanes of bright stars and ragged dark and bright clouds of dust and gas define their spiral-wound arms. The sense of rotation conveyed by the trailing arms and the oblate spheroidal shape of the nuclei is, happily, confirmed by Doppler shift.

Less numerous are ellipsoidal galaxies—brilliant nuclei without spiral arms. In somewhat larger number appear the irregular galaxies, the least irregular looking like spiral arms without nuclei. In the irresistible leap from taxonomy to evolution, an order of development was declared, from irregular through spiral to spheroidal. Excluded from such speculation were the numerous small and inchoate assemblages that also turned up on the astronomical plates.

Hubble took a large sample of the population of galaxies under study—not their portraits but the spectra of the radiation that portrays them. Reflected from a mirrored refraction grating at the Newtonian focus of the 100-inch telescope, the galactic spectra were captured by prolonged time exposure. Arthur Stanley Eddington well described their images on the astronomical plate: "torpedo-shaped black patches." Most readily recognized is the pair of emission lines in the spectrum of calcium. In the spectrum of Earth-bound calcium, these lines appear in the far ultraviolet. In the cooler outer atmosphere of billions of stars in a distant galaxy, the energy of their emission is absorbed. They register accordingly as absorption in lines in the image of the faint galactic radiance soaked into the plates [see illustration, page 31]. Hubble found these lines shifted into the visible spectrum toward the red end of the spectra. Attributing this "redshift" to the Doppler effect, Hubble concluded that the galaxies are moving away from the Earth.

The expanding universe

In 1929, Hubble announced a still more momentous discovery. His survey had established a redshift in all the galactic spectra. He had found, moreover, a high positive correlation between decreasing brightness of the galaxies and increase in their redshift. Brightness supplied, of course, indication of distance. The positive correlation of the two measurements showed that the velocity of the recession of the galaxies must increase with distance from the Earth.

The Hubble survey showed the universe is expanding. Recognition that light from distant galaxies reaches Earth from earlier times showed that the universe had a history. Its expansion had a beginning that required explanation and an end that invited contemplation.

In the correlation of redshift with distance, Hubble's law recognizes that light traveling at its constant velocity from a distant galaxy reaches Earth from an earlier universe. On an astronomical plate, the size of the universe at all times and distances is increased, therefore, by the same amount [see illustration, page 31]. The constant relation of redshift to distance would yield, with calibration of the brightness of the distant galaxies, the age along with the size of the universe.

Hubble first calculated the value of his constant at 558 kilometers per second increase in velocity per each megaparsec increase in distance. That made the history of the universe much too short: the faster the expansion, the shorter the time of its expansion to any measured size. The 558-kilometer estimate made the universe, at its then estimated size, younger than reliable estimates of the age of some stars and even the Earth itself. Measurement and debate now place the Hubble constant between 60 and 80 kilometers per second per megaparsec.

Cavendish experiment *in 1798 [top] set gravitational constant. Observing the deflection of a torsion bar bearing two small lead balls brought close to two massive balls, he set the force at 6.75 × 10⁻⁸ dyne [see page 51]. Eötvös, in 1908, proved the equivalence of gravity and inertia: two equal masses in the Earth's gravitational field were indistinguishable as inertial masses on orbit around the Sun [see page 102]. With laser light measuring the Earth-Moon distance, a super Eötvös experiment made the same proof [see page 177].*

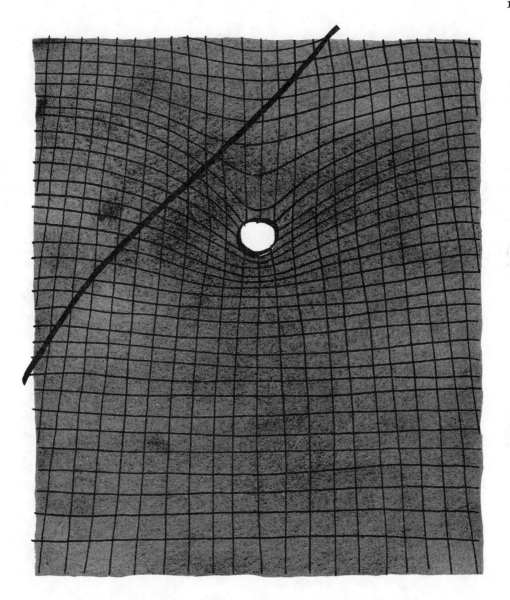

Curvature of spacetime *by the force of gravity was shown by Einstein to account for all observed effects of the force, including deflection of the path of starlight, an effect predicted for a different but analogous reason by Newton [see pages 54 and 105]. On the gradients of the curvature of spacetime, the velocities of masses increase and decrease; traveling at the ultimate velocity, radiant energy increases and decreases correspondingly in wavelength. Shift in apparent position of star in 1916 solar eclipse confirmed Einstein's prediction.*

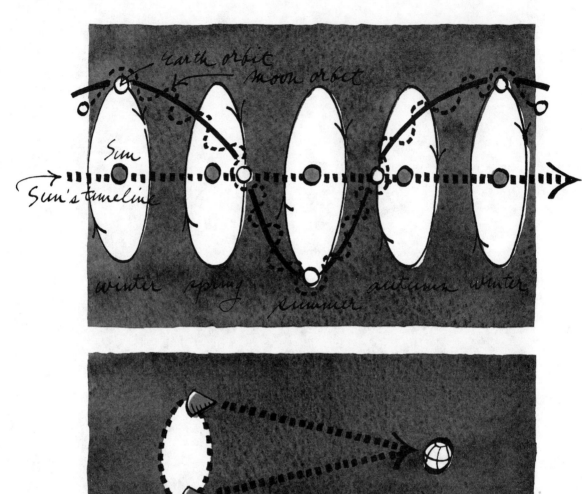

Visualization of spacetime *requires allocation of one dimension to time, here the horizontal. The Earth's motion in space relative to the Sun, along with the Moon's relative to Earth, is plotted through one year relative to the Sun's time-line, with Sun motionless in space [top]. The universe is observed in a spacetime cone fixed by the velocity of light [bottom]. If two events in the past [their space-time cones at left] were too far apart for light to travel between them, the universe might not be homogeneous [see page 236].*

For an expanding universe, General Relativity was on hand, ready to relate structure to dynamics but with the uncertainty that had troubled Einstein in 1917. His equations were consistent with either an expanding or contracting universe, but not with a static universe. Against expansion gravity would act as a negative force, slowing the acceleration down. To a contracting universe, however, gravity supplied a positive force, speeding up its collapse. A static universe required the addition of a negative—antigravity—term to the equations. Cosmological model-building did not have much to go on then, when the galactic grindstone was the observed universe. Moved by aesthetic preference for equilibrium and perhaps by the ancient and prevailing vision of a universe in eternal serenity, Einstein introduced the necessary corrective term into the equations published in 1916. This, some years later, Einstein confessed to Gamow and Wheeler, was "the biggest blunder of my entire life." The expansion of the universe would have been the most celebrated confirmation of General Relativity.

The cosmological term

Einstein's cosmological term has figured as the "cosmological constant" in the development of cosmology over the decades since. It has now come to the center of deliberations turning on the apparent present acceleration in the universal expansion. As the expression for the necessary force of repulsion, the cosmological constant is associated in the deliberations with the energy of the vacuum, source of the particle physicist's rest mass and much else. What is more, this force is finding a role in the Big Bang and in the continuing expansion since—all within the terms of General Relativity amended by the cosmological term.

For the static universe of his "corrected" General Relativity, Einstein had early reassurance. The Dutch astronomer Willem de Sitter elaborated an unchanging cosmos. To cosmology, the de Sitter model contributed the still strongly held principle of isotropy. This sees the cosmos as homogeneous. Appearing the same from everywhere, it offers no favored vantage point, such as the center people once thought they occupied. De Sitter held the universe to be isotropic in time as well. His was a steady-state universe, unchanging in density.

Alexander A. Friedmann recovered the expansion of the universe from the original General Relativity equations. In 1922, seven years before Hubble reported his observations, Friedmann showed that, in accordance with the equations, the apparent velocity of bodies in the cosmos must increase with distance from any point of observation.

When de Sitter published a correspondingly revised model of the cosmos, Einstein conceded, "If there is no quasi-static world, then away with the cosmological constant!" He therewith renounced claim to prediction of what may prove to be a fifth force at work in the cosmos and one embedded in his great achievement.

By 1930, Georges Lemaître, with the advantage of news from California, fashioned around General Relativity a full-blown model of an expanding, aging universe that had its beginning in a grand explosion from a primordial "atom."

Progress in the instrumentation has now made cosmology an observational science. Theory may be confronted with experience on the largest imaginable questions. The first question was the age of the universe. The latest calculations of the Hubble constant, setting it between 60 and 80 kilometers per second per megaparsec increase in distance from the Earth, secures a history of 13 to 15 billion years. That is long enough to yield the otherwise observed features of the universe.

For the age of the stars, there was evidence from Earth-bound physics as well as from astronomy. Early in the century, astronomers had begun classifying stars by their luminosity and spectra. They had established essentially two kinds of stars. Most nearby stars are "normal" stars, like the Sun; they were thought to have long lives. Stars of higher luminosity and greater mass—two or three or more, many more, solar masses—were thought to have proportionately shorter and more violent lifetimes. The natural history of stars is set out in the "H-R diagram." Originally composed in long-range collaboration by Ejnaar Hertzsprung in Denmark and Henry Norris Russell at Princeton University, it has been much elaborated by others [see illustration, opposite].

In the mid-1930s, with insights from nuclear physics, Subramanyan Chandrashekar at the University of Chicago and Hans Bethe were explaining how the Sun generates its light. Numbers could

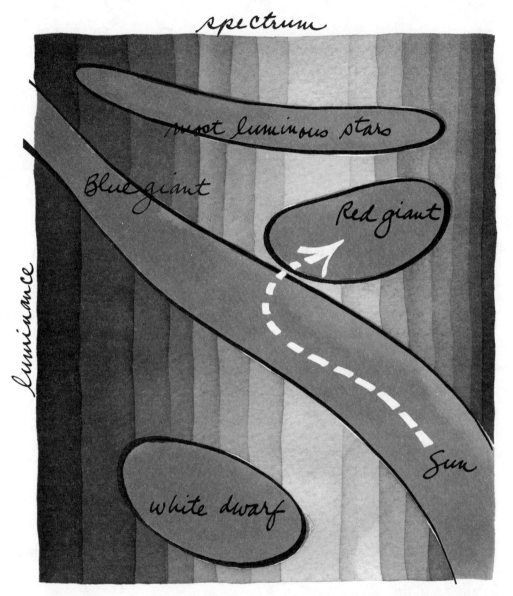

spectrum

most luminous stars

Blue giant

Red giant

luminance

Sun

white dwarf

Hertzsprung-Russell diagram *plots life histories of stars. The horizontal scale shows stellar surface temperature and color ascending right to left from red to blue; on the vertical scale stellar luminance ascends from bottom to top. The Sun is moving leftward in the "main sequence" band, from which it will swing up, after 5 billion years, into the region of Red Giants. Some of these evolve to Blue Giants in region above. Gravitational collapse brings most main sequence stars down to the White Dwarf region at bottom [see page 173].*

now be added to the H-R diagram. "Normal" stars, up to the mass of the Sun, move along "the main sequence" for their first 5 to 10 billion years. They shine with about the intrinsic luminosity of the Sun, at surface temperatures around 6,000 degrees K, until they have converted a tiny ration of their hydrogen to starlight and helium. At that turning point stars become larger, brighter and redder. Now they begin to burn the helium they have manufactured and, in a surrounding mantle, some additional hydrogen. They evolve for a few billion years at an accelerating rate up the luminosity scale, the more massive stars moving at a faster rate, to swell up and shine for a few million years as Red Giants. At last, in paroxysmal (i.e., in about 100,000 years) conversion, they become Blue Giants, with surface temperatures soaring to 12,000 degrees K. For shorter times, some oscillate between Red and Blue Giant states. These transformations give the average star a history of 10 to 15 billion years, depending upon its mass, before it collapses to a White Dwarf.

Stellar populations

In the 1940s, Hubble's colleague Walter Baade began a systematic survey of the stars in the distant Andromeda galaxy, which could now be resolved in sufficient number. At the end of the war, with the installation of the 200-inch telescope farther from city lights on Palomar Mountain, the Baade survey went into high gear. Comparing star fields on photographic plates selectively sensitive to the colors of the spectrum, Baade found a significant difference between spectra of the stars in the nucleus and those in the arms of the galaxy. In the arms he found many stars bluer and more massive than the Sun but most stars of the same brightness as the Sun, in progress along the main sequence. The population there resembled the stars overhead in what could be presumed to be an arm of the Galaxy.

Baade classified these as "Population I" stars. Their spectra in both galaxies were "solar" in that they showed lines, such as appear in the Sun's spectrum, of the heavier elements found on Earth. Baade took this to be evidence that these were younger stars incorporating elements synthesized in earlier generations of stars. In their collapse to White Dwarfs or neutron stars, the forerunners had blown off the products of nucleosynthesis into the clouds of interstellar space [see illustration, page 201]. Novae and supernovae supplied the carbon that catalyzes the fusion of hydrogen atoms to helium in the Sun and the heavier elements in its terrestrial satellites. As Harlow Shapley was then saying in the popular lectures for which he was famous: "We are made of star-stuff."

In the nucleus of Andromeda and in as much of the nucleus of our Galaxy as is visible from Palomar, Baade found a distinctly different population of stars. He saw that the 90 percent of the galactic population there was closer to the main sequence than the 10 percent in the spiral arms. The stars in the nucleus were redder and more luminous and farther along on the main sequence. He found the same population in the globular clusters around the nucleus and hanging in a halo outside the galactic plane. These stars were, accordingly, more ancient than the stars in the neighborhood of the Sun, perhaps as old as the galaxies themselves. The spectra of these stars showed little evidence of elements beyond helium. Baade designated the stars in the nuclei of the two galaxies as "Population II." His study of Population II clusters brought abrupt increase in the age of the universe. Finding that the intervening dust and gas in the Galaxy dimmed the brightness of those older clusters, Baade fixed their intrinsic brightness at double Shapley's calculation. That finding doubled the distance to the clusters in the Andromeda spiral. The observed cosmos, even with the Hubble constant close to his original estimate, now had an age closer to 5 billion years.

The steady-state universe

In the early 1950s, against the gathering evidence from observational cosmology for a dynamic, changing universe, Fred Hoyle, Thomas Gold and Herman Bondi at Cambridge University put forward a challenging "steady-state" model. Isotropic in space, it claimed as well the virtue of de Sitter's second isotropy: forever constant in density. To offset the observed expansion, Hoyle and his colleagues proposed that the cosmos holds its density constant by the continuous creation of new matter. For that proposal they could offer no physics; they could show, however, that it required the creation of no more than one atom of hydrogen every 1,000 years in a space the size of an ordinary room. For continuous creation they could argue the virtue of parsimony: it had no genesis or terminus to explain or contemplate.

They argued that the galaxies in reach of observation all appeared to be close to the age of the Galaxy. Galaxies farther out, when telescopes reached them, would prove to be constant in number per cubic volume of space. They would be older as well, befitting the longer travel time of their photons through spacetime.

Against the "Big Bang" universe—Hoyle's derogatory term has stuck—the steady-state cosmologists could point to a serious flaw in the evolutionary model at the very moment of the creation it required. Gamow had proposed that the earliest minutes brought the synthesis all of the elements in the table of elements. His scheme for the creation of hydrogen and helium persists in present understanding and is supported by the abundance of those two elements. Hydrogen and helium, as the immediate products of the Big Bang, constitute 76 and 24 percent of the rest mass of the observed universe, with all the other elements lost in the rounding of numbers as to mass as well as relative abundance.

Gamow envisioned the synthesis of the heavier elements by the process of neutron capture and the Rutherford beta decay that transforms a neutron to a proton [see page 148]. Neutron capture would build up the heavier elements from helium, with its weight of 4, adding one proton at a time. Given the 10-minute half-life of the free neutron, that would necessarily have been accomplished in the very first minutes after the precipitation of the proton and the neutron out of the cooling plasma of the Big Bang.

The nucleosynthesis of elements

As Hoyle was pleased to observe, however, Gamow's proposal stumbled at the first step. There is no stable element of weight 5 nor of weight 8. Their absence presented an insurmountable barrier to the Gamow buildup of elements beyond helium-4: the capture of neutrons, one at a time, could not proceed past helium-4, much less beryllium-8.

In favor of his steady-state model, Hoyle advanced findings from experiments conducted with the particle physicist William Fowler and the astrophysicists Geoffrey and Margaret Burbidge at California Institute of

Technology. Employing an accelerator in the low MeV range, the experiments reproduced conditions in the interior of stars at temperatures of 100 million degrees K. This is the temperature at which helium burns in the cores of stars on their way to the Red Giant stage. In the first stage, two helium nuclei fuse momentarily in a highly unstable beryllium-8 nucleus. Its short half-life helps to account for the rarity of elements heavier than helium. Some small percentage of the beryllium-8 nuclei do capture, however, a third helium nucleus to form carbon-12. Nucleosynthesis then proceeds by helium-capture up the scale from carbon through oxygen-16, neon-20 to magnesium-24. The main line of nucleosynthesis is marked by the larger abundance of the even-numbered nuclei over the odd-numbered that occur between them.

Exhaustion of helium brings collapse of the stellar core. The temperature soars in the gravitational collapse to 5 billion degrees K. Now fusion of heavier nuclei, in one combination or another, leads to synthesis of iron-56 and related members of the iron group. Prolonged ejection of much of the stellar mass into interstellar space attending the slow collapse of the Red Giant to White Dwarf—or abrupt ejection by a supernova—makes these products of stellar nucleosynthesis available to the next generation of stars. The addition, in these stars, of one proton at a time to heavier nuclei fills in what might be called the helium gaps to produce much of the rest of the table of elements. What remain are produced by other processes, off the main line, such as nuclear fission and the spallation, or chipping, of nucleons from atomic nuclei.

Hoyle's success as an astrophysicist did not sustain his cosmology. The 1950s brought a sudden, exponential increase in wealth of astronomical observation outside the visible spectrum. Observation in the invisible wavelengths of the radio spectrum brought contradiction and ultimately refutation of the steady-state hypothesis.

The number of galaxies observed per unit volume of space did not remain constant, but increased as befit the hypothesis of a universe smaller and denser when younger, with increasing evidence that the more distant galaxies are younger as well. The evolutionary model stood, corrected by what Hoyle, Fowler and the Burbidges had shown regarding the synthesis of the elements. Hoyle had thus usefully demon-

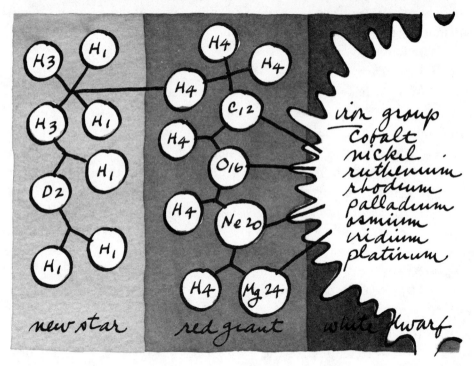

Thermonuclear synthesis of elements *generates the radiant energy of starlight and proceeds through successive generations of stars. Hydrogen and helium, formed in the Big Bang, are the starting material. Photons of radiant energy from thermonuclear fusion of hydrogen to helium in the core of a first-generation star [top] escape slowly through dense regions surrounding the core. In first-generation stars, successive fusions of protons (H) in deuterium (D), a hydrogen isotope holding a neutron, and then in helium-3 (H_3) yield helium-4.*

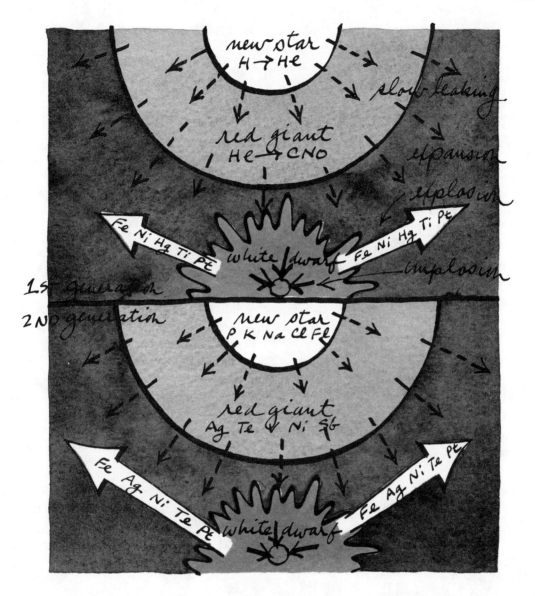

The end of hydrogen fusion brings inflation of the star to Red Giant diameters. Helium fusions then yield the even-numbered nuclei of carbon, oxygen, neon and magnesium. Collapse of the star to White Dwarf fuses these elements in iron and related elements and explodes products thus far of thermonuclear fusion into interstellar clouds. Second-generation stars, incorporating these elements, synthesize most of remaining elements, adding one proton at a time. Supernova explosions [see page 212] yield heaviest elements, some of them unstable.

strated that a failed hypothesis may make its contribution to under-standing along with a successful one.

Cosmic background radiation

Discovery of the cosmic background radiation settled the contest between the two cosmologies. There was no suggestion of it in Hoyle's steady-state theory and no explanation other than that supplied by an initial event followed by the expansion of spacetime. Arno Penzias and Robert Wilson made the discovery in 1965 at Bell Telephone Laboratories. They had hunted down all internal sources of static in a huge horn antenna designed to serve an early space-age international communication system. Despite chilling of its circuits down to super-conductivity, the antenna still put out a hiss that came from nowhere.

Coincidentally, at Princeton University, Robert H. Dicke and his colleagues were coping with the noise besetting a much smaller horn antenna. With this instrument they hoped to look for the predicted cos-mic background radiation. The signal they were looking for was 1,000 times weaker than the noise in their antenna. Hearing about the Princeton work, Penzias and Wilson reported their frustration to Dicke. The hiss in the Bell Laboratories antenna landed squarely on the predic-tion from theory. At 7 centimeters, the signal showed that the expand-ing universe had cooled to around 2.5 degrees K from the 10^{32} degrees K at which the standard theory starts the Big Bang.

The new radio astronomy that settled the cosmological question has since extended the radius of the universe three times beyond that set by the Baade measurement. Deep into the correspondingly lengthened past, radio telescopes have been observing, as current events, titanic transformations of matter-energy that have attended the expansion of the universe and the organization of its contents. Whole galaxies appear to be swept up and consumed in the headlong course of these transfor-mations. Inside the Galaxy and the nearer neighbor galaxies, radio tele-scopes have found scale models of these more distant events. The same processes consume mere stellar masses in more recent ages and nearer space. Evidencing the enormous forces they engage, these events make themselves known also on wavelengths other than radio.

The physical plant of radio astronomy built after the Second World War testifies to the wartime service of university physicists in the development of the technology of electromagnetism for service in the war. The "quasi-optical" behavior of the microwaves mastered for radar allows radio astronomers to collect faint emanation in these wavelengths from the sky on large parabolic mirrors and focus it on their antennae.

A typical .00002-inch light wave is 10^{-7} the diameter of the 200-inch Hale telescope mirror. The same ratio of wavelength to diameter for a radio telescope tuned to the important 21-centimeter band would require a reflector 2,100 kilometers in diameter. The very much smaller practicable radio telescopes proved huge enough, at 100 to 300 meters [right]. Radio astronomy made its debut in the 1950s as a necessarily Big Science. Only governments could supply its capital equipment. As governments did for the particle physi-

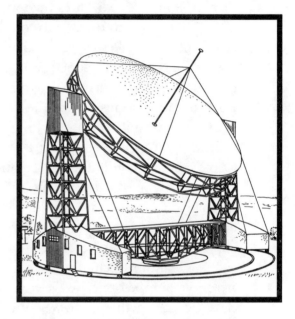

cists, so they did for the cosmologists: notably the 250-foot, steerable parabolic antenna at Jodrell Bank in England; the 140-foot steerable antenna at Green Bank, West Virginia, and the 1,000-foot fixed reflector at Arecibo, Puerto Rico.

On the 21-centimeter band—important because it is the frequency on which interstellar hydrogen radiates—the Arecibo telescope has a resolution of a few seconds of arc. That is the diameter of the smallest object it can measure at that wavelength. It is 1,000 times smaller than the 30 minutes of arc subtended by the Sun and Moon. Still higher resolution on radio wavelengths can be obtained by teaming up two or more antennae. They are made to serve, in effect, as the opposite edges of a single telescope mirror. With sufficiently long baselines separating the telescopes, resolution better than that of any optical telescope, down to fractions of a second of arc, can be attained. The Very Large Array of radio telescopes at Socorro, New Mexico, produces wonderfully detailed maps of diffuse radio sources. Supercomputers develop them from signals coming in from 27 linked, 25-foot steerable mirrors arrayed on a 21-mile baseline. Various other stratagems locate point sources of such radiation within sufficiently small circles of probable error and capture them in longtime exposures by optical telescopes.

For access to the cosmos on the short-wavelength, high-frequency side of the visible spectrum astronomers are indebted to the Cold War space programs of the superpowers. The opacity of the atmosphere to this high-energy radiation shelters life at the surface of the Earth. Rockets and balloons carried the first X-ray telescopes above the atmosphere and are still sent up for brief but productive looks around. Now instruments tuned to all the interesting frequencies on the far side of the visible conduct continuous observation aboard satellites.

Telescopes that see in the infrared, at the other end of the visible spectrum, also fly aboard satellites—above the CO_2 greenhouse. Earthbound infrared telescopes, with large gold-plated parabolic mirrors, do important service, however, at infrared wavelengths.

Quantum technologies have immensely increased the reach of astronomy in the optical wavelengths. Multimirror telescopes now outclass the light-gathering power of the 200-inch mirror of the Hale telescope. By electronic feedback from their joint image-making, the surfaces of four or six large mirrors are made to perform as a single mirror. The same feedback makes partial compensation for the boiling of the atmosphere that is the bane of ground-based astronomy. The two effective 10-meter mirrors of the Keck telescope on the summit of Mauna Kea on the island of Hawaii compare to the Hale telescope's 5 meters. A half dozen other instruments in operation and under construction exceed 8 meters in effective diameter. The largest of all, to be installed at the European Southern Observatory in Chile, will train an effective 16-meter mirror on the nucleus of the Galaxy.

In these and the other great telescopes, photographic film has yielded to "electro-optical detectors." The best astronomical film registers intelligibly only a small percentage of the light focused on it.

"Charge-coupled" devices respond to photons arriving one second apart and record objects far fainter than film can capture. Where photographic film becomes "overexposed," these devices respond linearly without limit. In arrays of 1,000 by 1,000, they have resolution as high as any film. They keep digital record of the incoming light, making it possible to process the data by computer in any desired fashion, for example to subtract light of given wavelength from an entire image.

What Galileo saw astronomers now deduce from third-, fourth- and fifth-hand records made by their instruments. Understanding of cosmic events at the scale of light-years approaches the intimacy, nonetheless, of the understanding of structures and processes gained at the targets in Earth-bound accelerators. To understanding won in the accelerators at the scale of 10^{-16} meter, astronomers have been able to refer the surprises brought in from the depths of space by the power of their new instrumentation.

The Galaxy

Over the last half century, the big radio telescopes have been mapping the major structural features of the Galaxy from what they are able to see in apparently empty interstellar space. Matter—hydrogen, principally, but ions, atoms, molecules and dust particles of other elements as well—is present in the plane of the Galaxy in an average density of a million atoms per cubic meter. That is a more perfect vacuum than can ever be drawn on Earth. In the 30 or 40 thousand light-years between Earth and the center of the Galaxy, however, there is enough of this material to hide from view the brilliance of the 90 billion stars in the galactic nucleus. Comprising about 5 percent of the observed mass of the system, it is in perpetual turnover through the stellar furnaces, at once the raw material and the debris of stars. The density of the interstellar atmosphere increases by 100 and 1,000 times near the nucleus of the Galaxy and along the spiral arms and by as much as 10,000 times in the dust clouds. That is still a 10^{-15} fraction of the density of the standard volume of gas at standard pressure on Earth [see page 56].

Every 1,000 years or so an atom of hydrogen (H) out there in interstellar space passes close enough to an errant racing electron to be excited to the state at which, perhaps a century or two later in accordance with the uncertainty principle, it emits a 21-centimeter photon. Molecular hydrogen (H_2) is detected by characteristic proxy radiation from carbon monoxide (CO), present in a tiny fraction of the abundance of hydrogen, that follows its rare collision with H_2. Yet, such is the depth of intragalactic space, the proxy radiation is continuous. The mapping by radio telescopes of the distribution of neutral atomic hydrogen and molecular hydrogen throughout the Galaxy (except for the region on the far side of the nucleus) has established the major features of the system, in particular the existence and layout of its spiral arms. To the unpracticed eye, it would look like Andromeda.

The maps show hydrogen in the plane of the galactic disk distributed through the region between the rim of the nucleus, at 10,000 light-years from the center, out to the vicinity of the Sun, at 40,000 light-years from the center and closer to the periphery of the Galaxy, near 50,000 light-years. Both atoms and molecules are at higher density in the arms. At the rim of the nucleus, the interstellar atmosphere gives the galactic disk a depth or thickness of 1,000 light-years. From there it tapers to a thickness of 500 light-years at the latitude of the Sun.

The ratio of depth, at 1,000 light-years and less, to the diameter of the entire system, 100,000 light-years, makes for a thin galactic disk. The 90 billion Population II stars swell the central nucleus to a thickness of around 10,000 light-years, a flattened spheroid 20,000 light-years in diameter. The Galaxy thus appears a fragile object.

The Doppler shift of the 21-centimeter line shows the Galaxy wheeling in counterclockwise rotation seen from "above" the "north pole" of its nucleus. Locally, the proper motion of the stars takes them in all directions. Whatever the direction and velocity of their motion with respect to one another and to the galactic coordinates, all objects in the system are entrained in the rotation of the whole. For the Sun, at its long distance from the center, the galactic year is 250 million Earth years. Closer to the center, the periods of rotation are shorter. In the course of the Sun's 20 or so galactic years, the configuration of the Galaxy has changed many times, and rearrangements of the constellations above the Earth have occurred still more often.

Around the nucleus of the Galaxy, radio telescopes see a ring of hot clouds of ionized hydrogen moving outward at 100 kilometers per second. Such material, simultaneously in rotation around the center, may constitute a nascent spiral arm. Mapping of the density of the interstellar atmosphere and its relative motion from region to region has brought into view at least four major spiral arms trailing the galactic rotation. They show the Galaxy to be more tightly wound than Andromeda. Along these arms the 10 billion Population I stars are strung [see page 196].

Stars made of stardust

Radio astronomy has given significance to the vast cool clouds that obscure the stars in the Milky Way. Time exposures yield wondrous displays on astronomical plates where the clouds are locally heated to ionization and incandescence by bright stars within. Densities of a billion molecules per cubic meter and higher prevail in these clouds. Principally composed of hydrogen, they are rich in carbon, nitrogen and oxygen, which constitute three-fourths of cosmic matter other than hydrogen and helium. CHON, for carbon, hydrogen, oxygen and nitrogen, respectively, may be recognized as a first-approximation formula for the substance of the living cell [see page 250]. Except for the abundance of oxygen in its crust, planet Earth, still the only known abode of life, has a relatively impoverished endowment of these elements. From their cosmic abundance, Fred Hoyle conjured a dust cloud imbued with life and wisdom in his science-fiction novel *The Black Cloud.*

In fact, more than 100 familiar "organic" compounds composed of these elements have been found in the dust clouds. Simple two-, three- and four-atom molecules, such as carbon monoxide, water and ammonia, abound, and synthesis proceeds beyond formaldehyde and methyl and ethyl alcohol to carbon chains of 10 and more atoms and even to the carbon ring compounds of coal-tar chemistry. Estimates show that the ethyl alcohol in a typical cloud would fill the volume of the Earth; it is tainted, however, by such molecules as hydrogen cyanide.

Dust plays a role in the accumulation of these chemical compounds and in the clouds. Dust screens the interior of the cloud from the high-energy radiation of the stars that promoted those chemical bonds, but disrupts them even more readily. The silicate, graphite and iron oxide dust motes also supply surfaces, apparently sometimes catalytic, on which atoms link up into larger and longer molecules.

The giant molecular complexes are among the most massive objects in the Galaxy. Typically, 100,000 solar masses of material are dispersed in one of these clouds, stretched along a spiral arm. Sample surveys of the more accessible regions of the Galaxy suggest there may be 4,000 of them. As their chemical diversity suggests, the material in these clouds has turned over in more than one generation of stars. The uniform relative abundance of elements in the composition of the Population I stars in the galactic arms testifies conclusively to the same effect.

Stars constantly return their substance to interstellar space. In quiescent loss of mass, the Sun blows a wind of particles at its satellites that now and then disrupts radio and telephone communication on Earth. Giant stars, ascending from the main sequence, expel huge masses of material as they switch from one thermonuclear cycle to the next. Collapsing to White Dwarf or neutron star density, exhausted stars give up still more of their mass to the next generation. For a while, perhaps 100,000 years, thereafter they present one of the most beautiful spectacles to be seen in the Galaxy: the expanding, glowing, spherical "planetary nebula," light-years across, of the material they have expelled. It is this material that gathers in the giant molecular complexes.

Birth of stars

The transparency of the giant molecular cloud complexes to radio wavelengths has made the formation of stars one subject in astrophysics in which observation has run ahead of theory. Ever since James Jeans subjected the question to classical analysis at the turn of the century, instabilities between inward and outward pressures have derailed any full accounting of the process. The difficulty may be expressed as the long scalar distance between a gas at 10^9 molecules per cubic meter and a commonplace stellar core at perhaps 10^{25} times that density.

The structure and contents of the spiral arms yielded important clues to one mode of star formation. The hottest, most massive stars (designated O and B) show up periodically along the trailing edges of the arms, where density in the molecular clouds is higher. It is thought that formation of these stars is triggered by the passage of density waves along the arms from mighty gravitational disturbances in the nucleus. Burning their fuel in 10 million years, the lifetime of O and B stars equates with only about two weeks in Earth's galactic year. They account for much of the turnover in the molecular clouds. The number of them that come and go in the course of a few Earth-galactic years may approach the number of senior, long-lived stars in the nucleus.

Elsewhere in the spiral arms, star formation gets under way by random processes. Fred Whipple of Harvard Observatory showed how photophoresis, the mechanical force exerted by photons of light [see page 92], moves molecules and dust particles about in the clouds to accumulate in regions of very much higher density. Self-gravitation condenses such concentrations then into Bok globules. At critical density, these globules go into gravitational collapse. They are named for Bart J. Bok, at Harvard Observatory, who first noted these objects and implicated them in star formation.

Now radio telescopes operating in millimeter wavelengths in tandem with telescopes operating in the infrared spectrum alongside have brought the apparent ignition of stars into view. In the constellation of Orion are notable bright clouds inside the local giant molecular complex. Penetration of these on radio wavelengths revealed the presence of apparent new stars deep within. They are enveloped by glowing clouds of ionized gas, some of it issuing apparently from the nascent stars.

In the early 1980s observation showed that the collapse of a Bok globule to a star must involve expulsion of some large portion of the material swept into it. Inside the giant molecular complexes, invisible on all but millimeter-radio and infrared wavelengths, are objects that are firing jets of material in 180-degree-opposite directions. In several cases it has been possible to identify disks of gas and dust spinning in the equatorial plane of these objects. The evidence is redshifted radiation that must come from the retreating edge and blueshifted radiation from the approaching edge of such disks. Pressure exerted by the density of the material in the disks, along with magnetic fields, may account for the ejection of the material at the poles of the collapsing sphere. Hundreds of such apparent nascent stars have been observed. They offer scale models of the spectacles presented, as will be seen, by stars collapsed to neutron stars in the local galaxy and by catastrophes engaging whole galaxies in the distant universe.

Some 0.08 of the mass of the Sun, theory says, is the minimum required to ignite and sustain starlight-generating thermonuclear reactions. Some number of smaller stars too small to shine must condense in the molecular clouds. The presence of these stars is also implied by the "local" missing mass required to account for the gravitational field of the Galaxy, but they are not yet visible even in the broadened spectrum of modern astronomy. These small stars, called Brown Dwarfs, must radiate the heat of their gravitational collapse, but it has gone undetected so far. They might also have magnetic fields awaiting detection. The great gassy planet Jupiter, of 0.01 solar mass, emits feeble infrared radiation and has a magnetic field. It presents a micro model of the many failed stars that may be concealed in the giant molecular complexes.

The detection of the disks spinning around nascent stars ratifies theoretical scenarios of the origin of solar systems. Rotation of a disk sorts out the dust and gas by weight, concentrating heavier particles and elements nearer to the star. Random aggregations of material here and there in the disk aggregate further in mass by gravitational attraction. Such aggregations go into runaway growth, sweeping up material from ever larger regions in the disk. Upon reaching a critical mass, around that of the Moon, the aggregation collapses around its center of gravity. The pressure and heat of the crushing infall brings the temperature to the melting point of the material. Now the material forms into a sphere, in expression of the spherical symmetry of the gravitational force. The heavier metallic elements, especially the relatively abundant iron, fall into the central core of the new planet; the lighter elements, such as aluminum, silicon and oxygen, rise toward the surface.

The circulation of metallic and stony asteroids in the solar system suggests that such protoplanets formed in numbers early in its history and shattered one another in collisions. Their planetesimal fragments have since been swept up, but not completely, and incorporated in the larger planets on orbit around the Sun. From an early protoplanet, it appears, the Earth had a narrow escape. A close encounter or even a low-angle collision with an object that thereupon went into terrestrial orbit supplies the best present explanation for the origin of the Moon.

Search for other solar systems has found planets on orbit around 24, at last count, stars that are close enough to Earth to be observed. The evidence—partial occultation of stars or displacement of their position in the gravitational field of the unseen planet—gives those planets the mass of Jupiter or greater. The romance of the possibility of another planet like ours awaits means for closer inspection.

The example of the Earth does strongly suggest that a planet of the right size and composition at the right distance from the right sort of star—the Sun, for example—gives rise inevitably to life. The course of natural selection that gave rise on Earth to conscious observers of the cosmos must be conceded, however, to be an exceedingly rare event. Yet, given the 100 billion stars in the Galaxy, the event is not excluded.

The profound implications of such a possibility have sustained an organized search for evidence of extraterrestrial intelligence over the past four decades. In 1959, Philip Morrison and Giuseppe Cocconi at Cornell University roused the community with the observation that, if signals from civilizations elsewhere in the Galaxy are present, "the means of detecting them is now in hand." They commended monitoring the 21-centimeter line of neutral hydrogen, so central to the mapping of the Galaxy that its importance "must be known to every observer in the universe." Today, screen-savers on computers in more than 2 million U.S. households, mobilized by the SETI Institute and programmed to find signals on the 21-centimeter and other strategically chosen lines, are processing the intake of the 1,000-foot Arecibo telescope and reporting continuously to central at the University of California at Berkeley.

Single-star candidates for realization of the romance in the Galaxy are in the minority. Stellar genesis, two-thirds of time, yields a binary star: two stars bound to rotation around the center of their combined, mutual gravitational field. Sometimes, the chance that governs the birth of stars yields three or even four stars so bound. A null result of the screening of radiation inside the Galaxy would not exclude the possibility of extragalactic intelligence in the 100 billion other galaxies. For detection of intergalactic signals, however, the means is not in hand.

Death of stars

People have had a longer acquaintance with the death of stars than with the little so far ascertained about their birth. Astronomers in China recorded the appearance of three "guest" stars in the years 1006, 1054 and 1181. The first gave as much light, they reported, as a half moon; it was bright enough, therefore, to cast shadows on Earth. No certain record of these events survives in the contemporary annals of European civilization, such as it was, in those centuries. The Chinese located them in their celestial maps accurately enough, however, to lead modern astronomers to their remnants in the sky today. Astronomers were on hand in Europe for the next two of these "supernovae" in 1572 and 1604, known as Tycho's and Kepler's, respectively. Observed or not, it is estimated that a supernova blows up somewhere in the universe every second and in the local Galaxy every 30 to 50 years.

To terminate as a supernova a star must have at least 1.4 solar mass or larger; estimates from debris have credited some with more than 10 solar masses. Adam Burrows of the University of Arizona has vividly described what, in present understanding, happens:

> Within one second, the core of a star that may have lived for 10 million years, cooking its hydrogen into progressively heavier elements, implodes from something the size of our planet to something the size of a city, achieving densities in excess of the atomic nucleus and velocities one fourth of the speed of light. At nuclear densities, matter is barely compressible and the core bounces, rebounding into the infalling mantle and, like a piston, generating a strong shock front that … overcomes the confining tamp of the infalling mantle … to launch a supernova explosion. The violent explosion disassembles the massive star, litters the interstellar medium with freshly synthesized heavy elements … blows a many parsec–sized hole in the surrounding galactic gas, and announces itself with a luminous display that can rival that of its parent galaxy for months.

Much of this description of the supernova event is owing to the detonation of SN1987A in the nearby larger Magellanic Cloud. This, in 1987, was the first supernova to be observed close up by astronomers equipped with modern instrumentation. The instrumentation included two traps set up for the detection of the elusive neutrino, one in a salt mine in Ohio and the other around the world in the deep underground Kamiokande neutrino observatory in Japan. They caught the same salvo of neutrinos from the explosion within the same 0.01 second. The neutrinos recorded were, of course, a tiny sample of those fleeting through the traps and a still tinier sample of all the neutrinos that flooded through the Earth. Calculation shows that the full flux of neutrinos carried energy, from 150,000 light-years away, equal to that simultaneously reaching Earth from the Sun, a mere 10^{-13} light-year away.

Given the angle subtended by Earth at the distance of 150,000 light-years, this was an infinitesimal fraction of the total flux of neutrinos from the SN1987A. Such eruption of neutrinos, it was thereby established, carries virtually all of the energy—99 percent of it—from a supernova. The collision of high-energy photons in the dense depths of the explosion gives rise to electron-positron, matter-antimatter pairs. Their mutual annihilation sends pairs of neutrinos in opposite directions on their endless voyage across the void and, without resistance, through any object they encounter.

The "IMB" neutrino trap in the Ohio salt mine—the joint venture of the University of California at Irvine, the University of Michigan and Brookaven National Laboratory—and the Kamiokande observatory have as a major objective the determination of the neutrino rest mass, if any. That is to be evidenced in a postulated exchange of identity among the electron, the mu and the tau, neutrinos [see illustration, page 167]. The neutrinos have standing in the search for the missing mass.

Inspection of the sites of the recorded supernovae, with modern physics to explain what is found, has now downgraded all but one from the status of supernova to that of nova. At the 1054 site, early astronomical photographs had revealed the glorious Crab Nebula. Not so long ago, radio astronomy found a neutron star inside. This ensured the identification of that nebula as the cloud of stellar substance ejected from a supernova and that was still expanding at high velocity. It is among the first of the many objects in the Galaxy now identified as sites of a supernova. J. Robert Oppenheimer, then holding joint appointments at the University of California, Berkeley, and California Institute of Technology, showed that a star large enough to explode as a supernova

yields a remnant neutron star, all in accordance with General Relativity. This was a stopping point on the way to his prediction of the collapse of a sufficiently large star into a black hole.

Such understanding has identified in SS433, in the constellation Aquila, the cinder of a supernova that shone perhaps before Homo *sapiens* appeared on Earth. The spinning neutron star, dragging an accretion disk from its companion star, generates energy at a million times the output of the Sun. From the infalling mass, it ejects two high-velocity jets of particles in 180-degree opposite directions from the accretion disk. Driven by the magnetic field of the star, inclined at 20 degrees to the axis of rotation, the jets carry shock waves thousands of light-years into the interstellar atmosphere, exciting much of the radiation that attracts the attention of its observers. The jets, rotated by the spin of the neutron star, send variable signals to Earth, blueshifted from the jet pointed this way and redshifted from the oppositely pointed jet [see illustration, page 216].

The neutron star in the Crab Nebula [above] is a pulsar; that is, it generates a sharply varying output of energy on all wavelengths. Its magnetic field, a trillion times stronger than the Sun's, is tilted at an angle to its axis of rotation. By the frequency of the consequent pulsation of the observed radiation, it is spinning at the rate of 30 times a second. The electromagnetic waves generated by this giant dynamo accelerate particles dragged in from the interstellar atmosphere to relativistic speeds. Much of the light from this source is, accordingly, synchrotron radiation shining on a continuous spectrum from X rays across the visible into radio wavelengths. Such radiation clearly distinguishes itself from the emission- and absorption-line spectra characterizing radiation from the thermonuclear and electromagnetic life processes of stars.

At the sites of the other two historically recorded events and two more recently discovered planetary nebulae, astronomers found no central object perceptible inside on any wavelength. By radio astronomy, a fifth such nebula was discovered in the constellation Cassiopeia, at the time the most powerful radio source in the Galaxy. Studies of large binary stars show the two stars may age on different schedules. One arrives at the White Dwarf stage, perhaps, while the other swells up to a Red Giant. The White Dwarf—with accelerated speed of rotation and amplification of its gravitational and magnetic fields that attends the shrinking of its diameter and the increase in its density—cannibalizes the substance of its companion and sweeps the material into an accretion disk.

From the accretion disk material flows on the magnetic lines of force down onto the poles of the White Dwarf. The accumulation of sufficient material on the surface of the White Dwarf blankets escaping heat from its interior. Increase in temperature then brings on a thermonuclear eruption. The White Dwarf shines for a few days or weeks as a nova—an event observed 10 times more frequently than a supernova. Some White Dwarfs go through this cycle repetitively.

A few White Dwarfs terminate their existence in an event that—at a tenth the supernova frequency—may outshine a supernova 10 times over. These are the White Dwarfs that leave behind an empty planetary nebula. They have been upgraded recently to the rank of Type 1a supernovae. Their redshifts supply the new cosmic yardstick that indicates the acceleration of the universal expansion.

Observation of these events in nearer galaxies showed them all reaching the same high intrinsic luminance. Further, it could be seen that the rise and fall of their luminance plotted a characteristic curve. This made it possible to identify with confidence much fainter 1a events in more distant galaxies. Their luminance at redshifts of 0.5 to 1 com-

Accretion disk *may take form at all scales from that of single stars to whole galaxies [see opposite and pages 209 and 228]. Here one of a pair of binary stars has collapsed to White Dwarf; the magnetic field of this spinning dynamo drags the substance of companion Red Giant into the accretion disk. Electromagnetic force ejects substance in jets from its poles. Accelerating to relativistic speeds, infalling particles emit synchrotron radiation [see page 174]. A new star [inset] may drag dust cloud material into an accretion disk [see page 210].*

Accretion disk jets *from Cygnus A—an elliptical galaxy spinning into a massive black hole at its center [see page 228]—reach out millions of light-years on either side. The hugest sources of radiation in the universe, quasars were first detected by radio telescopes. Mapping of the radiation from this object finds its most intense sources in the vast lobes at the ends of the jets, evidently entraining intergalactic as well as galactic material. Radio maps enable optical telescopes to find faint, because distant, galaxies at the center of these objects.*

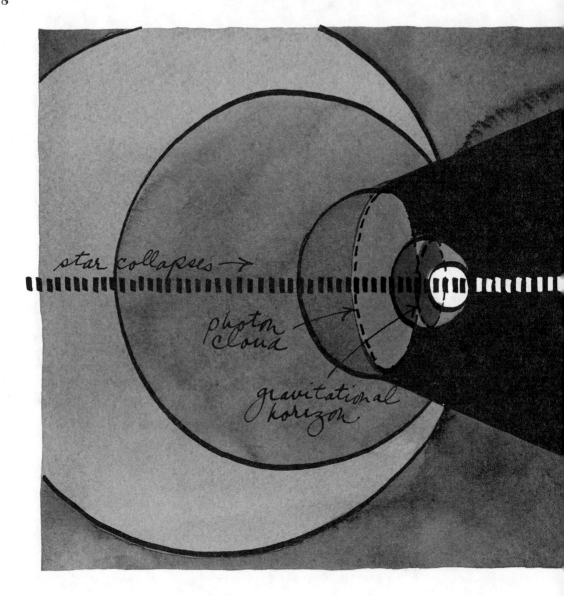

Collapse of a star into a black hole by the force of gravity is here depicted on the "timeline" [see illustration, page 192] of the event. Such collapse may overtake any assemblage of matter exceeding three solar masses [see pages 174 and 220]. A star blows up as a supernova upon exhaustion of the thermonuclear fuel that maintained the force of heat against gravity [see page 212]. The residual mass goes into accelerating collapse through two critical diameters. At the first, photons in trajectories tangent to its surface are caught in a cloud that thereafter

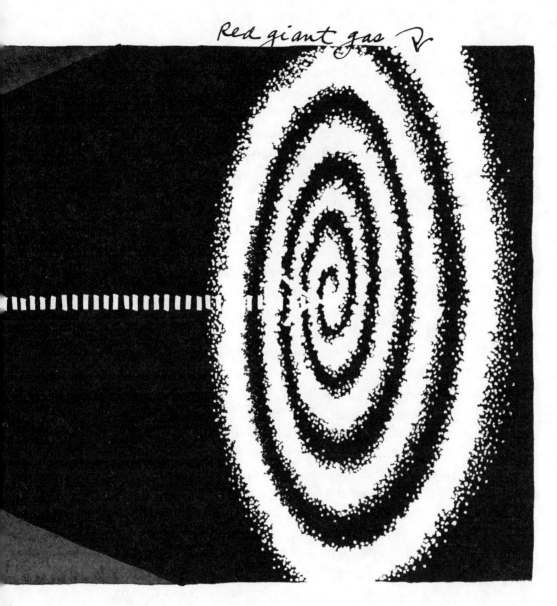

Red giant gas.

locates a black hole. At the second, it falls through the "gravitational" radius at which the escape velocity from its rising density reaches that of light itself; no photon can escape. (That is why a black hole is black.) At last, gravity overwhelms even the internal "pressure" of the substance of particles. What remains is their net charge and the angular momentum of their descent into the "singularity" of the black hole. This black hole has drawn the substance of its companion Red Giant into an accretion disk [see illustration, page 216].

pared to the intrinsic standard gave assured measurement of the distance to those galaxies. The distance proved to be greater than that measured by earlier, less assured, yardsticks. These 1- and 2-billion-year-old galaxies—their age fixed by their redshift—were receding at a faster rate than older galaxies out beyond them in spacetime.

This is the observation that has focused interest on the Einstein cosmological constant and compelled speculation about the play of a hitherto unknown "dark energy" in the cosmos.

Having established how stars up to three times the mass of the Sun expire, the search began for the black holes into which, theory predicts, stars of more than three solar masses must vanish. The idea first occurred in the 18th century to the Reverend John Michell. Impressed by Isaac Newton's prediction that starlight would be deflected in the gravitational field of the Sun, Michell essayed a calculation of the mass of a star sufficient to halt the escape of the corpuscles of light from its own surface. Not long after the equations of General Relativity were published, the German astronomer Karl Schwarzschild recognized that they predict the existence of black holes.

Oppenheimer and Hartland Snyder at the University of California, Berkeley, gave the theory its presently accepted statement in 1939. The increasing density of the collapsing star amplifies its gravitational field. At the critical density—when the mass has collapsed to what is called the Schwarzschild radius or gravitational horizon—the velocity of escape from the collapsing star and its field equals the velocity of light. For a body of the mass of the Sun, the radius would be 3 kilometers; of the mass of the Earth, 9 millimeters. While the existence of such "mini" black holes has been conjectured, the theory calls for mass sufficient to collapse through the counterpressure of quantum forces in the structure of matter, just above three solar masses. Upon collapse through its gravitational horizon, the mass goes on collapsing to the vanishing point. The gravitational horizon, however, remains.

Theory says that a black hole may manifest itself, on appropriate wavelengths, in radiation from matter under acceleration to disappearance across its gravitational horizon. On the model of the White Dwarf and neutron star binaries, therefore, astronomers have searched for evidence of a binary companion to a black hole. The best candidate for years has been an object in the constellation Cygnus. The flux of synchrotron and thermal radiation from this source strongly suggests the presence of an accretion disk from which a black hole is cannibalizing a companion star. Called Cygnus X at first, this object is now known as Cygnus X1 with the discovery of two other powerful sources of radiation in Cygnus. Cygnus X3 has proved to be a binary with a neutron star as its primary member. It not only generates synchrotron and thermal radiation but accelerates particles to energies beyond those hoped for from the Superconducting SuperCollider. No particles escape from Cygnus X1, strengthening its candidacy for status as a black hole.

Nucleus of the Galaxy

The most promising nearby site for proving the existence of a black hole is presented at the nucleus of the Galaxy where 90 percent of its mass is concentrated. This is but one aspect, however, of interest in the nucleus of the Galaxy that has brought the deployment of increasingly powerful instruments in the Southern Hemisphere, including the 16-meter multimirror telescope now being installed in Chile by a consortium of European observatories.

Over the past two decades, with telescopes tuned to the 21-centimeter radiation of hydrogen and to infrared wavelengths, observers have penetrated the 40,000 light-years of intervening stellar and interstellar matter to look inside the nucleus. The outermost feature, at 10,000 light-years from the center, is a ring of ionized hydrogen clouds, moving outward in the plane of the Galaxy at around 100 kilometers per second. Beyond this ring, within the dense nuclear congregation of stars, the interstellar atmosphere appears thin, as if what was there had been consumed in the condensation of those ancient Population II stars. At 5,000 light-years from the center, appears a disk of atomic and molecular hydrogen canted at an angle to the plane of the Galaxy. In this disk are hot spots of ionized hydrogen, at 10,000 degrees K, and blue-white supergiant stars of apparently recent origin. Then, at only 30 light-years from the center, another ring of ionized hydrogen is found at the cooler temperature of 5,000 degrees K [see illustration, pages 224–225].

The densest concentration of stars in the Galaxy is contained in this central sphere 10 light-years in diameter. Within, rotating about the center of the sphere, clouds of ionized hydrogen of about the mass of the Sun and three light-years in diameter have been detected. These clouds, it is thought, may be the remnants of collisions among the densely packed stars. They rotate in a period of 10,000 years, faster than the interior stellar mass, in gravitational thrall to whatever is there at the very center. That central body shows up as a bright source of infrared radiation. Within a diameter of 30 million kilometers—about one-sixth the diameter of the Earth's orbit—it appears to contain a mass of 50 million solar masses. Deep within may be a supermassive black hole.

If there is, this object remains presently quiescent. In action, it may, as the outmoving outer ring of ionized gas suggests, have a role in the genesis of the Galaxy's spiral arms.

The extended Galaxy

From the putative black hole at its center, the new instrumentation of astronomy has shown the gravitational field of the Galaxy to extend far into the volume of space around it. The distribution of globular clusters and of faint ancient stars outside the disk and bulge of the system had established the existence of a flattened spherical "halo" enveloping the Galaxy. It has a diameter of 120,000 light-years and a depth at the center of 100,000 light-years. To this halo, theory credits in part the stability of the Galaxy's thin extended disk. Rotation of the disk is ballasted by the enormous mass of the halo.

The motion of the Sun has been carefully measured with respect to the Magellanic Clouds, four nearby dwarf irregular galaxies, and 10 globular clusters located at great distances outside the halo, all within a sphere 300,000 light-years in diameter. These objects are entrained, though at lower rotational velocities, in the rotation of the Galaxy. In 1974, J. Einasto and his colleagues at the Tartu observatory in Estonia estimated that the gravitational field of the Galaxy must engage nearly 2 trillion solar masses distributed within this spherical "corona" enveloping the halo and the Galaxy inside [see illustration, page 226].

Einasto's calculations bring the local missing mass to seven times the visible. The stars in the clusters and in the dwarf irregular galaxies are all ancient and faint. Lone stars scattered throughout the vast corona may shine below visibility. Newer measurements raise the velocity of rotation in the outer reaches of the Galaxy proper. This correction supports the proposed extended reach of the gravitational field of the system throughout the corona. Moreover, seven nearby galaxies have been observed to exhibit similar internal rotation out to their peripheries, which suggests engagement of large unseen masses in their extended gravitational fields.

The shaping of the galaxies proper within their extended coronas must have proceeded by concordant action of the gravitational force and the conservation of angular momentum. As the material collapsed out of the slowly rotating coronas into highest density at their very centers, accelerating rotation spun the galaxies into flattened wheels.

The motion of the Galaxy with respect to its neighbor galaxies suggests a common gravitational field. Galaxies, the building blocks of the universe, tend to swarm in clusters; clusters, in superclusters. The clusters, too, are self-gravitating systems. Radio sources in the interior of some clusters signal ionization of the thin intergalactic atmosphere, suggesting the action of immense gravitational-tidal forces among those galaxies. Deep in some superclusters, galaxies are observed in coalescence into supergalaxies.

Quasars and the distant universe

Surveys of the distribution of galaxies show superclusters lining up in strings, hanging in curtains and forming bubbles around vast empty caverns. Thus, locally, the cosmic principle of homogeneity does not hold. When the distribution of galaxies is mapped in sufficiently large volumes of spacetime, however, homogeneity is restored.

The cosmos in reach of such galactic mapping has a radius of 5 billion light-years. Reach into far greater depths of spacetime has been accomplished with the discovery and close observation of the steadily increasing number of objects called "quasars" (from "quasi-stellar radio sources"). Their intense output of synchrotron radiation identified them as a new order of cosmic object. With no emission or absorption

The nucleus of the Galaxy *has been brought into view by the concentration of observing power in the Southern Hemisphere on all wavelengths [see page 221]. The outermost feature, at 10,000 light-years from the center, is a ring of ionized hydrogen clouds moving out in the plane of the Galaxy at around 100 kilometers per second—possibly a nascent arm. Beyond is a dense congregation of ancient Population II stars. At 5,000 light-years from the center appears a disk of atomic and molecular hydrogen canted at an angle to the plane of the Galaxy, with*

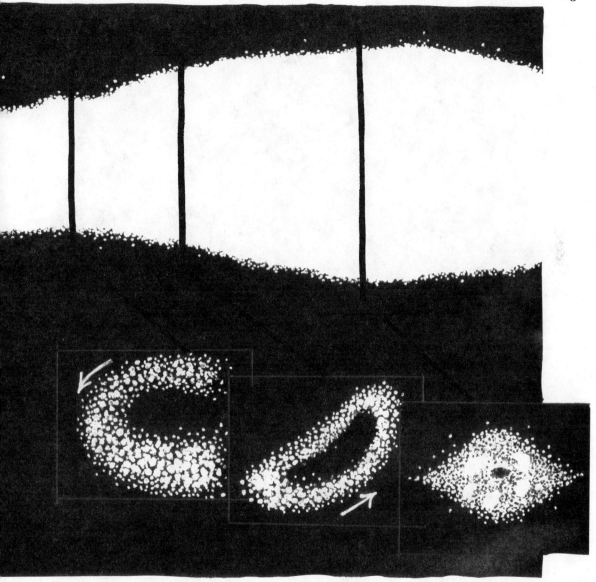

hot spots of ionized hydrogen and blue-white supergiant stars of apparently recent origin. Within 10 light-years of the center is the densest concentration of stars in the Galaxy. Collisions among them apparently yield clouds of ionized hydrogen in rotation around a bright source of infrared radiation at the very center. No more than 30 million kilometers in diameter—one-sixth that of the Earth's orbit—it appears to contain a mass of 50 million solar masses. Deep within there may be a supermassive black hole.

Spherical "corona," 300,000 light-years in diameter, evidences the reach of the Galaxy's gravitational field into nearby space [see page 222]. Objects within are entrained in rotation with the Galaxy. Principal are the Magellanic Clouds; from the larger cloud a tenuous stream of material appears to be drawn toward the Galactic nucleus. Closer in, the Galaxy is enclosed in "halo," principally defined by globular clusters. The gravitational mass in the corona is estimated at 2 trillion solar masses, far more than is visible.

lines in that radiation, redshift could not be established. It could not be decided, therefore, whether they were inside the Galaxy, and so "stellar," or outside, perhaps a new kind of galaxy.

The first of these intense radio emitters was observed in 1946 at the Royal Radar Establishment in Malvern, England. Adapting military gear to the uses of peace, observers there discovered this source in the constellation Cygnus (in that constellation again, by coincidence). A few years later, Baade and Fritz Zwicky were able to match the radio object to a barely visible one in a Hale-telescope star field. The huge output of energy suggested to them that it might be an intergalactic collision. Zwicky had observed elevated energy output from such collisions plainly portrayed, as this object was not, in photographic images.

In their immensity, galactic collisions were the hugest generators of energy for which there was evidence at that time. Given the ratio of galactic diameter to the typical spacing of these objects, only 1/1,000, collisions between them must be reckoned as infrequent but not rare. Collisions between stars, occupying space more than 10^{25} times their own volume, are rare.

During the 1950s, radio telescopes detected many more point sources shining brightly in the radio spectrum. Many of them came in on energy comparable to that of Cygnus A. If these sources were outside the Galaxy, their absolute radio luminosity would be orders of magnitude higher than that of a number of well-identified, hyperactive, elliptical galaxies that had meanwhile been classified as radio galaxies. Some of the sources, and of some of the radio galaxies, as well, varied periodically in output. That put an outside limit on the size of the generators of their radiation. To be seen as a single source, an object cannot be larger than the distance light travels in the interval of variation. Periodicity of weeks, days and hours showed the generators in these radio sources to be compact enough to fit inside a galaxy.

Radio interferometers, reducing the uncertainty of location to an arc-second, made it practical to search for these radio sources in photographs of crowded star fields. The matching visible objects proved invariably to be giant elliptical galaxies. Their spectra, secured by as many as seven hours of exposure, presented the next problem. The emission and absorption lines showed no correspondence to the spectra of other objects. In 1963, Maarten Schmidt, at the Palomar Mt. Wilson Observatory, discovered that emission lines in the near ultraviolet of one powerful radio source could be matched to lines in the far ultraviolet that identified hydrogen in highly excited states. This large redshift placed the quasar at a distance of 2 to 3 billion light-years from Earth. Quasars thus must lie outside the Galaxy, not stars but kin to galaxies.

Source of quasar brightness

The radiation of all but a few quasars seemed to be coming from comparably remote distances. To sustain what must be their high absolute luminosity even for the brief cosmic time of a million years, they had to be transforming a cumulative 100 million to a billion solar masses totally to energy. What process could so rapidly transform such immense amounts of matter? A closer look at nearer quasars by the Very Large Array of radio interferometers at Socorro, New Mexico, supplied the spectacular answer.

Computer images from precise measurement of energy gradients from a cluster of sources in Cygnus A showed its radiation spread across millions of light-years in intergalactic space. The radio energy that first registered the existence of this object comes principally from two sources, separated by tens of million of light-years, each ballooning into thousands of cubic light-years. About midway between, a compact radio source coincides with an elliptical galaxy, the brightest and biggest in its local galactic cluster. From this galaxy in one direction extends a jet, sharply defined at radio wavelengths, of particles moving at near-light velocity. What must be the oppositely pointed jet has not been thus directly observed. It is evident, however, in that two jets have bored through millions of light-years of intergalactic space in opposite directions on each side of the galaxy. In the vast radio-loud regions, they have accumulated clouds of excited particles of their own substance and shocked and excited particles swept up from the intergalactic atmosphere [see illustration, page 217].

The general explanation for this and for the similar spectacles presented by other quasars is a supermassive black hole at the center of these galaxies.   Rotating at considerable speed, it is thought, the black hole drags the substance of its galaxy into an accretion disk shaped by its grav-itational field.   Around the axis of spin, powerful magnetic fields gener-ated by fast-moving charged particles in the accretion disk eject materi-al into the high-velocity jets.   In this activity, some considerable fraction of the galactic rest mass is converted to the energy of radiation.   Much the larger substance of the galaxy disappears across the gravitational hori-zon into the ever more supermassive black hole.   The quasar-galaxy is consuming itself.   The lifetime of this now recognized class of objects cannot exceed much more than 100 million years.

Edge of the observed universe

Maarten Schmidt's redshift measurement establishing that quasars are distant objects showed also, of course, that their radiation comes from long ago.   His colleague Alan Sandage instituted an imme-diate search for objects not first heralded by radiation in the radio spec-trum but with spectra and redshifts.   His sky surveys, ingeniously designed to detect these faint objects by their spectra even in crowded star fields, did find them, in greater numbers than the radio-strong quasars.   Their spectra, showing highest intensity at lines in the ultravi-olet of the standard spectrum, evidence high absolute luminosity dimmed by large redshifts, often to the middle of the visible spectrum. Their radiation issuing as from a point source establishes these objects as a distinct class of galaxy, classified now with quasars.   Their nuclei are involved in hugely energetic activity.   No force but gravity could gener-ate that energy, most probably by the agency of a black hole.   The improbable black hole thus gains a leading role among the generators of luminous energy in the cosmos.

In the late 1960s, the number of quasars, both radio-loud and radio-quiet, approached 1,000; it then became possible to consider them statistically.   Schmidt found most numerous the quasars 13 billion light-years and more distant from Earth and dated, therefore, to 13 billion and more years in the past.   Per cubic volume of space, they outnumber the nearer and more nearly contemporary quasars, such as Cygnus A, 1,000 times.   A decade later, when more than 1,500 quasars had been detected, Schmidt and Sandage reported few quasars shining from more than 15 billion years ago.

Surveys of the southern sky from the Inter-American observatory at Cerro Tololo, Chile, conducted by Malcolm G. Smith and Patrick Osmer in the late 1970s confirmed the finding that few objects older than 13 billion years could be detected. In one small patch of sky, the surveys located nearly 150 hitherto unrecorded quasars, whose redshifts gave nearly all of them almost the same age; that is, around 13 billion years. The most distant of these objects (about one-fourth of the total) were not more than 15 billion years old.

These observations suggest the possibility of a "quasar era" in the history of the cosmos. Here it needs to be recalled that the nearer universe is just as old as the distant and presents the same spectacle from its early history to observers out there. In the nearer universe, the self-cannibalizing quasars of 13 billion years ago would have vanished, perhaps into their black holes. Alternatively, there may have been a quasar phase in the history of some or all galaxies. If so, some of the quasars may persist, transformed to more stable elliptical galaxies. The search for intermediate "primitive" galaxies nearby has begun, but with uncertainty as to how they are to be identified.

Radiation from the postulated quasar era brings no evidence for the existence of "normal" galaxies, similar to the Galaxy, back then. The younger Population II stars in their nuclei would have been moving along the main sequence, shining more brightly than they do today. But their principal output of radiation, in the visible spectrum, would be redshifted into the infrared and radio spectrum below present detection.

Recognition of a quasar era has thus taken observational cosmology deep into the past close to the beginning of the grandeur so recently brought under observation. The extreme redshifts under study in this work have an instructive example in a representative object reported from the Sloan Digital Sky Survey, which is conducted by a consortium of observatories with headquarters at Apache, Arizona. The Lyman Alpha line of hydrogen, prominent in the faint radiation from the distant quasars, has on Earth a wavelength of 1.216×10^{-5} cm, in the far ultraviolet. That line in the representative quasar is shifted across the entire visible spectrum to 8.3×10^{-5} cm in the infrared. Long division (8,300 by 1,216) yields a redshift of 6.82 times. At the presently agreed-upon Hubble constant—60 to 80 kilometers per second per megaparsec—the universe has an age of 14 billion years plus or minus a billion years. The redshift sets the time at which the radiation started from the quasar, accordingly, at the end of the first billion years of universal history.

Echo of the Big Bang

Into the still deeper past, observational cosmology reaches the echo of the Big Bang itself. This is the universal background radiation predicted by Alpher and Hermann and discovered by Penzias and Wilson in 1965. Satellite survey around the sky of the intensity of this radiation across the spectrum of its wavelengths, between 0.2 millimeter and 80 centimeters, set its temperature—in accordance with the black-body equation—at 2.73 degrees above absolute zero K. Calculation of its redshift—1,500 times!—from the far-off spacetime whence it comes gives the radiation back then a temperature of around 3,000 degrees K. In sight is the universe at 300,000 years. Its expansion has carried it to a radius of 45 million light-years. The density of radiation has fallen below that of matter. Protons and alpha particles have cooled and slowed down enough to permit the electromagnetic force to attach electrons and give them their identity as hydrogen and helium. Photons of radiant energy have started on their endless journey, a few toward that tiny space in the expanding universe occupied by the solar system and its planet Earth, where they are today detected so near absolute zero K.

The implosion of the universe into the past, thus traced by observational cosmology, has become more apparently the story of creation, told backward. This version of creation can be told backward because experience—observation and experiment—connects the present universe continuously to the past. Observational cosmology has carried the story back to the end of the first 300,000 years. From thence, theoretical extrapolation of the trends of increase in temperature and density of the universe and decrease in its radius reaches the first 100 seconds. There theory overlays and is affirmed by experimental evidence from the particle accelerators. The increasing power of the accelerators over the past half century has taken the story deep into the first second of time. Thence theory—extension of theory that is sustained by accelerator experiment—takes the Big Bang to its beginning.

Big Bang: standard model

This improbable story of creation responded, it should be recalled, to the question plainly posed by the recognition in the 1930s that the universe is expanding: from what did the expansion start? George Gamow's bold answer carried the testable proposition that the hot beginning must have an echo. The discovery of the universal background radiation in 1965 drew others into refinement and elaboration of the Big Bang. The story of creation may now be told forward.

In the standard model of the story, held by the consensus until the early 1980s, the Big Bang begins abruptly at 10^{-37} second [see illustration, pages 234–235]. That stratagem obviates consideration of the state of the universe at 0.0 second, where known physics fails.

At 10^{-45} second, the universe wears that ghostly symmetry. Otherwise featureless, it is in size below naked-eye visibility, hot, 10^{37} degrees K and dense: 10^{100} grams or, equivalently, 10^{120} ergs per cubic centimeter, of which volume it is an infinitesimal fraction. For this unimaginable entity, the universe at present offers scale models in the neutron star and the black hole.

As the Big Bang gets under way, a phase transition—as water comes to a boil or freezes to ice—breaks the featureless symmetry, disengaging the gravitational force from the quantum forces. The radius reaches 10^{-3} centimeter, still below visibility; the density falls to 10^{91} grams per cubic centimeter and the temperature to 10^{32} degrees K. This corresponds to the energy of 10^{19} GeV, at least a dozen orders of magnitude beyond the reach of the most ambitious accelerator technology. The GUT phase transition follows at 10^{-37} second, disengaging the strong force from the electroweak force. The radius has just passed 1 centimeter; the density of energy has fallen to 10^{79} grams per cubic centimeter and the temperature to 10^{29} degrees K, corresponding to 10^{16} GeV, still well out of reach of experiment.

These numbers are, even in theory, uncertain: as to radius, temperature and energy, by about one order of magnitude; time, by two orders of magnitude and density, by four orders of magnitude. They are extrapolations, however, from evidence from the accelerators.

The next phase transition is coming within reach of experiment. At 10^{-12} second, with the radius at 100 million kilometers, the temperature, fallen steeply to 10^{16} degrees K, corresponds to the energy of 10^3 GeV, first attained in the Batavia accelerator. The steep fall in density has taken it to 10^{29} grams per cubic centimeter. This is the electroweak phase transition, the break in symmetry that separates the electromagnetic force from the weak. The quarks and leptons and the heavy quanta take their rest mass from the dense Higgs field energy that charges the then universe.

Big Bang in reach of accelerators

Evidence for this event was established over the years from 1973 to 1984 in the accelerators at Batavia and CERN [see page 159]. The event was to have been certified by the appearance of the Higgs particle in the target zone of the Superconducting SuperCollider. The "God particle" may make its appearance before 2005 in either the Batavia accelerator, now being boosted to a multiple of its 10^3 GeV (1 TeV) energy, or the 27-kilometer ring of the Large Hadron Collider now under construction at CERN.

At 10^{-6} second, in the expansion of the universe to the radius of 100 billion kilometers or 0.1 light-year, the QCD—quantum chromodynamics—phase transition gives particles take their identity as leptons and quarks. This brings the universe into the state explored by the accelerators through the 1950s and 1960s. The matter-antimatter annihilation in the moment preceding had, for a briefer moment, spiked the decline in temperature to 10^{13} degrees K. The tiny surviving excess of matter goes through the QCD transition to combine in the exotic nuclear particles composed of excited quarks that erupted from the accelerators when they reached the MeV energy range.

From the first second on into the first 100 seconds, the decelerating expansion takes the universe to the radius of 10,000 light-years. Temperature and density approach that of the interior of stars. The

234

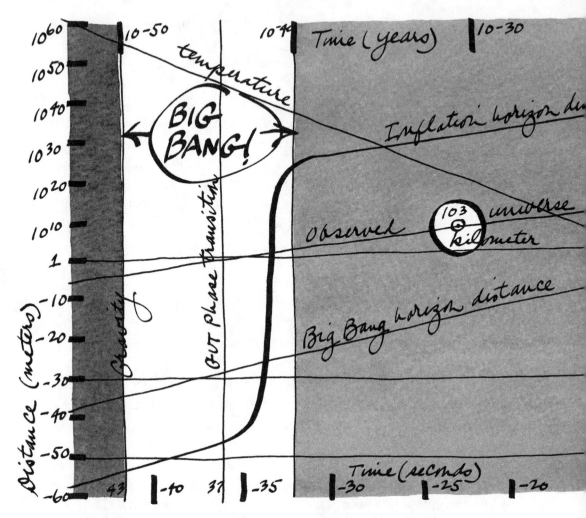

Big Bang genesis of the universe is charted here in order-of-magnitude scales. In the original Big Bang scenario, the universe at 10^{-45} second is a dense, hot, otherwise featureless entity below visibility in diameter [see pages 232]. "Phase transitions" attending the abrupt expansion and cooling of the universe bring the four forces and four particles and the atoms they constitute [see illustration, pages 166–167] into being in the first second. At 10^{-43} second, gravity separates from the quantum forces; at 10^{-37} second, the strong force from the electroweak. These events remain 10 orders of magnitude beyond reach of experiment. At 10^{-12} second, the electroweak force separates into the weak and electromagnetic forces, and matter separates from energy. Now the temperature of the universe is in the TeV accelerator energy range; experiments addressed to this event are

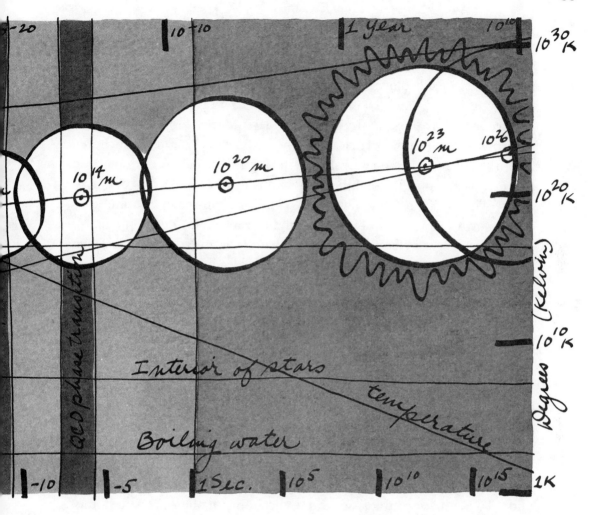

under way. In the QCD phase transition, particles of matter assume identity as leptons and quarks. Accelerators have reenacted this event, generating the two generations of high-energy particles from which ordinary matter "descends" [see pages 118 and 164]. Circles compare expansion at crucial intervals. The "fireball," at 10^{23} meters and a temperature of 3,000 degrees K, is observed in the cosmic background radiation redshifted to 2.725 degrees K. "Black body" uniformity of that radiation supports the "inflationary" amendment of the Big Bang scenario [see page 237]. Inflation from less than 10^{-50} meter to 10^{25} meters secures the observed homogeneity of the universe and brings all parts of it well within the "horizon distance," the distance traveled by light in the 15-or-so-billion-year lifetime of the universe [see illustration, page 192].

cooling-down has long since bound the up- and down-quarks in protons and neutrons. In the still-high universal energy, some of the primordial protons enter the first level of the self-organization that is to structure the diversity of substance observed in the universe today. Some bind by the strong force to neutrons, forming the nuclei of deuterium, the stable isotope of hydrogen. In pairs, a larger number bind with pairs of neutrons to form alpha particles. The neutrons thus bound are sheltered from the decay. The rest transform, through the 10.3-minute neutron half-life, to protons; the number of protons nearly doubles.

Through the next 300,000 years the universe continues its expansion to a radius of 45 million light-years. The dense plasma of electrically charged electrons, protons, deuterons and alpha particles has been opaque to radiation. Now, with the decline in temperature to 3,000 degrees K, the density of radiation falls below that of matter. Binding to electrons, the protons and alpha particles assume their identities as hydrogen and helium. The electrically neutral gas of hydrogen and helium is transparent to radiation. Photons with energy of 3,000 degrees K light the universe. Subsequent expansion, by two orders of magnitude, has red-shifted the cosmic background radiation to 2.725 degrees K.

There are things wrong with the standard Big Bang model— quite apart from the difficulty that recently arrived observers must find in imagining the unimaginable compacting of spacetime and the correspondingly unimaginable density and temperature at the first 10^{-45} second. To begin with, the story cannot account for the observed homogeneity of the universe. If the observed universe is homogeneous, its most distant parts must once have been in contact. The standard Big Bang model has the universe exploding faster than light—to a radius of 10^{14} meters in a millionth of the first second. Regions in the early universe would necessarily have been separated by "horizon distances," beyond which light could not have traveled even in the then-brief lifetime of the universe [see illustration, page 192]. Such velocity of expansion violates no law. Light has its ultimate velocity, but the expansion of space need observe no such limit. Accordingly, even within the tiny radius from which the universe entered the Big Bang, not all regions could have been subject to the same forces at the same time. Expansion since, moreover, would have carried the presently most distant regions too far away for light ever to report on them to observers in this corner of the universe or on us to them in their corner.

Inflation

To cure this and other shortfalls in the Big Bang, Alan Guth, then at Stanford Linear Accelerator Center and now at MIT, ventured deeper into the first second, back behind the first 10^{-43} fraction of it. He perceived that a peculiar form of matter predicted by supersymmetry versions of GUT theory would exert negative gravity: a force of repulsion. Such a force has standing, of course, in the cosmological constant of the 1917 edition of General Relativity. If the early universe contained just a tiny patch of the peculiar form of matter, this patch would be stretched exponentially by the gravitational repulsion, inflating it to encompass the entire observable universe.

In the inflationary model of genesis that Guth now proposed, the expansion of the universe begins slowly and then rapidly accelerates. That contrasts with the expansion in the standard model that starts instantaneously at high speed and then slows rapidly under the force of gravity. The universal inflation starts slower than the velocity of light from a radius of less than 10^{-50} meter, allowing the tiny universe to become fully homogeneous. Then the force of repulsion takes over. Driving through the first phase transitions to 10^{-33} second, it inflates the universe to a radius exceeding 10^{20} meters.

In this proposal, published in January 1981, Guth acknowledged an implicit problem: the runaway inflation might be stopped by disorder akin to the boiling of water. Random bubble formation would negate the isotropy sought in the proposed inflation. Within about a year Andre Linde, then at the Lebedev Institute in Moscow and now at Stanford University, proposed a solution to this problem, soon after affirmed by Paul Steinhardt and Andreas Albrecht at the University of Pennsylvania. They showed that not all phase transitions proceed like the boiling of water. Given specific assumptions about the underlying particle physics, the inflationary phase transition could proceed smoothly and uniformly. At the end of the inflationary era, around 10^{-35} second, all continues according to Big Bang scenario, supported by experimental physics from soon after the first trillionth of the first second.

This inflationary preliminary detonation was welcomed as curing the defects found in the "simple" Big Bang. It is now strongly supported by the measured isotropy of the background radiation. The Cosmic Background Explorer satellite, through four years beginning in November 1989, made precision measurement of the wavelengths of the radiation around the universe open to observation above and below the plane of the galaxy. At no wavelength, from 0.5 millimeter to 0.5 centimeter, does observation depart by more than 50 parts in a million from the ideal black body curve at the temperature of 2.725 degrees K. The glow on the horizon of the universe proves to be the "smoothest," "flattest," most featureless of all things observed. The best explanation for such pure isotropy is the sudden, extreme inflation of the nascent universe.

Interest has turned now to the 50-parts-in-a-million departures from isotropy in the map of the background radiation. Technology developed in the last decade permits accurate contour mapping of these modulations of the background radiation in the regions where they appear. The stakes are high. These regions, it is thought, contain the seeds for the evolution of the galaxies and the larger-scale structures of the universe. Regions of tiny variation in density in the dawning universe condensed, it is thought, in those now-observed aggregations of mass as the universe cooled down and declined in density. The data coming in from high-resolution ground-based instruments so far find excellent agreement with prediction from inflationary theory. It is anticipated that readings by satellite-borne instruments on expeditions in the planning stage will do the same.

Particle physicists and cosmologists have thus converged upon a unitary picture of the universe and its history from the beginning. Much unfinished business troubles satisfaction in the work accomplished. Foremost is the nature of the missing mass. It could be of the same kind as the observed luminous matter but radiating below the reach of present instrumentation or lost in black holes. If the elusive neutrinos can be shown to possess the tiniest rest mass, they might supply a substantial fraction of the missing mass. Partially in this interest, the big accelerators at Batavia and Geneva are firing salvos of neutrinos straight through the rock, on the chord of the arc of the Earth surface, at detectors located around the surface of the planet in Ontario and Italy. The missing mass sought from classical considerations in General Relativity may prove to be of an unexpected kind, exerting the antigravitational force of the cosmological term that should have secured a static universe. The new physics in supersymmetry and string theory offers a variety of particles of "dark matter" that could meet such specification. In the eruption of particles from the experiments designed to trap the Higgs particle may be found the first evidence for the presence of superpartners of the particles of matter and antimatter proposed by supersymmetry.

Dark energy

Perhaps ahead of the nature of the missing mass comes the nature of the force of repulsion. The acceleration of the universal expansion is forcibly indicated by the precision measurements made with the Type 1a–supernova yardstick. The "dark energy" exerting this force could be another manifestation of the energy of the vacuum from which the particles of matter took their rest mass and with which they exchange virtual particles. This force may always have been at work, from the beginning in the inflation and the Big Bang and in the expansion of the universe that now, it appears, is accelerating.

Resolution of these new questions may hang on—or it may secure—the closing of the conceptual gap between the quantum theory of the three polar forces and the field theory of gravitation that otherwise work so well together. Einstein's own stubborn effort could not bring the forces together in a single field theory. All attempts by imagination, observation and experiment have failed to bring gravity into a quantum-theoretical statement that yields propositions open to test. Comprehension of reality in "a conceptual system built on premises of great simplicity" remains the unreached goal of inquiry.

Nor is it settled that physics is finite. The "inflationary universe," now so credibly established, opens to speculation a cosmos very much larger than the universe accessible to observation. The inflation of the primordial kernel places the horizon distance perhaps 20 orders of magnitude beyond the reach of observation. Within the inflated universe the presently observed universe assumes the status to which observation reduced the Galaxy in the course of the 20th century. Inflation of the primordial universe by 10^{20} times makes room for countless other universes. Those other universes may proceed through their life histories on different schedules and even under different physical laws. The objects of such speculation lie, so far, beyond any conceivable experience and so outside of consideration here.

Of still another objective imputed to cosmic inquiry, Steven Weinberg has observed, "As we have discovered more and more fundamental physical principles, they seem to have less and less to do with us."

The century's work has shown that human existence is confined to a tiny, momentary sanctuary in a universe that is everywhere the scene of hitherto inconceivably gigantic overturns of mass and energy attending a history that began long before we were here to observe it, and will continue long after the Sun, Earth and we are gone. If the fundamental physical principles have anything to do with us, it is in the narrow exceptions to the generally prevailing conditions that permit our existence. In lieu of hope to find the purpose of existence in the starry heavens, the new understanding counsels the cherishing of life on planet Earth. People owe their existence to pre-*sapiens* forerunners who found such purpose in the dawning consciousness of their fleeting existence. To life and the origin of the human species this book now turns.

5

The Living Cell

*We must therefore not be discouraged by the difficulty
of interpreting life by the ordinary laws of physics.*
Erwin Schrödinger

C oming forward from the Big Bang, this account of the new story of creation reached a time when the radius of the expanding universe was passing 10 million light-years. Corresponding decline in the density of energy then freed photons of electromagnetic radiation to illuminate what was there. At that epoch, there was nothing there but energy, still at high density, and hydrogen and helium. Substantially all the atoms in the universe today are hydrogen, about 93 percent, and helium, about 7 percent.

"Substantially all" leaves some fraction of the atoms in the universe to be accounted for. That fraction is the ash of all the starlight radiated since the gravitational force first collapsed the stars. It is hydrogen and helium fused in the nuclei of all the heavier atoms in the table of elements. These heavier atoms constitute no more than 10^{-3} (0.001) of the mass of all the known matter in the universe. Starlight, over the eons, has carried off a mere 10^{-4} (0.0001) of the primordial mass. Generating starlight, thermonuclear reactions have been taking the naked nuclei of hydrogen and helium through the first steps in the organization of matter toward the complexity that, in this corner of the universe, has come to have conscious observers. At around 80 protons and 120 neutrons in a heavy atomic nucleus, the assembly begins to exceed the capacity of the strong and weak forces to contain complexity. Heavier atoms fall apart in radioactivity.

The matter thus assembled in the stars thereafter realizes its further potential by action of the electromagnetic force, which engages the external electrons of intact atoms. In the vast regions of interstellar space within the galaxies, at least 5 percent of the galactic mass—principally hydrogen, of course—exists in the intact atomic and molecular state. The atoms and molecules form chemical compounds and aggregate in the solid particles of the dust clouds that obscure starlight in the arms of the galaxies. Those clouds hold a lengthening list of organic compounds. They may conceal additional huge aggregations of matter collapsed in Brown Dwarfs, stars not big enough to shine. These stars may afford hospitable ranges of temperature and pressure for inorganic and organic chemistry within their interiors and in their vicinity.

The model of our local solar system suggests that the electromagnetic force best displays its virtuosity in satellite systems attending single stars. Robot satellites and Earth-bound radar have disclosed wondrous landscapes on the other terrestrial planets and the terrestrial satellites of Jupiter and Saturn. The atomic organizer of these spectacles is oxygen. Oxygen is the most abundant element built up from hydrogen and helium in the stars. In compound principally with four of the 89 other elements present in relative abundance in the solar system—silicon, aluminum, manganese and iron—oxygen constitutes the solid-rock substance of these satellites of the Sun.

On Earth—and possibly on one or more of those other rocky satellites of the Sun—oxygen combines with hydrogen and the two next most abundant elements, carbon and nitrogen, to carry the organization of matter to another level of complexity. Life, it turns out, is as natural as starlight. It is electromagnetism that connects hydrogen, oxygen, nitrogen and carbon to one another in intricate configurations by their respectively one, two, three and four external electrons. In the laboratory, outside the living cell, these four elements may spontaneously combine to form molecular components of the cell. Early in the history of the planet such molecules assembled in self-reproducing living cells.

Thence came the conscious observers. The French astrophysicist

Pierre Auger saw life as a "standing wave" in the downhill dissipation of energy and order in the cosmos: a ripple of higher organization of matter and energy—of lower entropy—through which runs the universal increase of entropy prescribed in the second law of thermodynamics.

A wonderful coincidence in the history of science steered inquiry into life on its present course at the very beginning of the 20th century. In 1900, three botanists simultaneously and independently discovered that heredity conveyed by sexual reproduction is quantized. A trait from one parent appears pristine in the offspring; the trait is not blended with the corresponding trait from the other parent. We are genetic compounds, not blends or mixtures, of our forebears.

Carl Correns in Germany, Erich Tschermack in Austria and Hugo de Vries in Holland, searching the literature before they went to press, were the more astonished to make another simultaneous and independent discovery. Someone had published their discovery 34 years before.

Mendel's gene

In a garden of the Augustinian monastery at Brünn (now Brno, in the Czech Republic), Gregor Mendel had conducted long series of experiments in the hybridizing of peas. He found that peas in the daughter pod are either green or yellow, not greenish yellow or yellowish green; smooth or wrinkled, not less wrinkled. The evidence is strong that Mendel designed his experiments to test his hunch that a trait is carried thus intact from one generation to the next by a factor that his successors have called the gene. The gene has proved as sound in concept as the atom, and as real.

In proposing the gene, Gregor Mendel posed questions that engaged the life sciences throughout the 20th century: How in the parent does the gene express the trait it transmits to offspring? How is the gene carried intact from the parent to the offspring? The concept of the gene laid the trail from animals and plants to the cells of which they are composed, to the large molecules of the cell and on to the "organelles," the little organs, of the cell. These array the molecules for the ordered, sequential biochemical activity that is the life of the cell, all as encoded in the molecular structure of the genes.

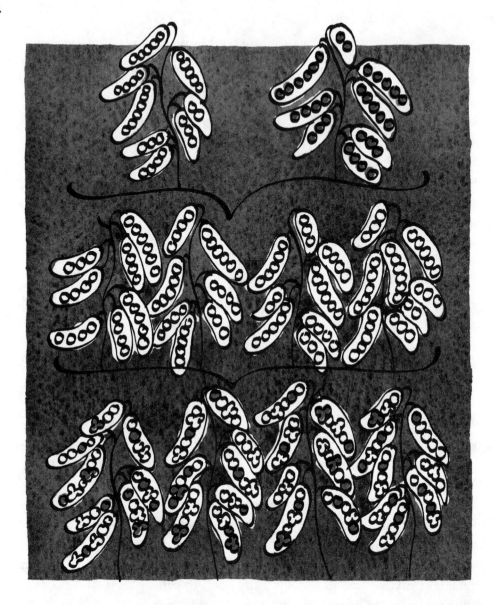

Inheritance is by combination, not blending, *of genetic traits [see preceding page]. Genes for smooth and yellow in peas [top left] are dominant; genes for green and wrinkled [top right] are recessive. In first-generation crossing, the dominant genes prevail: all peas are smooth-yellow. In second [bottom] half the peas are smooth-yellow; a fifth each carry one recessive gene, green and smooth or wrinkled and yellow, and a tenth carry both recessive genes, wrinkled-green. Genes keep traits in circulation against environmental change that might favor them.*

The cell is the irreducible unit of matter organized in the living state. It is the minimal organism capable of self-assembly from energy and matter in its environment and, then, of self-reproduction. The earliest cells gave rise to at least two persisting major kinds of cell. Closer to the original and constituting more than half of the tonnage of living matter on the planet are the bacteria. They are the smaller and anatomically the simpler of the two; some bacterial organisms are multicellular. Bacteria are also called prokaryotes, which means "before" or "not yet" nucleated. Cells with a nucleus, the "well" nucleated eukaryotes, constitute virtually all of the visible organisms, animals, plants and fungi, plus the vast kingdom of protoctists, single-celled and even multicellular eukaryotic organisms not recognized as animal, plant or fungal. By "living cell," the eukaryotic human being usually means a eukaryote.

The locus of Mendel's gene

Early in the 20th century, inquiry recognized the self-replicating nucleus in the eukaryotic cell as the locus of Mendel's gene. At mid-century the DNA molecule incarnated the gene. The way then opened to understanding of how hereditary experience is conveyed from the parent cell, how it is expressed in manufacture of the cell's substance and how it is replicated for transmission to the next generation.

Through its enveloping membrane, no mere container but itself an active organelle, the cell draws incessantly and selectively from the environment the raw material of its substance. Another organelle manufactured in expression of the cell's heredity supplies the energy necessary to sustain this activity. Still another organelle, most likely in another cell, prokaryotic or eukaryotic, captured that energy from the Sun, one photon at a time. Each electron boosted by a photon into higher orbit thereafter cascades downward, one quantum jump at a time, in the ladders of large molecules arrayed in the organelles to yield the energy that brings air, water and earth into the assembly and activation of living organisms. Transformed by the 10^{14} cells in an active human body, for example, solar photons yield the mechanical energy equivalent of 150 kilowatt-hours a year—plus a great deal more energy as heat.

The organelles and large molecules of the cell have been shown to carry on their vital functions in isolation from the cell. Their sub-structures and molecular components have been shown to self-assemble on their own. Ancient technologies and significant industries depend upon these capacities of "organic" matter.

How, on the other hand, the large molecules and organelles of the cell concert their activity in the life of the cell itself and of the multicelled organism beggars understanding. The present turnover of energy in living cells got its start, the geological record indicates, more than 3.5 billion years ago, just as soon as the planet cooled down enough to keep water in the liquid state. Enumeration of the smaller and smaller number of features that all now-living varieties of cells have in common is bringing into sight what must have been the rudimentary organisms that started it all. From that beginning, the genetic record celebrates the capacity of the cell to respond opportunistically to new resources encountered in its environs.

Most of the mystery of life is concealed, therefore, in the first nearly 3 billion years of its history, a sufficiently long time to accumulate experience. That is time measured at the human pace. At the atomic and molecular reaction time of 10^{-15} second, the time available for the accumulation of experience may be more appropriately reckoned in the 10^{17} seconds of Earth history. Incorporating new elements from the Earth's crust in its substance and drawing energy from the Sun, the cell made itself at home in every latitude on Earth. Before multicelled eukaryotic organisms appeared, perhaps 700 million years ago, prokaryotes had engaged the atmosphere, the water and the crust of the Earth in the creation of the biosphere. They thereby remade the planetary environment to suit their flourishing existence. Since then, the cell has realized ever greater variety, higher order of organization and increasing autonomy in its multicelled offspring. All the while, the bacterial pioneers have persisted. They secure the planetary environment for their offspring and set for their conscious observers the example of survival.

Self-replication, the transformation of energy and the incorporation of new experience in the gene together, therefore, drive evolution. Darwin drew upon the baleful vision of Thomas Malthus in assigning the paramount role in the origin of species to the struggle for the means of subsistence and the selection thereby of the fittest. The source of variation that placed new organisms in contention was a mystery when Darwin published *The Origin of Species,* seven years before Mendel reported his findings and more than a half century before they were recognized.

The work started by Mendel has put evolution in a new light. The dichotomy of variation and selection—variation proposing and natural selection disposing—does not suffice. Most of life's first experiments and inventions must have failed on their own. Even at the outset, the agent of selection—the physical environment—was undergoing change in its occupation by life-forms. New species pioneered new niches and themselves supplied new niches to new species. Variation does not "solve problems" of survival presented by the environment, as the population geneticist Richard Lewontin has observed. Life itself created the environment of life on Earth today.

The fittest are now more neutrally defined as the bearers of surviving progeny. The genesis of every species is a special case to reward investigation. To transform energy, to self-replicate and to incorporate new experience in the self-replicating gene: together, these capacities are unique to living matter. Only living organisms evolve. The rest of the universe has history.

Before 1800, with little more than a bead of glass as his microscope, Antonie van Leeuwenhoek discovered that "animalcules a-swimming" will be found in "water taken out of ditches and runnels." That observation, not now surprising, was the Moon landing of its time. Christopher Pitt, in England, caught the moment:

> Thro' whose small convex a new world we spy,
> Ne'er seen before but by a Seraph's eye!
> So long in darkness shut from human kind
> Lay half God's wonders to a point confin'd.

A century later, with microscopes that improved upon Leeuwenhoek's "small convex," observers could see the multiplication of cells by the division of each cell into two cells. The botanist Matthias Jakob and the zoologist Theodor Schwann saw that cells compose the tissues of plants and animals. Tissue cells—no free-swimming animalcules—could be seen bound together in variously specialized, interdependent services to the organism. Rudolf Virchow then stated the cell theory: Cells come only from cells—*omnis cellula e cellula.*

A few serious observers still argued the folklore of spontaneous generation of life. Maggots and flies came from offal; fermentation corrupted brews and broths left standing. Louis Pasteur settled this question by a simple experiment: exposed to air, broths spoiled; covered over, they did not. To meet the objection that covers stifled spontaneous generation, he supplied air through a pipe twisted to exclude the air borne single-celled organisms he held to be the agents.

Such spoiling and the fermentation that he monitored for the wine industry, Pasteur declared, could proceed only in the presence of living cells. With Virchow, Pasteur thus posed for investigation the question of the origin of life at some remote time in the past.

To the understanding of life at the molecular level, Pasteur made a quite different contribution. From incomplete ferments and from some wines, crystals of tartaric acid often precipitate. Pasteur noted that these precipitates weakly rotate polarized light, sometimes to the right and sometimes to the left. Managing to separate them, he found two sorts of crystal mixed in the precipitates. One rotates the light strongly to the right; the other, strongly to the left. On this evidence he proposed that the molecules of tartaric acid have three-dimensional structure, twisted or coiled to the right or to the left, which has a role in their chemical and biological activity.

The cell and its nucleus

Under the microscopes in use at the turn of the 20th century the cell—the eukaryotic cell—appeared a transparent globule of "protoplasm" contained, apparently, in a membrane. In its nearly featureless interior, a darker small speck would draw the first interest of an observer. This "nucleus" was at the center of the signs of life [see frontispiece].

In due time, the nucleus would be seen to grow somewhat in size. Then it would come into the reach of translucent rays that fanned out toward it through the protoplasm (now "cytoplasm"; *cyto* = cell) from points on opposite sides of the cell. The nucleus would extend in length at a right angle to the axis of the rays. Thereupon, it would disassemble into tiny strings that would move in equal numbers to the opposite poles drawn, apparently, by the now-retracting rays. The rays of the "mitotic spindle" would vanish; the two sets of strings would bundle into separate nuclei. The approximate equator of the cell would cinch inward toward a point. The cell was soon two cells.

An observer had to wonder what vital force ordered these events. The cell that was there was gone. Its life now animated two cells. Upon the death of a cell or a multicelled organism, everything that was there before is still there, except life.

In his luminous treatise on *Growth and Form* published in 1917, the British biologist D'Arcy Thompson observed, "the principal forces … and the principal properties of matter with which our subject obliges us to deal … have, to say the least of it, a close analogy with known physical phenomena." He cited the small size to which single-celled life is constrained. Surface tension sets limits beyond which cells as well as raindrops cannot grow. Small size brings surface tension and intermolecular forces into contention with gravity in sustaining the cell's third dimension. The tiny forces of diffusion and osmosis conduct the cell's transactions with the world outside. Yet it had to be seen that "there are actions visible as well as invisible taking place within living cells which our knowledge does not permit us to ascribe with certainty to any known physical force."

Some sought in life a force outside physics. Vitalism stayed in contention to the mid-20th century.

Physicians' concern with infectious disease brought recognition in the 19th century of the bacterium as another kind of living cell. The menacing of other organisms, it is now known, occupies the tiniest percentage of the increasing variety of identified bacteria. They thrive in mutually sustaining communities wherever there is water or just enough moisture, manufacturing their substance ultimately from air, water and rock. The prokaryote genetic material is strung on a hooped "nucleoid" and floats naked in the cytoplasm. All but a few prokaryotes are encased in a tough polysaccharide capsule; therein they survive hostile environments. Bacteria include, in all their variety, ancient immediate descendants of the first assemblers of living matter, that dwell in environments that approximate conditions on Earth at the origin of life.

At the end of the 19th century, physicians and cell biologists were troubled by evidence, in communicable illnesses, for a still tinier organism. The "filterable virus" escaped through the pores of filters fine enough to strain out the tiniest bacterium.

Constitution of living matter

By the mid-19th century, chemists had a rough assay of the constitution of living matter. This they learned from the nutritional requirements of multicelled mammals, people and their domestic animals. Believing that only living organisms could make such matter, they called this new branch of their science organic chemistry. Organic molecules had not yet been detected in interstellar space. On Earth, such molecules were shown to be made principally of elements of air and water.

These elements join up in three familiar organic compounds. Carbon, from carbon dioxide in the air, and hydrogen, from water, together make the fatty hydrocarbons. The starchy carbohydrates contain oxygen, from water, as well as carbon and hydrogen in their somewhat larger and more elaborate molecules. Largest and infinitely various in architecture are the proteins, with the most abundant atmospheric gas, nitrogen, in their composition. Sulfur turns up as well in many proteins, supplying strong bonds here and there between atoms in their intricate structure. These three organic molecules are made, in short, of CH, CHO, CHON and CHONS, respectively.

In 1869, three years after Mendel published his work, Friedrich Miescher added a fourth organic compound and a sixth element to the list. From cell nuclei, he isolated the element phosphorus, associated, he thought, with protein. Miescher called the compound "nuclein." The significance of this discovery had to await rediscovery of Mendel's work and the events that followed. CHONSP, then.

A more detailed inventory by Preston Cloud, geologist and geohistorian at the University of California, Santa Barbara, shows, "The reader, thoroughly desiccated, would be [by weight] about 48.4 percent carbon, 23.7 percent oxygen, 13 percent nitrogen, 7 percent hydrogen,

3.5 percent calcium, 1.6 percent phosphorus, 1.6 percent sulfur, and less than 1.5 percent a dozen other things." The calcium comes from the skeleton and, among the "dozen other things" are sodium, potassium, magnesium, iron, manganese, cobalt, boron, copper and zinc.

It is carbon, with its four electrons, or valence of four, that gives organic molecules their infinite diversity. The organic chemists found that the elemental constitution of the molecules—thus, sugar: $C_6H_{12}O_6$; alcohol: C_2H_6O; vinegar: $C_2H_4O_2$—does not begin to tell the story. It is important to know which atom is linked to which. Carbon links them up not only in linear chains but with two-handed oxygen and three-cornered nitrogen to form square, pentagonal, hexagonal and still other, two- and three-dimensional, figures.

The chemical versatility of such compounds had an unanticipated demonstration in 1897. The brothers Eduard and Hans Buchner extracted the juice of a yeast culture in the expectation of finding medical uses for it. They added sugar to this extract as a preservative. To their surprise, and contrary to Pasteur's dictum, the combination went into vigorous fermentation, producing alcohol.

Here were organic compounds, outside the cell, exercising a competence of the living cell. The Buchners' "zymase" (Greek: "yeast") engaged the next generation of organic chemists in separating and identifying its active ingredients.

For all the interest that attended study of the cell and advances in organic chemistry, talented young biologists were also asking questions about the whole animal. At Columbia University in 1900, Thomas Hunt Morgan was engaged in study of the regeneration of severed body parts—a faculty possessed notably by arthropods, insects and other animals with external skeletons and by a few endoskeletal lizards. The discovery of Mendel's work diverted Morgan's attention and talent to genetics. He soon demonstrated that heredity in the ubiquitous fruit fly, Drosophila *melanogaster,* obeys Mendelian law—and so, presumably enough, in other animals as well as in plants.

To the fruit fly is owing much of present understanding of genetics. Drosophila offers itself rather ideally for such service. Its generation time of 12 days yields results more quickly than Mendel's plants. Moreover, the chromosomes in the salivary glands of Drosophila swell up when meta-morphosis brings on a molting phase. Inspection in a light microscope is thereby facilitated.

In 1903, Walter S. Sutton, an alert graduate student (who went on to a career in surgery) called Morgan's attention to a correlation between the fruit fly's inheritance of certain genetic traits and details in its chromosomes. By 1911, such coincidence persuaded Morgan

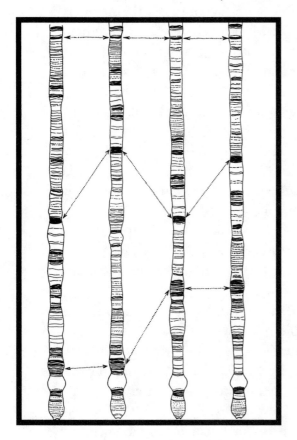

that genes might be organized in linear order on chromosomes. Improvement in microscopy fully confirmed Sutton's prescience at mid-century. Traits manipulated in breeding experiments could be mapped on the ribbonlike D. *melanogaster* chromosomes [left].

In 1927, Morgan's pupil Herman Muller succeeded in inducing mutation in familiar traits of Drosophila by exposing the flies to mild X rays. It was known that high-energy radiation could injure tissue cells. Incautious early investi-gators of this radiation had suffered skin burns, local cancerous tumors and leukemia. There could then be no doubt that the abstract concept of the gene has its reality in mole-cules of some sort lined up in the chromosomes for their work in the cell's moment to moment life and ultimate replication. In nonlethal muta-tion, moreover, could be seen a source of genetic variation to consider, along with the assortment and recombination of genes demonstrated by Mendel.

Organic chemists had meanwhile found the phosphorus of Miescher's nuclein associated not with protein, but with a quite different compound that they called deoxyribonucleic acid (DNA for short). This DNA had an elaborate structure, somehow involving a 5-carbon sugar (deoxyribose), a phosphoric acid and two pairs of nitrogen-containing compounds. In the 1920s, Robert Feulgen, a German biochemist, using a brilliant fuschia dye specific for DNA, reported that both dye and DNA are confined to the nuclei of plant and animal cells. This suggested strongly that DNA might be the substance of the gene.

No function, however, had yet been demonstrated for DNA. Meanwhile, proteins—the known constituents of muscle, bone, tendon, skin and hair—had abruptly emerged as important actors in cell chemistry. Analysis of the Buchners' zymase had yielded 14 intermediate products between sugar and the end-product alcohol and by-product carbon dioxide. Study of the succession of chemical reactions that produced these 14 intermediates led to identification of 20 or more "enzymes," proteins functioning as catalysts.

Like inorganic catalysts, enzymes enter into and accelerate—catalyze—chemical reactions between other molecules, retaining their own identity throughout. The enzymes, the chemists found, have energetic bonds located at specific sites in their structure. Binding their "substrate" into chemical combination with themselves, they then bond the substrate to its destined partner. Thereby, enzymes accelerate the reactions between organic molecules by a million to a trillion times. By such orders of magnitude they improve upon the chance that the molecules will encounter one another in the right orientation at the right velocity to effect their binding. In wonderfully coordinated sequence, it was found, the zymase enzymes carry the fermentation of sugar through the synthesis and breakdown of the 14 intermediates to alcohol and CO_2.

In the late 1920s, James B. Sumner at Cornell University and John Northrop at the Rockefeller Institute for Medical Research succeeded in isolating enzymes in such pure form as to precipitate them as crystals. Crystallization proved their molecules identical in size and shape and so to fit snugly together in crystals. Chemical analysis showed them to be proteins. They were large proteins, of 10^3 to 10^4 molecular weight ["gram molecular weight" or "mole"; see page 71].

The diversity of function found in proteins gave new significance to understanding of their structure that had now begun. Proteins were known to be composed of subunit "amino acids." From 100 to 1,000 of them link up by "peptide bonds" to join in a "peptide" chain. Some 20 amino acids, out of the many known and possible, are identified as the constituents of the proteins of the living cell. They vary in molecular weight from glycine, at 75, to tryptophane, at 204.

The amino acids are distinguished from one another by "side groups" of different weight and configuration. The linking-up of their identical amino-acid groups sheds a molecule of water. In effect, an oxygen, O, weakly connected at the "far end" of one amino acid, joins up with two hydrogens at the "near end" of the other amino acid to form H_2O. The peptide bond fixed in this reaction joins a carbon thus uncovered at the "head" end to a nitrogen uncovered at the "foot" end [see illustration, page 262]. This crucial bond is broken by pepsin, the enzyme that the chemists found promoting the digestion—the breakdown—of proteins in the stomach. About the linkage of identical "monomers," such as amino acids, into "polymer" chains, chemists had been learning from experience with the synthesis of plastics. The side groups that festoon a protein chain along its length distinguish one protein from another and endow them with their activity.

To the investigators who established this knowledge, it was apparent that the specificity and diversity of function attributable to proteins, especially to enzymes, must be related to the effectively infinite number of possible combinations of the 20 amino acids. That infinity begat another infinity with the recognition that the function of an enzyme must be related to the sequence in which the amino acids are linked in the length of its chain—just as the sequence of elements establishes the identity of a chemical compound. Proteins strongly commanded consideration as the substance of the gene.

The tobacco mosaic virus

Crucial episodes in determination of the substance of the gene now transpired at the Rockefeller Institute. William Henry Welsh of Johns Hopkins University and Simon Flexner of the University of Pennsylvania had persuaded John D. Rockefeller to establish, in 1901,

an institute devoted to inquiry into the nature of life. From understanding at the fundamental level would surely come answers to the most
intractable questions then confronting the science and practice of medicine. Such a question was presented by the filterable viruses.

At Princeton, New Jersey, the Institute set up a greenhouse-lab-
oratory for the study of plant viruses as models of their kind. There,
Wendell M. Stanley had the tobacco mosaic virus under study. In 1935,
employing techniques devised by his colleague Northrop to crystallize
the enzyme proteins, he managed the first isolation of a filterable virus.
Stanley showed tobacco mosaic virus to be constituted of much the same
combination of protein and nucleic acid as the chromosomes of the cell.
In this virus, however, the nucleic acid was not DNA but ribonucleic
acid, or RNA, containing the sugar ribose in place of deoxyribose.
Stanley was fortunate in his choice of virus. The rod-shaped tobacco
mosaic virus is long enough, 3×10^{-7} meter, to have registered its image
in the first-generation electron microscope then available. The thread of
the virus RNA, Stanley showed, is encased in protein.

Upon separation of the virus from leaf extract and subsequent
purification, he could precipitate it in crystals. When he inoculated solutions of this purified nucleo-protein in a tobacco leaf, however, it came
"alive." It took over the genetic machinery of the leaf cells to manufacture itself in quantity and show up in the characteristic mosaic pattern of
wilt in the leaf. Stanley found it, occasionally, inactivated in crystal
"inclusion bodies" in a cell in a wilted leaf. Having killed the cell and
wilted the leaf, the tobacco mosaic virus in more than one serial passage
in Stanley's laboratory, showed itself to be a gene without a cell, ready
to replicate in the cells of the leaves of the next tobacco plant. Interest
in nucleic acids as the likely substance of the gene revived.

Stanley noted another implication of this work. The viruses, he
said, make it "difficult, if not impossible, to place a sharp line separating
living from non-living things ... the principle of the vital phenomenon
does not come into existence suddenly but is inherent in all matter."

In 1955, Stanley came as close as anyone has to assembling a living organism. Then at the University of California, Berkeley, he and his
colleagues disassembled the virus into its component parts, its protein

overcoat and thread of RNA. They then reassembled it—that is to say, arranged the setting for the self-reassembly of the component parts. The reconstructed virus duly infected a tobacco plant.

DNA: the genetic molecule

Chemists by now had a better understanding of the DNA structure. The 5-carbon sugar plus phosphoric acid groups link up into long chains, shedding molecules of water, just as the amino-acid peptide groups do in proteins. To each link is attached one of the four nitrogen-containing compounds, forming a "nucleotide," just as side groups characterize the amino acids. The four nitrogen-containing compounds, called "bases," are of two kinds: a pair of hexagonal "pyrimidines" and a pair of larger, geometrically more elaborate "purines." It was a structure sufficiently formidable to carry a significant function.

In 1945, again at the Rockefeller Institute, an experiment by Oswald Avery and two young colleagues, Colin MacLeod and Maclyn McCarty, tipped the scale in favor of DNA. Starting in the 1930s, Avery had undertaken to develop an "antiserum" against the pneumococcus, on the precedent set earlier in the century by Flexner, who had developed an antiserum against the meningococcus. Neutralizing the toxin generated by the bacterium, Flexner's antiserum had significantly reduced mortality in the then-widespread cases of meningitis. The pneumococcus exhibited, however, a frustrating genetic variability. A pneumococcus strain would outflank an at-first-effective antiserum by mutating to production of the toxin of a different strain.

The British bacteriologist Fred Griffith had shown in 1928 that such change in a pneumococcus might be induced by exposure to another strain. His subject was a naked pneumococcus, not enclosed in the usual bacterial cell wall. He injected mice with this pneumococcus in a medium containing the macerated debris of a virulent strain. Infection from pneumococci of the virulent strain soon killed the mice.

Turning to the genetics of the pneumococcus, Avery and his colleagues instilled proteins from a donor strain, including protein separated from the DNA in its nucleus, in a culture of a target strain. This induced no change in the target-strain toxin. The DNA from the donor strain, however, brought genetic transformation of the target strain; its descendants were soon producing the toxin of the donor strain.

This historic experiment also suggested another mode of genetic variation. By more or less accidental exchange of DNA, it was later realized, bacteria in an ecosystem pool their genetic resources.

From work in other laboratories, spurred by the Avery publication, evidence soon established DNA as the genetic molecule. At Strasbourg, André Boivin and Roger Vendrely showed that the average DNA content in the eukaryotic cells of the tissues of mammals, birds and fish differs from one species to the next, but is constant for a given species. Alfred Mirsky and Hans Ris at the Rockefeller Institute, with a technique to establish the DNA content of single cells, got the same confirming result from assay of the tissue cells of fish and mammals.

Boivin and Vendrely and Mirsky and Ris then showed, further, that the DNA content of the sperm or the egg is half that of the tissue cells of the same animal. The result comported with the knowledge that the sex cells carry half the number of the organism's chromosomes [see illustration, page 305]. At the Cold Spring Harbor Laboratory, Alfred Hershey and Martha Chase showed that a phage—a bacteria-infecting virus—injects 80 percent of its DNA into a bacterium and leaves 80 percent of its protein outside.

Chargaff's rule

In 1949, Erwin Chargaff at Columbia University made a finding that strongly implicated the two pairs of bases in DNA, the two pyrimidines and two purines, in the transmission of genetic information. He showed that in all DNA samples the concentration of thymine, one of the pyrimidines, always matched that of adenine, one of the purines, and that the concentration of the other pyrimidine, cytosine, always matched that of the other purine, guanine. He found, too, that the ratio of the thymine-adenine combination to the cytosine-guanine combination invariably differs in DNA samples from different organisms. "Chargaff's rule" was to point the way to the storing of the genetic code.

Even as the protein enzymes were displaced as the substance of the gene, it was shown that it is a major function of DNA to generate protein enzymes.   The red bread mold *Neurospora* can be grown on a culture consisting exclusively of known chemical compounds, nitrates, phosphates, other inorganic compounds and nothing more than sugar and one of the B vitamins supplied by other living cells.   Almost from the ground up, therefore, the *Neurospora* genes secure the manufacture of its entire substance—including the 20 amino acids as well as the protein enzymes made of them that in turn catalyze their synthesis.

To George W. Beadle and Edward L. Tatum at the California Institute of Technology, *Neurospora* presented a broad genetic target for experiment in mutation induced by high-energy radiation.   *Neurospora* commended itself for such experiment because it multiplies profusely in its asexual spore phase, when it carries but one set of chromosomes.   A mutation induced in this phase may then be isolated as a single unit of inheritance in descendants from the crossing of the mutated line with another in the sexual phase.   The nutrient that must be supplied to the otherwise barren culture medium to sustain the crippled strain identifies the gene deleted by radiation.   By this procedure, Beadle and Tatum established the "one gene, one enzyme" rule.   One gene presides at the synthesis of one enzyme that catalyzes one essential chemical reaction.

Molecular architecture

In confirmation of Pasteur's conjecture about the significance of three-dimensional structure in biological molecules, chemists were finding that the biochemical activity of nucleic acids and proteins is related to that structure.   Imaginative chemists began to speak of enzymes as fitting their substrates as a key fits its lock.   This "secondary" structure of the molecules, however, is a physical rather than a chemical attribute.   Its resolution called for tools from physics.   Those tools and physicists seized by questions in the life sciences led the next advances in understanding of the gene and its action in the synthesis of proteins.

This development was accelerated, without doubt, by disbursement of Rockefeller Foundation funds budgeted for science.   At Warren Weaver's instigation, those funds went to "application of the whole range of scientific tools and techniques ... which had been so superbly developed in the physical sciences, to the problems of living matter."

When Weaver retired in 1958, molecular biology was a recognized discipline. The four Nobel prizes in chemistry and in physiology and medicine that year went to recipients of early grants under his administration.

Funding a revolution

The Weaver grants had gone not only to the investigators—Pauling, Beadle, Monod and others—but to the fashioners of their tools. A principal task was the separation of closely similar molecules. In 1906, a young Russian botanist named Mikhail Tswett discovered that "If ... a solution of chlorophyll is filtered through a column of an adsorbent ... tamped firmly into a narrow glass tube, then the pigments, according to the adsorption sequence, are resolved from top to bottom in various colored zones." Called "chromatography," the technique underwent refinement, on Weaver grants, as the principal tool for separating proteins and their amino acids, the nucleotides, and all the other molecular actors at the center of interest in the life sciences.

In one early innovation, the edge of a sheet of filter paper is dipped in a solution of unknown substances. As the solution migrates into the fibers of the paper, it lets go its burden of molecules in the descending order of their mass. The sheet may then be hung at a right angle to the first trail of separations to let a different solvent separate the separations. Electrophoresis promotes such differential migration of the molecules by flow of electric current. The separated molecules may be clipped out of the paper for analysis or identified *in situ* by tagging with radioactive "tracer" elements. Today, scaled up and automatized, chromatography is a full-fledged technology. Its essential principles are employed in sequencing genes, at a continually declining cost per base, and separating molecules in quantity in manufacture of therapeutics.

At Cambridge University in the 1930s, the physicist J. D. Bernal adapted X-ray diffraction for study of the structure of inorganic crystals to tease out the structure of proteins. With X-ray crystallography, Linus Pauling at California Institute of Technology established the "alpha helix" as the key feature of that structure. Pauling's visual imagination saw that any asymmetric figure linked to replicas of itself, head to foot and at the same angle, must form a helix [see illustration, page 262]. He recognized the helix as the simplest structure in which amino acids might self-organize. Looking for the helix in the pattern of dots and streaks offered by an X-ray diffraction photograph of the crystal of a protein, he found it. In each turn of the helix, he counted 3.6 amino acids.

In collagen—the most abundant protein in the tissue of animals, the helix is triple. Three left-handed alpha helixes, composed of three amino acids, intertwine to form a right-handed superhelix [see illustration, page 262]. Coiled in coiled coils, as demonstrated by Max Perutz at Cambridge University, collagen fibrils supply the fibrous substance of skin, connective tissue, bone, tendon and ligament; in all, 40 percent of the tissue in a human body. Waning production of collagen, especially in the connective tissue, is the surest sign of advancing biological age.

The alpha helix in the enzyme proteins may run straight for short stretches, fold on itself, unwind for stretches and be cross-linked at strategic points by sulfur bonds, to form an approximate sphere, the shape of most enzymes. The sequence and folding present the right amino-acid side group or combination of groups at its "active site" for contact with the reactive site of the enzyme's substrate.

Like one of Pasteur's tartaric acid crystals, the alpha helix rotates polarized light invariably to the left. That is because the asymmetry of its constituent amino acids points left. Laboratory synthesis can produce a right-rotating amino acid to match every "natural" left-rotating one. Synthesis can produce, moreover, many more amino acids, both right- and left-handed, than the 20 engaged in the structure of the proteins of the cell. Early in the history of life, it is evident, some decisive event favored the left-handed over the right-handed amino acids.

Function and structure

In the first resolution of the secondary structure of a protein, in 1959 at Cambridge University, John Kendrew and his colleagues resolved also the "tertiary" structure involved in the action of some few of them. Their protein was myoglobin, the molecule in muscle that temporarily stores the oxygen delivered by the hemoglobin of the blood. They were able to precisely fix the relative position of each of this molecule's 2,600 atoms—in 150 amino acids—in three-dimensional space.

The polypeptide chain maintains the alpha helix configuration over about three-quarters of its total length as it folds around to hold the iron-clasping heme group at its four corners [see illustration, page 263].

The heme group displays tertiary structure. A "porphyrin" ring of carbon atoms holds in place an inner ring of four nitrogen atoms. In one configuration, the nitrogens hold the single molecule of iron oxide; in the other, the iron oxide has released the oxygen. Such metal-clasping porphyrin structures characterize all molecules involved in energy-exchange transactions in the cell. To those transactions, the atom of iron or other metal brings one or more electrons ready for action.

Encryption of the code

The understanding that enzymes catalyze the manufacture of the substance of the cell pointed the generalized notion of heredity at a specific biochemical function. The genetic apparatus, as Beadle and Tatum had shown, must convey the design of the whole concert of enzymes. The understanding, in turn, that the nature of the enzyme depends upon the sequence in which its component amino acids are linked correspondingly suggested how the genetic apparatus might convey an enzyme's design. The gene, it was clear, must somehow incorporate that sequence in its structure, secure its expression in the synthesis of the enzyme and replicate the sequence to pass on to the next generation of cell.

Max Delbrück at California Institute of Technology gave this consensual conjecture its formal statement in the simplicity of the Morse code. Two signs, a dot and a dash, in clusters of up to four, suffice to encode the English alphabet, with combinations to spare. Some analogous organization of substructures in the DNA polymer would serve to encode the sequence of amino acids, only 20 of them, in a protein.

The resolution of the structure of DNA molecule—in *The Double Helix* by James D. Watson—made one of the all-time best-selling books ever written by a scientist. Watson had his doctorate from the University of Indiana, where Herman Muller presided over study of genetics. He joined Francis H. C. Crick, trained in X-ray crystallography in the Bragg laboratory, in this historic discovery. They deduced correctly that the purine and pyrimidine bases in the nucleotides are joined to one another

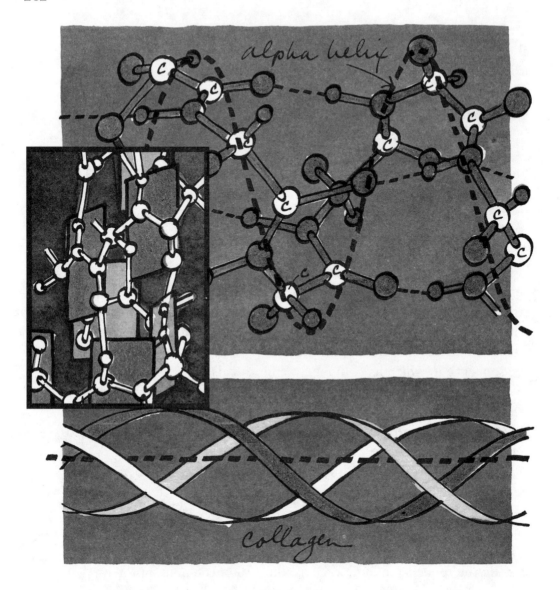

Highly structured large molecules, here in four representative examples, conduct life processes in the living cell. Attachment of subunit amino acids to each other, "head to foot" at the same angle [inset], generates a helix, in proteins the "alpha helix," the minimal arrangement by which such subunits can build three-dimensional structures [see page 254]. Collagen, constituting 40 percent of mammalian tissue, is composed of three nested alpha helixes coiled in a right-handed helix [see page 260]. The porphyrin structure is common to molecules

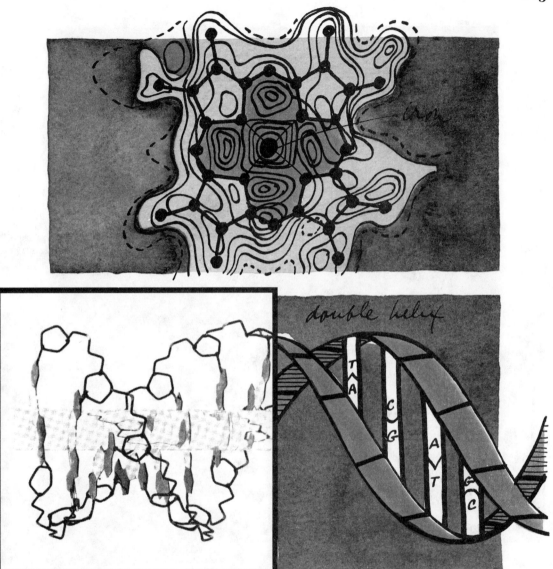

engaged in energy transactions. Four nitrogen atoms form a claw that holds an atom of a metal, of iron in hemoglobin and of magnesium in chlorophyll [see pages 261 and 280]. The metal atoms contribute free electrons to engage in the energy transactions of photosynthesis and respiration. The double helix (Crick-Watson model at left) encodes on one helix the sequence of amino acids in a protein, each cluster of three (out of four) bases constituting a "codon" encoding an amino acid. The second helix carries the anticodons [see illustration, next two pages].

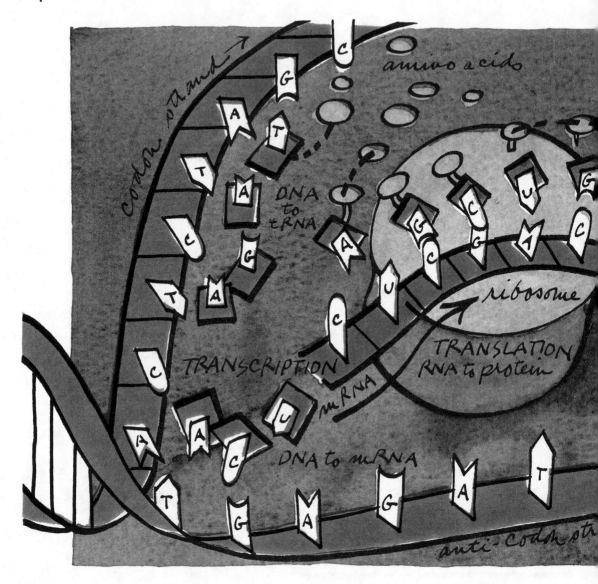

Transcription, translation and replication of genetic information encoded in the double helix of DNA are illustrated here. The 20 amino acids that go into the structure of proteins are each encoded in the sequence of three of the four DNA "bases" [see page 266]. The sequence of these base triplets in the codon helix encodes the order of the amino acids in one of the countless different proteins. Such sequences are transcribed to long messenger RNA (mRNA) strands from the order of their complementary base triplets in the anticodon helix.

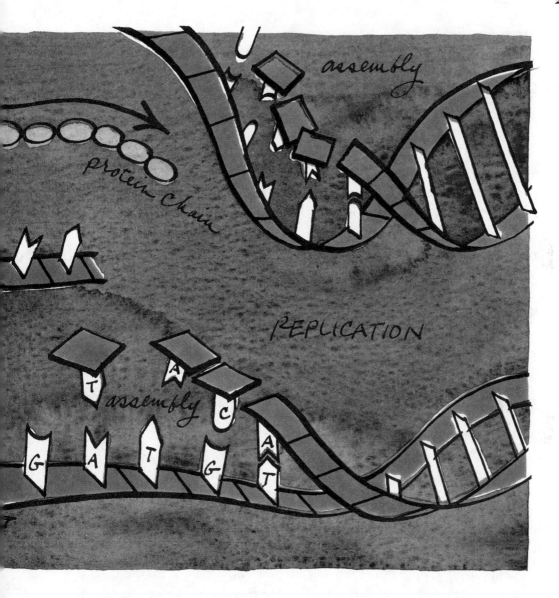

Simultaneously, triplets of bases in the DNA codon strand are matched, one at a time, to their anticodon triplets to form transfer RNA (tRNA) molecules [see page 268]. At the ribosomes, the transcription is translated to protein by the matching of tRNA triplets, each carrying the amino acid for which it is encoded, to complementary mRNA triplets in the order encoding the protein [see page 269]. The double helix is replicated as single bases are matched to their complementary bases in the codon and anticodon strands, making two double helixes.

to form a winding staircase that crosses the interior well described by the two helical phosphoric acid-sugar chains. Their work was facilitated, as Watson confessed in his book, by the exacting crystallographic portrait of the DNA molecule secured by Rosalind Franklin, a Bernal protégé, and Maurice H. Wilkins, also at Cambridge University. "Confessed," because the picture had not been published by its authors. On the other hand, the authors had not yet recognized its significance when Crick and Watson went to press in 1953.

The Watson-Crick model of DNA fulfilled all expectations. The DNA strand encodes the amino acid sequence of a protein in the linear order of its bases. In the double helix was found the explanation, as well, of how the code is replicated for conveyance to the next generation.

The genetic code

George Gamow made one of the early tries at cracking the genetic code. He observed that if the four bases—linked, as by the Chargaff rule, spades to clubs and hearts to diamonds in his playful metaphor— are clustered in groups of three, with the sequence of entities in each triplet disregarded, then 20 is the number of possible combinations. Arrayed in triplets, four bases were enough to line up 20 different amino acids in unique sequences in protein chains of infinite variety.

Nature proved at once more liberal and conservative than Gamow. More liberal: if sequence is regarded, the possible number of triplets becomes 4^3, or 64. More conservative: 64 triplets afford redundancy against accident in the encoding of the amino acids that are incorporated in larger relative abundance in proteins. Of the 64 possible triplet "codons," 61 specify one or another of the 20 amino acids. The three left over supply punctuation, "stop codons," to locate the ends of a gene in the long chain of DNA. The DNA codons thus constitute the amino-acid alphabet. The sequence in which the codons line up the amino acids on the DNA chain then spells a protein word.

The pairing of the bases conveys heredity by the exclusive partnerships of the pyrimidines and purines. The pyrimidine thymine, on one helix, binds only with adenine, its purine partner, on the other helix. Correspondingly, the pyrimidine cytosine binds only with the purine guanine. This is more easily written and read as the affinity of T to A and C to G. The sequence of bases on one helix will thus line up its complement of bases on the other: the codons TAG and ACT on one side of the double helix will match to ATC and TGA, their "anticodons," on the other. The sequences thus run in opposite sense on the two strands.

In replication, the tertiary structure of DNA comes into play. The two strands come apart, permitting the codon chain to bind on anticodons, and the anticodon chain to bind on codons. Each strand of the parent DNA thus replicates the genetic endowment in a double helix for conveyance of that endowment to two members of the next generation.

The crucial role of the sequence of bases in expression of heredity has compelling demonstration in the mechanism of mutation. Rearrangement of the sequence ACT must call up a different—or no—amino acid, disabling the enzyme. In the early 1960s, Leo Szilard, then at the University of Chicago, put a graduate student to work on the question: does the rate of replication affect the rate of mutation in a DNA chain? Replication would seem to afford occasion for accidental substitutions and deletions along the length of the chain.

By feeding and starving, warming and chilling a culture of bacteria, Aaron Novick was able to race and to slow the generation-cycle time of the culture. The rate of mutation turned out to be constant; that is, independent of reproduction rate. The close to error-free result depends upon active error-correction processes, only recently unraveled, in the transcription of the critical base sequences from DNA to RNA. Speeding up the rate of replication simply speeds up the proofreading as well.

The background radiation from radioactive elements in the crust of the Earth amplifies the net accidental rate of mutation by about 10 percent. To the rate of error represented by the incidence of cancer, human behavior contributes a great deal more.

DNA transcribed to RNA

Molecular biologists are grateful that the DNA alphabet proved so legible. There is nothing simple in what follows. It is not to be thought that DNA directs activities in a straight line to the enzymes and the reactions they catalyze. The cell is more approximately visualized as

a system of interdependent variables and instantaneous quantum interactions of thousands of molecular entities coordinated by interlacing
feedback loops. A more likely organizer of activity is presented not in
the DNA double helix, but in the single-helix RNA. Genetic information encoded in DNA is transcribed first to RNA. Between the genetic
plan encoded in DNA and the realization of the plan in the biosynthet-
ic activity of the cell, RNA conducts almost all transactions. Replication
of the DNA double helix itself requires mediation by RNA. DNA is the
archive, and RNA the executor of the genetic processes.

In transcription of a gene to RNA, a specialized enzyme, RNA
polymerase—encoded, of course, in DNA and manufactured in a prior
round with the mediation of RNA—first locks on to a base sequence that
signals the location of one end of the gene. A length of the double helix
thereupon uncoils, exposing a longer or shorter sequence of bases. To
the "coding" strand, which carries the anticodons of the gene, the
enzyme RNA-polymerase draws RNA nucleotides that are to compose a
molecule of messenger RNA, or mRNA. Moving along the strand, propelled by differential intermolecular forces, it attaches the nucleotides to
their complementary anticodon DNA bases. In the long thread of
mRNA (with the pyrimidine uracil taking the place of thymine) the
entire gene sequence is transcribed in less than a minute. To the DNA
codon strand, meanwhile, the RNA polymerases draw the nucleotides
for the assembly of a huge number of RNA molecules, each only three
bases in length. Each of these transfer RNA molecules, or tRNAs, registers the anticodon for a single amino acid.

In DNA of the eukaryote, it was discovered simultaneously by
Philip A. Sharp at MIT and Richard Roberts at Cold Spring Harbor
Laboratories in 1977, long stretches of meaningless sequences interrupt
stretches of meaningful codons. It is suspected that these are fossils of
viruses survived by early generations of eukaryotes. The transcription to
mRNA picks up the meaningless "introns" along with the "exons" that
encode the gene. James E. Darnell, then at MIT, showed in 1963 that
the just-transcribed mRNA inside the nucleus contains about 5,000
bases compared to the average 1,000 bases required to specify a protein.
Some sort of editing process inside the nucleus cleaves the introns out of

the mRNA molecules and splices the exon fragments together in correct order. The nuclear membrane segregates the processing of raw mRNA in a subcompartment of the nucleus called the nucleolus.

With respect to some RNA transcripts at least, Thomas Cech at the University of Colorado discovered in the early 1980s that RNA may edit itself. Isolated from protein, this mRNA spontaneously excised its introns and spliced the meaningful exons in correct order. RNA was thus shown to exercise enzymatic function, displacing protein from exclusive command in that sector. Cech and others soon thereafter observed RNA engaged in self-replication without benefit of enzymes. These discoveries, as will be seen, placed RNA at the center of inquiry into the origin of life.

RNA translated to protein

In the eukaryotic cell, upon completion of the editing process, the functional mRNA is extruded through the nuclear membrane into the cytoplasm. The mRNA thread now finds its way to one of the clusters of ribosomes scattered, by the 100,000, over the vast surface afforded by folds in the endoplasmic reticulum. Here the codon sequence transcribed in RNA nucleotides is translated to the amino-acid sequence of the protein they encode.

A tiny organelle, the ribosome, conducts the translation. It is composed, about half and half, of protein and yet another type of RNA, ribosomal RNA or rRNA. Under a high-magnification electron-microscope, the 10^{-24}–cubic meter ribosome appears as an aggregate of a larger and a smaller particle. The particles are called "50s" and "30s," respectively, after their rates of sedimentation in the ultracentrifuge. Resolution of their structures at 10^{-8} meter—the atomic order of magnitude—by X-ray crystallography has recently resolved the respective roles of these particles in the translation process.

On the mRNA thread clamped between the particles, the rRNA "acceptor" site in the 30s particle activates the next mRNA codon. That attracts the complementary anticodon arm of a tRNA carrying the appropriate amino acid. The enzyme-reactive arm of the tRNA positions the amino acid in the "peptide" site of the 50s particle. There an rRNA catalyzes the peptide bond attaching the amino acid to the emergent protein chain. The "exit" site releases the tRNA to find another molecule of its amino acid for attachment to the next protein chain.

On a single mRNA chain, 10 or more ribosomes may be thus engaged at the same time in assembly of as many copies of the protein. Some million ribosomes thus call in 20 million tRNA molecules to maintain the pace of protein synthesis in the cell at the rate of a million peptide bonds per second [see illustration, pages 264–265].

Transcription and translation of genetic information proceed similarly in the prokaryote, but with no need to edit mRNA chains. No introns punctuate the average bacterium's 10 million base pairs. The DNA chain strung in the prokaryote nucleoid is anchored at two places to the membrane inside the cell wall. Transcription of DNA to mRNA begins at one anchor and proceeds simultaneously along both half hoops to the other anchor. While transcription proceeds, ribosomes mount the unfurling mRNA chain and bring in the tRNAs to translate the mRNA transcript to protein. Fragments of the nucleoid will function floating free in the bacterial cytoplasm. Such "plasmids" of many bacteria will function when transferred to other cells, relatively indifferent to the genetic company they keep [see illustration, page 305].

Order from probability

The reliability of the genetic process poses the question about the nature of life in a new and prospectively productive way. Order in the physical world turns on the statistics of the jiggling of an Avogadro's number—6.02×10^{23}—of atoms. In *What Is Life?* a decade before *The Double Helix,* Erwin Schrödinger asked a "naive physicist's" question:

> How can we, from the point of view of statistical physics, reconcile the facts that the gene structure seems to involve only a comparatively small number of atoms (of the order of 1,000 and possibly much less), and that nevertheless it displays the most regular and lawful activity— with durability and permanence that borders on the miraculous?

In illustration of "durability and permanence," the Viennese physicist cited the Habsburg lip. This "peculiar disfigurement of the lower lip" makes its first appearance in family portraits from the 16th century and identifies Habsburgs living in the 21st century. In tentative explanation of the reliability of the underlying physics, Schrödinger observed that the large molecules of the cell have been shown to be "aperiodic crystals"—in a sense, solids. He cited the grandfather clock. For this indubitable solid, he said, room temperature is equivalent to absolute zero; quantum uncertainty perturbs its timekeeping not at all.

The molecular physiology of the cell proceeds in the realm of space and time between the Newtonian world of the grandfather clock and the indeterminate probabilistic world where QED governs. Here, the hundreds and thousands of atoms in the strands of DNA, RNA and the other large molecules are "jiggling, jiggling" [see page 111]. The chemical bonds tying them together and connecting them to one another are vibrating in all planes, lengthening and shortening around their mean lengths. Electrical charges, positive and negative, differentially distributed along their lengths engage reciprocally with those of other molecules in their range. Propelled by these forces, the polymerase enzymes "read" the anticodon strand of the DNA and assemble the complementary codon strands of the mRNA, all without error that might have terminated the story long ago. The random activity is undoubtedly constrained, and perhaps organized, by the "aperiodic" structuring of the large molecules, about which so much has been learned since Schrödinger wrote. Given the 10^{-21}–second reaction time of the electromagnetic force, there is tolerance for much molecular writhing and many a false intermolecular contact antecedent to the sealing of each of a mere 10^6 peptide bonds per second. Quantum biology is just now framing its first questions for confrontation by experiment.

Cells thus have anatomy—molecular anatomy, to begin with—to be understood along with physiology. Plainly, interdependent and sequential steps in long series of reactions cannot be conducted with their enzymes mixed up in solution. The all-but-empty 1900 globule of the eukaryotic cell is transformed now to a dense, seething assembly of membranes, ligaments, tubes, valves, pumps and organelles of more intricate structure and less apparent function. In molecular cell biology, cytology and biochemistry have converged. Molecular anatomists are discerning

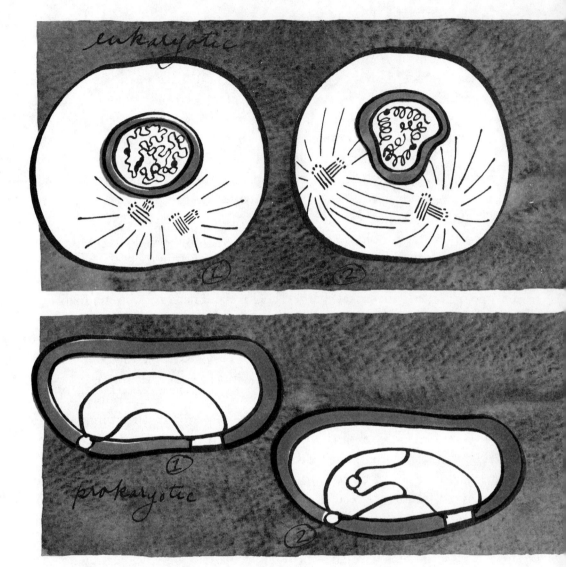

Replication of eukaryotic and prokaryotic cells *proceeds on different but parallel courses, reflecting one of the major innovations in evolution [see page 298]. A single, circular "nucleoid," anchored at two points to interior cell membrane (1), carries the prokaryote genes; replication begins at one anchor (2) and proceeds simultaneously on both plasmid stands. Replicated plasmid finds new anchorage (3); common anchor point divides, followed by division of cell (4). Prokaryote genes are also carried in small free-floating "plasmids." Eukaryote*

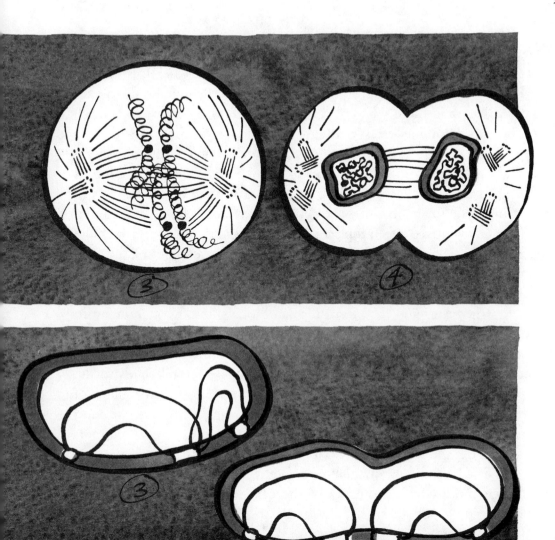

genes, strung on linear, free-ended chromosomes [see page 252], replicate inside nucleus. At the same time, the centrosome [see page 275] replicates outside the nucleus (1). The two offspring centrosomes move to opposite sides of the nucleus (2). The nuclear membrane then evanesces as the tubule rays of the "mitotic" spindle radiate from the centrosomes and attach to "centromeres" on the replicated, now separating chromosomes (3), which are enclosed in nuclear membrane as the cell itself at last divides (4).

in those organelles the arrays in which enzymes are ordered for their sequential contributions to a cycle of reactions.

The cell membrane, a mere envelope as late as 1950, is now an active organelle. The electron microscope shows the membrane folding inward here and there. The membrane thus faces the inside of the cell to the outside over a correspondingly more extensive area and brings the outside deep into the seeming interior. Joined to and continuous with the inward-folding membrane, sheets of membrane inside function as "endoplasmic reticulum." The total surface of membrane, inside and out, multiplies by many times the apparent surface of the cell. Scattered on the vast surface of endoplasmic reticulum, the ribosomes in huge number carry on their work. What the electron microscope still life does not suggest is the constant rippling, folding and unfolding, generation and evanescence of membrane that attend its function.

Chemistry and the electron microscope show the membrane to be constituted of two layers of cylindrical "lipid," fatty acid, hydrocarbon molecules. They butt hydrophobic ends together at the midline of the membrane cross section and face hydrophilic, wettable ends to the watery fluids inside and outside the cell. From 50 to 100 million of them compose the membrane of one animal cell. Free to move around one another in the undulating plane of the membrane, these paired molecules give it the structure of a two-dimensional fluid.

The membrane maintains nonetheless cleanly the difference between the inside and outside of the cell. It resists the pressure of osmosis from both sides. It upholds electrical potential differences, set up by difference in the concentration of ions between inside and out. Valves and pumps made of protein, in the hundreds and of a dozen different kinds, conduct passive and active transport across the membrane in both directions. The two-dimensional fluidity of the membrane permits these organelles to move swiftly to where their service is invited. The membrane itself invaginates and pinches off inside to bring particulate matter in from the outside and performs the topologically reverse operation on particles to be expelled to the outside.

The reproduction of a eukaryotic cell engages the first organelle after the nucleus to attract the attention of observers. This is the mitotic spindle. Supplying the spindle is one function of the mesh and matrix of microtubules that constitute the skeleton of the cell. Muscle to the metaphorical skeleton is supplied by actin and myosin. These same proteins together produce the contractile force in mammalian muscle. Interpenetrating the membrane folds of the endoplasmic reticulum, this musculoskeletal system supports and modulates the shape of the cell.

The mitotic spindle

Microtubules are constituted of two kinds of tubulin protein that pair up in a "dimer." In laboratory glassware, the dimers self-organize spontaneously in sheets. They then roll into tubules, 13 dimers in each turn, in the inevitable helix [see illustration, pages 276–277]. With an outside diameter of 2.5×10^{-9} meter and an inside diameter of 1.5×10^{-9} meter, microtubules may extend 10^{-6} meter, 1,000 times their thickness, in length. In service as cytoskeleton, tubules come and go, exhibiting the rigidity of rods or the *souplesse* of ropes. They function also as pipes in the transport of material, especially in the axons of nerve cells that may extend a foot or more in length.

At the reproduction of the cell, it might appear that microtubules are in charge. Alongside the nucleus, as the chromosomes are replicating, a tiny organelle, the centrosome, proceeds to replicate itself. Its two component centrioles—each a stubby cylinder with a stubby cylinder protruding at a right angle from it—become four. These stubby cylinders are composed of 27 microtubules fused in triplets arrayed, in cross section, in a circle of nine. Upon replication, the offspring centrosomes migrate to opposite sides of the cell. Microtubules, sprouting from the centrosomes, then form the mitotic spindle that attaches one member of each complementary pair of replicated chromosomes symmetrically to the oppositely positioned centrosomes. Along the microtubule scaffolding, the chromosomes move apart. The drama suggests what the student of cell evolution Lynn Margulis has called "a system of heredity independent of DNA."

The nine-triplet structure of the centrosomes does duty also in the structure of the kinetosomes. From them, under the membrane of the many motile eukaryotic cells, protrude microtubules arrayed in a

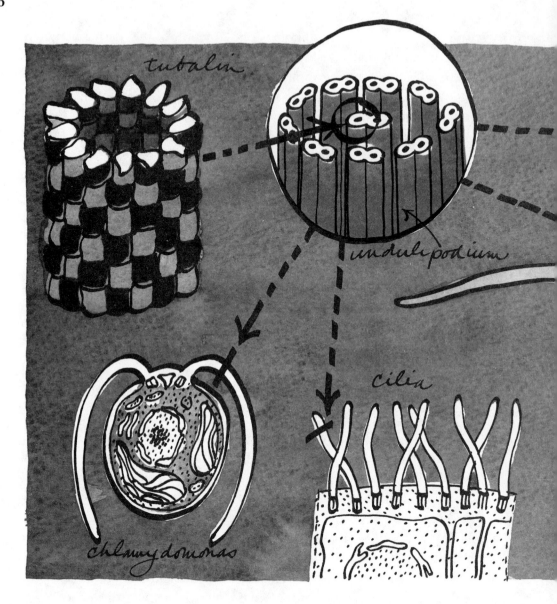

tubulin

undulipodium

cilia

chlamydomonas

Mechanical function in eukaryotic cells *is served universally by the same molecular apparatus. Two kinds of tubulin molecule form a "dimer." Dimers self-assemble in a tubule, with just 13 tubulin molecules in its circumference. Tubules may reach 1,000 and more times their 2.5×10^{-9}–meter diameter [see page 275]. In the undulipodium, the organelle of motility in many protoctista, 20 tubules fused in pairs in a circumference of nine pairs surround a central pair. The same assembly of tubules forms the whiskery cilia on many eukaryotic cells,*

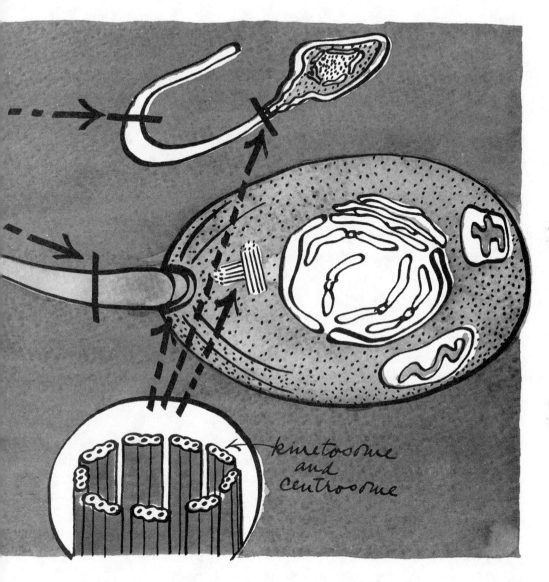

kinetosome
and
centrosome

including those in the mammalian mucosa. The motion of undulipodia and cilia is governed by tubules in another configuration: 27 tubules fused in nine triplets forming a squat cylinder, the kinetosome. The same configuration of tubules makes the centrosomes from which the mitotic spindle radiates from the opposite sides of the eukaryotic cell and engages the separating chromosomes in the course of eukaryotic cell division. This is strong evidence for the symbiotic origin of eukaryotes [see illustration, pages 300–301].

quite different cross section: nine fused pairs of tubules surrounding a single fused pair [see illustration, page 276]. Bundles of these assemblies form the whiplike undulipodia—undulatory "feet"—that propel single-celled eukaryotes through the fluid media in which they live. A single undulipodium propels the mammalian sperm. On stationary tissue cells, as in the respiratory system, whipping cilia of the same cross section keep the surrounding fluid medium in motion. In the labyrinth of the inner ear, cilia protruding into the interior fluid set off nerve impulses that signal the orientation of the head in the three dimensions of Newtonian space [see inset, page 375].

ATP: energy transformation in the cell

In the life of the cell, the transformation of energy must be reckoned a process as vital as the replication of the gene and the accumulation of experience in the concert of genes. Cells transform chemical-bond energy to energy in every other form. Delivered to actin and myosin in the muscle, chemical-bond energy transforms to mechanical energy. In the nerve impulse, it transforms to electric energy. Across the membrane it sustains osmotic pressure and drives active transport. From the firefly and from "fox-fire" mold in rotting wood, it radiates as light. One ubiquitous molecule captures and delivers chemical-bond energy to the sites of its transformation in the cell.

Adenosinetriphosphate, or ATP, is composed of adenine, one of the purines in the DNA and RNA nucletotides, and ribose (together: adenosine), tied to a chain of three phosphate groups. Losing one phosphate group, which occurs in its energy-transformation activities, ATP becomes ADP, adenosine*di*phosphate. Minus two phosphate groups, it becomes adenosine*mono*phosphate, or AMP, not surprisingly one of the four RNA nucleotides engaged in genetic biosynthesis.

The energy required to attach each phosphate group to AMP increases from the first to the third. It takes 1 eV to attach the third phosphate group to a molecule of ADP, transforming it to ATP. This is the energy that ATP delivers to the energy-requiring processes in the cell. Losing the third phosphate in the process, ATP falls back to ADP, ready for recharging to ATP again.

Karl Lohman at the University of Heidelberg first isolated ATP from muscle in 1930. In the test tube the breaking of the third phosphate bond liberated heat. In the cell, this energy lifts electrons in reacting molecules to higher energy states with no loss of energy to heat. Such efficiency in the movement of electrons suggests the superconductivity attained by bringing metals close to absolute zero in temperature.

Photosynthesis and respiration

The "phosphorylation" that attaches the third phosphate group and charges ATP with its energy has proved, in the last half century, to be the pivotal reaction around which turns the energy economy of life on Earth. Photosynthesis captures the Sun's radiant energy to manufacture organic molecules from inorganic carbon dioxide and water and releases oxygen to the air. It first transforms that energy to charge up molecules of ATP. Cellular respiration, in turn, brings in oxygen to burn organic molecules to obtain their chemical energy, producing water and carbon dioxide in the process. The energy derived goes first to attach the third phosphate to ATP. These, the complementary processes by which plant and animal (and prokaryotic) life sustain their mutual existence set up the great planetary turnover of oxygen and carbon dioxide in the atmosphere [see illustration, pages 346–347].

Respiration and photosynthesis, it can be seen, are physiological modes of the two fundamental chemical-energy exchange reactions: oxidation and reduction. Oxidation-reduction reactions, or "redox" for short, bring the transfer of one or more electrons from one reactant to another. Oxidation is the energy-yielding donation of electrons from one reactant; reduction, the energy-consuming addition of electrons to the other. Because an electron transfer requires both an electron donor and an acceptor, oxidation and reduction always go together.

When the complementary cycles of photosynthesis and respiration were first recognized nearly two centuries ago, the primary carbohydrate formula CH_2O suggested that photosynthesis was a matter of attaching carbon from carbon dioxide, CO_2, to water, H_2O. In the early 1930s, C. B. van Niel at Hopkins Marine Laboratory in California corrected this misapprehension in a historic study of the photosynthesis conducted by blue-green and purple bacteria. These bacteria, he showed, engage hydrogen sulfide, H_2S, to bring electrons (attached to the hydrogens) into the reaction cycle in their photosynthesis. Both sulfur and oxygen have two valences; they turn up interchangeably in many styles of chemical reaction. Sulfur, not oxygen, is the waste product of this bacterial photosynthesis. Plainly H_2S serves for H_2O in purple bacteria photosynthesis donating the essential hydrogens to the reaction. With the $_{18}O$ isotope to mark the water molecule, Martin Kamen at the University of California, San Diego, left no doubt that photosynthesis in plants engages H_2O to supply the two electron-bearing hydrogens to CH_2O. The $_{18}O$ isotope was found in the air, not in the CH_2O.

The oxygen that lets go hydrogen in the energy-requiring reaction and accepts hydrogen in the energy-yielding reaction shuttles therefore between water and air. Carbon shuttles from and to the air: captured from atmospheric CO_2, reduced to CH_2O by photosynthesis and released by respiration to the atmosphere in CO_2 again.

In the phosphorylation of ADP to ATP, photosynthesis and respiration engage many of the same enzymes in the same reactions. Their molecular agents for the energy-moving transfer of electrons are metal-grasping porphyrins and sugar-phosphate coenzymes that are, like AMP, nucleotides related to those involved in genetics. These enzymes and agents are arrayed for their highly ordered sequential action in the two most elaborately structured—after the nucleus itself—organelles of the cell. Of significance to the evolution of the eukaryotic cell, as will be seen, is the fact that each of these organelles carries its own vestigial DNA and accessory RNA and enzyme complement that function in its replication inside the cell and at the reproduction of the cell.

The organelle of photosynthesis

Inside the organelle of photosynthesis [see illustration, page 282], the chloroplast, the chlorophyll molecule, captures solar photons. Chlorophyll is another structured entity. A flat porphyrin ring, like that of the heme group in myoglobin, holds in its nitrogen talons an atom of magnesium; to the ring is attached a long carbon-chain tail. The molecule is so tuned that one photon in the yellow-red wavelengths—the peak wavelengths of the energy of sunlight—kicks one electron loose from the magnesium atom.

From a test-tube solution of chlorophyll, the electrons release solar energy in a flash of fluorescence. In the cell, the chloroplast secures controlled release of energy from light-activated electrons in an organized molecular matrix. The chlorophyll molecules are embedded in membranes that the electron microscope shows stacked in crystalline order. The hot electrons are taken on the leash at once by the first of a succession of cytochromes. These too are porphyrins, with iron in their claws. They transport the electrons, without loss of energy, to sites where the chloroplast has raw materials and the right enzymes and coenzymes in place. There each electron yields its energy to chemical bonds. In reactions that go on in the light, the energy goes first to make molecules of ATP. Then, in the dark, a cycle of reactions takes the energy from ATP to more stable bonds in the sugar, glucose. This is the starting material for manufacture of much of the substance of the cell and the source of the energy required.

On good evidence, the light reactions convert 75 percent of the photon energy impinging on the chlorophyll molecule to chemical energy in the bonds of ATP. That is twice the rate at which central-station prime movers convert the energy of fossil fuels to electricity. It is well that the molecules function so efficiently. By the estimate of G. E. Hutchinson of Yale University, a five-layer depth of leaves on land and a 1-millimeter depth of phytoplankton in the ocean establish the tenuous connection between life on Earth and the source of its energy in the Sun. Where they function, they capture 0.1 percent of the 1 kilowatt per square meter of solar energy that reaches the Earth's surface.

In the 1950s, Daniel Arnon and his colleagues at the University of California, Berkeley, dissected out the light reactions. They did so literally; they made chloroplasts conduct the full cycle of photosynthesis outside plant cells in laboratory glassware. Their work established that photosynthesis proceeds, in the light reactions, as photophosphorylation. The capture of solar energy in the third phosphate bond of ATP proceeds on two routes. The routes depart from two different chlorophylls. The hot electron kicked from chlorophyll*a* goes to phosphory-

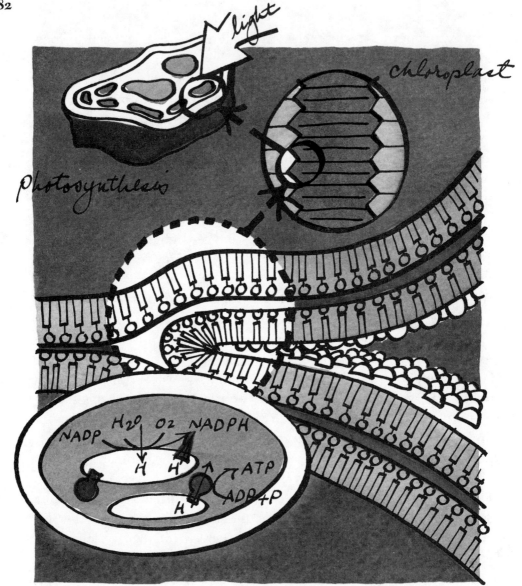

light

chloroplast

photosynthesis

NADP H₂O O₂ NADPH

Photosynthesis transforms solar energy to the energy of chemical bonds. A single solar photon kicks an electron loosely bound to the magnesium atom in a chlorophyll molecule into a higher energy state. Chlorophyll molecules arrayed in crystalline order in membranes send the excited electrons on two cycles, aerobic and anaerobic, to fixing of their energy in the energetic chemical bond that attaches phosphorus atoms to ATP and NADP [see page 280]. Molecular mediators of these cycles are arrayed in the intermembrane space.

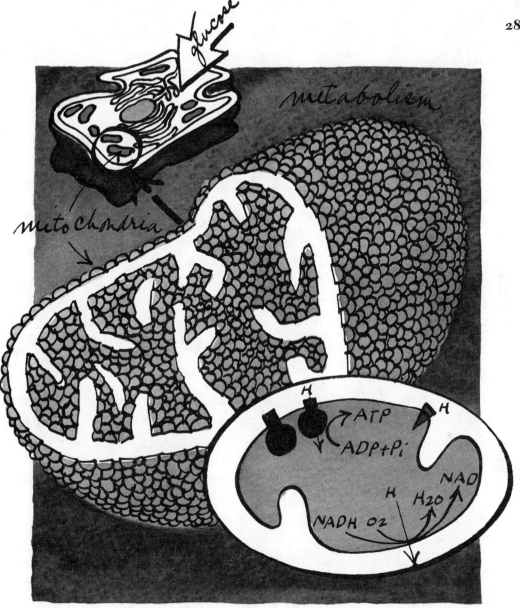

Respiration transforms chemical energy in glucose to the bond attaching the third phosphorus to ATP, which energy-carrier makes that energy available for conversion to mechanical, electrical and all other forms of energy generated by life [see page 285]. Molecular mediators, arrayed on the outer membrane of the mitochondrion, extract the glucose energy on anaerobic and aerobic cycles; the energy goes to synthesis of ATP on inner mitochondrial membrane [see page 286]. Most ATP energy goes to protein synthesis on outer membrane.

late a molecule of ATP and to reduce one of the electron carriers, the pyridine nucleotide PN, to PNH_2. This cycle of reactions, "Photosystem II," captures the two hydrogen atoms from water and releases the residual oxygen to the atmosphere. Before there were plants, this mode of photosynthesis—and aerobic respiration—in the vast family of Cyanobacteria began the reconstitution of the Earth's atmosphere.

The green leaf and one group of bacteria capture additional solar energy in electrons that travel the anaerobic route of Photosystem I, conducted by chlorophyll*b*. The hot electron is handed on in a closed cycle from cytochrome to cytochrome and returned to neutralize the positive charge it left at the chlorophyll home base. On the way, with vitamin K and one of the B-complex vitamins as coenzymes in the catalysis of the reactions, it surrenders its excitation energy stepwise to the charging of two ATP molecules.

PNH_2 delivers the H_2 to the uphill dark reactions that force the two electrons into combination with carbon. The three ATPs charged by the two photosystems deliver the energy necessary for this reduction. Melvin Calvin, also at the University of California, Berkeley, and also in the 1950s, worked out the dark reactions in elegant detail. He showed that the assembly of the six-carbon glucose chain proceeds through a succession of intermediate compounds by the turning of a many-stepped cyclic mill, each reaction facilitated by its specific enzyme or enzymes. Many of these reactions and their enzymes, he found, play their part in the reverse turning of the mill that disassembles the glucose chain in the "citric acid cycle" of oxidative phosphorylation.

Calvin and his associates discovered that glucose is not the only product of what is called the Calvin cycle. In their guinea-pig plant, a single-celled eukaryotic alga, they found that 30 percent of the carbon goes directly to synthesis of some of the amino acids and of fatty acids as well. The sequenced complexity of the 20- to 30-step cycle, depending on which end-product it yields, suggests the high spatial order in which the enzymes and electron carriers must be arrayed in the chloroplast for this end-phase of photosynthesis.

An organelle no less elaborate than the chloroplast conducts the oxidative phosphorylation of the glucose manufactured by photophosphorylation [see illustration, page 283]. The mitochondrion conducts respiration only in, but not in all, eukaryotic cells. By oxidative phosphorylation, plants burn about 50 percent of the glucose they manufacture. The other 50 percent supports the animal and fungi kingdoms.

The organelle of respiration

The cycle begins with the anaerobic—nonoxidative—splitting of six-carbon glucose to two three-carbon molecules of lactic acid. This operation proceeds through six intermediate products at the cost of two ATP. It yields four ATP, however, for a net gain of two. Thermodynamic bookkeeping conducted by the gram molecule shows the two net moles of ATP carrying 24,000 calories. The efficiency of this transaction is, by engineering standards, a respectable 43 percent.

This preliminary splitting of glucose stands alone as the fermentation of glucose to lactic acid. The energy it yields suffices to sustain the existence of much of the prokaryote kingdom. In the human organism, the accumulation of lactic acid in the muscles brings on the early stress that is relieved by "second wind." That signifies the turning-on of oxidative phosphorylation proper.

The aerobic, truly "respiratory," phase of oxidative phosphorylation engages oxygen in the complete oxidation of the two lactic acids and recovers the starting materials, carbon dioxide and water. Also called the "citric acid cycle" because the subtraction of the first carbons turns the two lactic acids into citric acid, its understanding represents a major milestone in biochemistry. Honoring the scientist who planted that milestone, it is more often called the "Krebs" cycle. With this work, Hans Krebs at Oxford University, in the 1930s, completed the inquiry begun by the Buchner brothers 40 years earlier.

The aerobic phase of oxidative phosphorylation secures the oxidation of all six carbons in the molecule of glucose. The energy yielded from the bonds by which photosynthesis attached two hydrogens to each of the carbons will ultimately yield 36 molecules of ATP. From the 690,000 calories yielded by the complete combustion of a mole of glucose, this constitutes a considerable capture. The energy invested in the third phosphate bond in a mole of ATP is known to be 12,000 calories.

Multiplying 12,000 by 36 yields 432,000 calories, an efficiency of more than 60 percent. Adding the energy gained in the two ATP derived from the first anaerobic splitting of glucose brings the overall efficiency of oxidative phosphorylation to 66 percent. That is 66 percent of the 75 percent of the radiant energy of the Sun first fixed by photophosphorylation. The net capture of solar energy in ATP by the two phosphorylations in tandem, therefore, exceeds 40 percent.

While Krebs was at work, Herman M. Kalckar in Denmark and V. A. Belitzer in the Soviet Union simultaneously and independently took note of a chemical event that seemed to attend respiration in animal tissue. To suspensions of ground muscle or kidney they supplied glucose and oxygen. They observed that phosphate, in solution from the tissue, disappeared as respiration proceeded in the ground tissue. They found the missing phosphate incorporated in various organic molecules, but especially in what is now familiar as ATP. Kalckar postulated correctly that the phosphorylation of ADP to ATP was coupled to the process of respiration. Present understanding of the universal role of ATP has since developed necessarily from many different lines of investigation. Thus, the Hungarian physiologist Albert Szent György established that ATP supplies the chemical energy to the interaction of myosin and actin that produces the mechanical energy of muscle contraction.

In the electron microscope, the mitochondrion appears a blimp-shaped body. Cross sections show it is composed of two membranes; highly convoluted infoldings of the inner membrane greatly increase its surface area [see illustration, page 283]. Throughout the membranes, enzymes that do the work deploy in ordered relation to one another. A mammalian tissue cell may have 50 or so mitochondria and 500 when, as in the liver, it is engaged in manufacturing service to the organism.

David E. Green and his colleagues at the University of Wisconsin showed in the 1960s that the oxidative reactions proceed in the outer membrane. Aboard NPH_2 molecules, the electrons captured by these reactions are ferried inside. There the surface area of the inner membrane, enlarged by infolding, allows the charging of ADP to ATP in great and greater volume. The ATP molecules are at once transported to the outer membrane. Then and there, most of them deliver their energy to the synthesis of proteins. The rest find their way to energy-requiring operations at other sites in the cell.

The anatomy and physiology of the eukaryotic cell constitute the ground plan of nearly all organisms visible to the naked eye and a great many of those too small to be seen. By variation on this ground plan, single-celled, free-living eukaryotes exhibit the range of appearance and behavior that first enchanted Leeuwenhoek. By selective shutdown of such omnicompetence, eukaryotes construct multicelled organisms, committing each cell to some specialized contribution to the whole.

Of the much that remains to be learned about the living cell, no topic is greater than the generation of the multicelled organism from a single cell. Long ago, investigators observed, in the early embryonic developments of organisms in the animal kingdom, much the same series of events occurred. Around the tenth division of the fertilized egg cell, the several thousand daughter cells form a hollow ball, called the "blastula." Down one side, then, the cells fold inward to form a hollow sac, the "gastrula," within blastula. From this point development diverges on the 33 different pathways that lead to end points as different as sponges from arthropods, mollusks from vertebrates.

Differentiation of cells soon after the infolding of the gastrula begins their dedication to the development of tissue types and end organs. Cells transplanted from one "imago" to another in the early embryo of the fruit fly, for example, will fashion a leg on the adult's head instead of an antenna. Cloning demonstrates that tissue cells retain the full complement of their species' genes. Transfer of the nucleus of a tissue cell to an enucleated egg awakens the full competence of the genes.

Molecular biology poses the question of embryonic development in a new way: how does the linear information contained in DNA generate a specific three-dimensional organism? Walter J. Gehring at the University of Basel, who framed this question, made a start on its answer. In a segment of the fruit fly DNA, he identified a cluster of genes that control the expression of a great many other genes in the spatial organization of the organ to which they give rise. A mutation in this cluster brings the sprouting of a leg in place of an antenna on the fruit fly's head. The "homeobox" cluster has now been found in every organism in which it has been sought, from worms to human beings. Homeobox genes regulate, apparently, by synthesis of proteins that repress and release genes at other sites in the chromosomes, somehow on schedule.

The genetic code and evolution

Every cell carries in its genes the common heritage of all organisms, as different to start with as eukaryote and prokaryote. The same DNA and RNA call up the 20 amino acids by the same alphabet. The RNA codons UUA, UUG, CUU, CUC, CUA or CUG redundantly encode the leucine, the "e" in the alphabet. No less vitally, nucleotides in the genetic dictionary, notably AMP, engage in the energy transformations that activate it. As spelled out in the dictionary, the same structural and enzyme proteins do the same service in diversely different organisms. The simpler hydrocarbons and carbohydrates serve their humbler functions in structures and as fuels.

Here, surely, is evidence of a common history and a common origin. Such observation, in a few years, will be commonplace. By the mid-1970s what had been first-time laboratory triumphs establishing the sequence of bases in a DNA or RNA chain were routinized. Mechanization of the routines then made it possible to conceive of sequencing the bases in the entire genome of an organism. At the turn of the millennium, the sequencing of the bases in six genomes had been accomplished. The millennium turned with the publication of two incomplete "drafts" of the genome of Homo *sapiens.* One is the product of the intrusion of commercial interest (more than 20,000 patents on DNA sequences have been applied for and nearly 800 issued at the U.S. patent office) and pharmaceutical fortunes to be made. The other is the product of an international consortium of laboratories financed, beginning in the early 1970s, by public funds lately supplemented by foundation funds motivated by concern to keep the human genome in the public domain and ensure immediate and open publication of all findings. The "drafts" hold the sequencing for most of the 3 billion bases and project strategies for attack on the rest. Beyond must come the discrimination of intron from exon, the identification of the genes and then the identification of the genes with their function. By no means do all of them encode proteins. Of interest and complexity yet to be imagined are those committed, as in the homeobox, to the realization of three-dimensional organisms from the design encoded in one dimension on the DNA chain.

This Big Science enterprise has yielded its first significant contribution to understanding. It is now possible to make reliable estimates of the number of genes in the human genome: around 30,000. That is larger than the genome of D. *melanogaster.* It is but a fraction, however, of the number required, it was thought, to specify Homo *sapiens.*

As economic interest in this development suggests, it has already sired its technology. "Genomics" carries the most awful of all the power objective knowledge has placed in human hands. People are urgently challenged to frame the meaning and purpose of their place in the order of nature. They are challenged, perhaps more immediately, to keep this power in the public control of open, self-governing societies.

The origin of species

No such evidence for evolution was at hand in the three agonized decades Charles Darwin gave to his statement of natural selection. The age of the Earth then postulated by geologists scarcely afforded the time necessary to have carried existing species down their diverging and converging evolutionary paths. The fossil record offered little more than proof that whole troops of species had gone before. That record went silent in rocks older than the Cambrian formations in Wales, the then earliest known fossil-bearing rock. In an often-cited passage Darwin himself set what he found to be the most difficult challenge to his vision of the history of life:

> To suppose that the eye, with all its inimitable contrivances for adjusting the focus to different distances, for admitting different amounts of light, and for the correction of spherical and chromatic aberration, could have been formed by natural selection, seems, I freely confess, absurd in the highest degree.

The long history of the living cell has answered that riddle. The geological record now affords the time not available to Darwin for the diversification and innovation that stemmed from the protean living cell.

The earliest fossils of recognizable living organisms carry the origin of life back to the end of and possibly into of the first billion years of Earth history, well named the Hadean period. The planet's gravitational field was still sweeping up debris left over from the collapse of the Sun and the planets from their natal dust cloud [see pages 208 and 335].

Oxygen is excluded from any possible first atmosphere. Its high chemical reactivity bound it tightly to other elements in the crust, of which it constitutes nearly half the number of atoms. In the absence of the ozone layer now sustained in the upper atmosphere, high-energy photons of sunlight reached the surface with undiminished energy. In this sterile environment life began.

Experimental inquiry

The Russian geochemist A. I. Oparin persuaded his colleagues in the 1920s that enough had then been learned about the abiotic chemistry of organic molecules to bring the question of life's origin into scientific inquiry. As a first necessity, he saw "the existence of a certain boundary ... a membrane ... an envelope" separating what was to become the protocell from its surroundings. He observed that lipids afloat on water spontaneously form hollow vesicles-bubbles. In England, J. B. S. Haldane and J. D. Bernal embraced Oparin's enterprise and saw persuasive possibilities in a presumed wealth of chemistry in the "hot dilute soup" of the primordial ocean. What the ocean might fail to brew would be supplied from extraterrestrial sources. The meteoritic bombardment brought in numerous carbonaceous chondrites. In their carbonaceous fraction are found organic molecules synthesized in the solar-system dust cloud before its gravitational collapse.

Experimental inquiry into the origin of life started with a project instigated by Harold Urey. With Urey looking on in the early 1950s at the University of Chicago, Stanley L. Miller circulated the presumed ingredients of the primordial atmosphere—water vapor, methane (CH_4) and ammonia (NH_3)—in a quartz tube through an electric spark gap. After a week the water in the reservoir, from which it was boiled into the quartz-enclosed atmosphere, turned pink. Analysis found five amino acids in the solution along with other organic compounds.

Experiment excited by Miller's work continues today in many laboratories, and the catalogue of the monomers of life produced abiotically continues to grow. These compounds issue from other hypothetical first atmospheres, consisting principally of carbon dioxide and nitrogen, regarded as more likely than Urey's hydrogen-rich mixture. Unshielded solar radiation is reckoned as the principal source of energy promoting such synthesis. The then-higher radioactivity in the crust of the Earth, the warmer surface temperature and lightning bolts in the turbulent, hot and humid atmosphere would have supplied energy to the same effect. Curiously, the first atmosphere, thin but rich in carbon dioxide, provided a greenhouse to hoard warmth from the fainter Sun.

Experimental work in this prebiotic chemical phase of evolution has not advanced much beyond the amino-acid monomers and sub-assemblies of the nucleotides. The peptide chain is vulnerable to hydrolysis; that is, dissolution of the peptide bond by water. Still more vulnerable to water is the well-established sequence of reactions by which five molecules of formaldehyde (a compound known in interstellar space) combine to form one molecule of ribose, the link in the RNA chain. Oparin's vesicles would have a function here, enclosing the vulnerable molecules in a more friendly solution. So also would reduction, in the culinary sense, of the dilute organic solutions by evaporation in estuarine and other shallows. Sidney Fox at the University of Florida showed in the 1950s that heated and dried-out solutions of amino acids yield polymerized chains as much as 50 units long. Polymerization and stability of the bond might also have been promoted by adherence to mineral surfaces. Such surfaces would enhance the self-organizing propensity of the amino acids. John Maynard Smith at the University of Sussex proposes a primordial "pizza" in place of Haldane's soup.

RNA and "ribo-organisms"

Demonstration that RNA may catalyze its own replication and the engagement of RNA nucleotides in energy transformation as well as genetics marked them as the ancestral molecules. "In a primeval soup appropriately stocked with building blocks, sources of energy, and catalysts," Christian de Duve declared, "RNA could be reached in a matter of years." He took note that "information may have entered by way of energy." John Maynard Smith had the RNA nucleotides organizing what he calls "ribo-organisms." Later, but soon, they met up with the "iron-sulfur world" and instituted protein chemistry.

In the first proto-organisms, however they originated, the genetic code could have been simple enough. As few as 12, and even as few as 4, amino acids have been argued as adequate to carry the transition of fully self-sustaining life processes. These are all available from abiotic synthesis, as contrasted with larger, elaborate amino acids thought to issue only from cell chemistry. The amino acid arginine is redundantly specified by mRNA codons in which the third base appears redundant: CGU, CGA, CGC, CGG. The middle base suffices, with one of the others, to specify the dozen amino acids most often called in. It has been estimated that 50 genes might carry the protocols for the primary metabolism of a minimal cell. The cell would have evolved to considerably higher complexity before increasing virtuosity in organic chemistry and natural selection called in DNA to manage information storage. DNA serves as an archive of experience, thanks to the high stability of its phosphate bonds. That virtue, however, disqualifies it for the diverse functions served by RNA.

Experiment has thus far failed to replicate the prebiotic synthesis of the RNA nucleotides. Leslie Orgel, for many years engaged in the effort at the Salk Institute, conceded: "The formation of sugars [i.e., ribose and deoxyribose] in plausible conditions and their incorporation in nucleosides have not been achieved.... The origin of nucleosides and nucleotides remains one of the major problems in prebiotic synthesis."

Around this impasse other authors find the way by assigning the nucleotides the status of latecomer products of biotic synthesis. Behind the common ancestor, they look for the protocell from which it evolved. They propose other starting materials for the proto-organism that anteceded the common ribo-ancestor. If only as counter to the nucleotide consensus, Robert Shapiro at New York University offers for consideration an all-protein protocell. Replication is managed by something like the little "two-handed" proteins that catalyze the connection between tRNA molecules and the amino acids they encode. In the protocell, these facilitators would need only to recognize the same amino acid in a protein chain undergoing replication.

Harold J. Morowitz starts with an Oparin fatty acid vesicle. A "primitive pigment" in the membrane would engage the energy of sunlight in synthesis and concentration of more membrane, to start with, and therewith the multiplication of vesicles when their size outgrew surface tension. The membrane barrier would allow compositional difference to develop between interior and exterior. Over time, these reaction chambers could contain the start of protein or nucleotide chemistry. By natural selection among, perhaps, many such start-ups under different environmental conditions, the common ancestor would take off.

If the origin of life remains elusive, Shapiro urges his colleagues to prize the joy of the quest. "But" he adds, "we may be closer to the answer than we think."

Whether in "a matter of years" or millennia, life indubitably originated. Either way, the origin transpired toward the end of an interval no more than 300,000 to 500,000 years. The geological record establishes that this was the interval between the time the earth cooled down enough to hold liquid water and the first sign of life in the rock.

Earliest fossil cells

Elso Barghoorn of Harvard University and J. William Schopf now at the University of California at Los Angeles rolled back the history of life with the finding, in the late 1960s, of fossil cells in ancient yet unmetamorphosed sedimentary rock in the geologically well-charted Fig Tree mining district at the boundary between the Republic of South Africa and Swaziland. The deposition of these sediments is reliably dated to more than 3.2 billion years ago, perhaps as early as 3.36 billion years ago. In microthin transparent sections of this rock, they made positive identification of rod-shaped and spherical prokaryotes, along with threadlike filaments of possibly biotic material. Chemical analysis of these microscopic fossils showed selective sorting of isotopes of carbon and oxygen associated with biotic, as contrasted with abiotic, chemistry.

In Australia in 1980, near a desert hot spot called North Pole in the northwest wilderness, Schopf found evidence of photosynthesizing bacteria in another rare find of ancient sedimentary rock dated, with some uncertainty, to 3.5 billion years ago. While the date of crystallization of the grains of rock can be determined with confidence, the date of deposition of sediments cannot. Embedded in the layers are what appear to be stromatolites, filamentous mats formed by myriad photosynthesizing Cyanobacteria. Such stromatolites pile up in mounds today in tidal pools not invaded by grazing slugs and mollusks.

The fossil evidence thus reaches back to the beginning of the second billion years of Earth history. Confirmation of the North Pole dating would strongly ratify Barghoorn's tentative identification of the spherical Fig Tree bacteria as photosynthesizing "blue-green algae" or Cyanobacteria. This evidence argues that evolution had already taken a major hurdle at that early time. Living cells could live on air and water.

Harnessing the Sun

Because photosynthesis engages elaborate molecular apparatus, the first organisms were thought to be, like the latest human organisms, "heterotrophs," dependent on a ready-made supply of sustenance. The hot dilute soup in which they supposedly originated was there to nourish them. With success, the demand of this heterotrophic way of life would soon have exceeded the supply of sustenance. Natural selection would then have favored the arrival of "autotrophs" capable of self-sustenance. Now, it appears, autotrophs may have arrived in advance of such Malthusian crisis. They may even have arrived first.

Living fossils supply the best insights into this early phase of evolution. Remote ancestor-prokaryotes survive today in nooks and crannies in the community of life and persist in whole regions of the planet where conditions approximate the environment in which they first appeared. Many of these Archaebacteria manufacture their entire substance in environments barren of any ready-made organic compounds. One group, the methanogens, secures the necessary energy from hydrogen—available in the primordial atmosphere in greater abundance before its escape from the planet's gravitational field. By something like the Calvin cycle, the energy of ATP goes then to supplying glucose to their fermentative metabolism. Such "chemoautotrophs" abound today and make their contribution to the planetary ecosystem in the intestines of animals, in sewage and in the sediments of the ocean, in lakes and marshlands, where their methane ignites as will-o'-the-wisps, and, perhaps most numerously, in warm damp fissures in the deep crust of the Earth.

Another group, the Archaebacteria, lives in hot salt springs on the continents and the seafloor. Some are known to thrive at 70 to 75 degrees C—nearly twice mammalian body temperature—and to survive at temperatures as high as 88 degrees C, but "freeze" to death at 55 degrees C. These heat lovers secure energy from sulfur to reduce CO_2 to the starting molecules of their substance. The release of H_2S, in place of H_2O, from their metabolism adds the smell of rotten eggs to emanations from marshlands. As demonstrated by these descendants, the first autotrophs found ready sources of energy in inorganic compounds.

The founding bacteria became fully autotrophic when they could harness solar energy to fix carbon from CO_2 in the atmosphere. One photon-trapping pigment found today in salt-loving bacteria is related to rhodopsin in the vertebrate retina; it can be seen, a pinkish hue, staring skyward from salt-evaporation pans. The arrival of the Cyanobacteria advanced the community of life a long step toward liberation from local supply of ready-made necessities, whether organic or inorganic. An early chlorophyll in the photosynthetic apparatus of their living descendants conducts photophosphorylation just as plants do, by Photosystem II. The first pioneers would have engaged H_2S, instead of H_2O, to supply the necessary electrons. Still flourishing are descendants of these innovators. They switch to anoxygenic photosynthesis in sulfur-rich and oxygen-poor environments.

When at last they turned over air and water with Photosystem II, Cyanobacteria could occupy prospectively the entire planet. From their output of oxygen, they were at first protected by capacious oxygen sinks, such as sulfide gases and ferrous iron, with which the oxygen avidly combined in the atmosphere and in the ocean water. Over time, they developed enzymatic traps for free oxygen in their own cytoplasm. Other organisms, under pressure of oxygen from them, followed suit. In the course of such adaptation, some went on to aerobic metabolism of the product of their anaerobic fermentative metabolism. Prokaryotes today practice a diversity of aerobic metabolisms and, of course, an even more varied range of anaerobic metabolisms.

With the Krebs cycle yielding 36 ATP per gram molecule of glucose, instead of 2, the aerobic metabolizers took over the world. Their anaerobic forerunners vanished in perhaps the most complete of all episodes of extinction, or retreated to the niches they now occupy. Over the next billion years, capturing and transforming the energy of the Sun, aerobes—autotrophs and heterotrophs—enveloped the planet in its first biosphere. On all the continents, Cyanobacteria and their stromatolites appear in sedimentary rock dated to 2.5 billion years ago.

Symbiosis in evolution

Living fossils, again, tell the story of the next immense transition in evolution. These fossils are the two critical energy-transforming organelles that, along with the nucleus itself, distinguish the eukaryotic cell. The chloroplast and the mitochondrion each have their own vestigial but active genetic apparatus: DNA, RNA, ribosomes, porphyrin electron-transfer molecules and transcription and translation enzymes. In each case, the retained genetic apparatus has its role in the replication of the organelle inside its "host" cell. The organelle offspring are inherited by the next generation upon division of the cell.

In the 1890s, long before compelling evidence came from molecular biology, a few cell biologists were surmising that the chloroplast and mitochondrion might once have been free-living organisms. It was enough that these organelles were packaged in their own membranes.

This evidence suggested that the evolution of the eukaryote had proceeded on shortcuts around the tedious course of genetic mutation. It was proposed that these organelles had found their way into what became the eukaryotic cell through prolonged, mutually sustaining symbiotic association of ancestral prokaryotes. Such association, involving many different kinds of interdependency, is observed among prokaryotes today in communities of soil bacteria and in out-of-the-way places like the alimentary tracts of arthropods and vertebrates. Eukaryotes are found as well in intimate symbiotic association with prokaryotes.

To the proposition that eukaryotic cells are chimeras—like the

Sphinx or the centaur, the fusion of two or more different organisms—Lynn Margulis, long at Boston University and now at the University of Massachusetts in Amherst, has made a perceptive and persuasive contribution. "The serial endosymbiotic theory of the origin of cells with nuclei" has engaged all of her professional life. No one knows more about the natural history of the unseen world of the prokaryotes and the protoctista, a catchall category of single- and multicelled eukaryotes. In their respective and combined communities, she identified the actors and developed scenarios of symbiotic association that have now secured the consensus in favor of this acceleration of evolution.

Margulis points to the lichen: a photosynthetic Cyanobacterium or green alga, a eukaryote, takes up company with threads of fungal cells; together they form a new and more complex organism. Up to 10,000 species of lichen testify to the frequency as well as advantage of this association. Many more such associations thrive—or have until recently—in the coral reefs. There, coral polyps, animals in symbiosis with photosynthetic protoctists, sustain whole marine communities.

Of pivotal importance to life on Earth today is the symbiotic relationship between the Rhizobium family of soil bacteria and the legumes. Along with many Cyanobacteria and some Clostridia, Rhizobia fix nitrogen, N_2, from the air in organic compounds. No eukaryote is capable of such enzymatic chemistry. Industrial fixation of nitrogen from the atmosphere requires pressure at 300 atmospheres and temperature in the range of 500 degrees C. Such is the potency of the Rhizobia enzyme system. The legume root hairs attract Rhizobia from the soil into their cell walls to form an "infected" nodule. The Rhizobia transform into large nitrogen-fixing cells, supplying the nitrogen compounds to the whole plant and, through the soil, to plants incapable of symbiotic association with Rhizobia. Legumes and other symbiotic associations of plants and nitrogen-fixing bacteria, and bacterial nitrogen-fixers alone underpin every ecosystem: no organically fixed nitrogen, no protein!

In the tiny community of the hindgut of the termite, Margulis found a direct illustration of the sort of symbiotic association that brought the evolution of the mitochrondrion. There, inside a myriad of different protoctista, nucleated but lacking mitochondria, can be seen whole communities of different sorts of bacteria. On first impression, some of these guests look just like mitochondria and some of them help their hosts to survive the low local supply of oxygen. *Pelomyxa palustris,* a large swamp-dwelling, multinucleated eukaryote, lives by symbiosis with three different kinds of internalized bacteria, two of them methanogens. An antibiotic that kills the symbionts kills this proctista. Margulis cites one association so close that it was thought to be a single bacterium and had a name, *Methanobacillus omelianski.* It is now known to be an intimate association of a bacterium that exhausts hydrogen and carbon dioxide as waste products and a methanobacillus that engages carbon dioxide and hydrogen in its metabolism, exhausting methane.

Origin of the nucleus

It is, of course, the membrane-enclosed nucleus and the apparatus engaged in its replication—most elaborately in the mitotic spindle in plant and animal cells—that first distinguished the eukaryotes from prokaryotes. For the origin of this organellar system, no ready model symbiotic association offered itself. Margulis and her colleagues have only recently identified the living fossils that demonstrate the scenario at its crucial turns. They propose that the symbiotic fusion of an Archaebacterium and a Eubacterium founded the eukaryotic line.

The distinction between the two kinds of bacteria follows from the sequencing of the bases of the ribosomal RNAs of hosts of bacteria, conducted by Carl R. Woese at the University of Illinois. He has shown that these two prokaryote lineages are "each no more like the other than they are like eukaryotes." "Archaebacteria" is his term, embracing, among others, the notably autotrophic methanogenic and salt- and acid-loving bacteria persisting today in environments that approximate the Hadean world in which they originated. From this splitting of the bacteria family, Woese goes on to declare that "there are actually three, not two, primary phylogenetic groupings of organisms on this planet." Just as firmly, Margulis insists that there are two: the prokaryotes and their offspring eukaryotes that gave rise to four new kingdoms, giving the planet five kingdoms of life in all. She concedes to Archaebacteria the status of subkingdom of the ancestral bacteria.

Accepting Woese's Archaebacterial-Eubacterial distinction, Margulis proposes that the eukaryotes inherit genes for the enzymes of protein synthesis from the Archaebacteria and the microtubular internal and external motility system—in plant and animal cells notably the mitotic spindle—from the Eubacteria [see illustration, pages 300–301]. The contribution from the Archaebacterium is fully documented by the work of R. S. Gupta of McMaster University in Ontario on amino-acid sequences in its proteins. The Gupta sequences furthermore confirm contribution from the Eubacterial side, although he does not endorse Margulis's nominee from the family of motile, undulating spirochetes.

The Archaebacterium, in the Margulis scenario, had no wall armoring its pliable membrane. A versatile autotroph, it lived—as do its descendants today—in warm, acidic and sulfurous waters. There it could engage either sulfur or oxygen (providing local concentration did not exceed 5 percent) as terminal electron acceptor, to form either H_2S or H_2O. The Eubacterial spirochete, an anaerobic heterotroph, required H_2S and protection from oxygen. Its fermentative metabolism generated in return carbon-rich products and elemental sulfur required by its partner. To the partnership the undulations of the spirochete provided motility away from oxygen in higher concentration and toward supply of carbon. The fusion of the two organisms, over time, was facilitated by absence of the cell wall in the Archaebacterium. When the DNA of the partners merged in the formation of the nucleus, the spirochete disappeared. The undulipodia and the cilia of so many eukaryotic cells and, of course, the mitotic spindle are the best evidence of its former existence.

Today, from microbial colonies in at least six mineral springs around the world, from interior Siberia to the Kurile Islands to New Guinea, specimens of *Thiodendron latens* have been harvested. The name "thiodendron," calling attention to the long stringy filaments of its propulsion system, implied this was a single organism. Upon closer inspection, it turned out that the species had to be delisted. It is the symbiotic association of a *Desulfobacter,* taking the place in Margulis's scheme of the Archaebacterium, and a spirochete huge enough to be visible to the unaided eye. In more forbidding environments, Margulis seeks the association of an Archaebacterium in such a partnership.

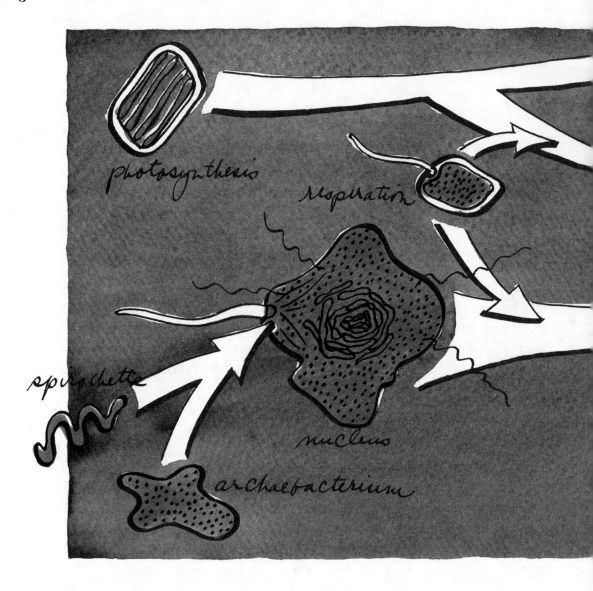

Eukaryotes originated in symbiosis of prokaryotes *beginning [lower left] with symbiotic association of an Archaebacterium with a spirochete, to which the first supplied sustenance and the second motility. Prolonged association brought the fusion of their genetic apparatus in the nucleus, with the spirochete contributing the mitotic spindle that coordinates replication of chromosomes, as well as the undulipodium, the organ of motility, in some and the cilia in many eukaryotic cells. Such association is observed in living organisms at many stages*

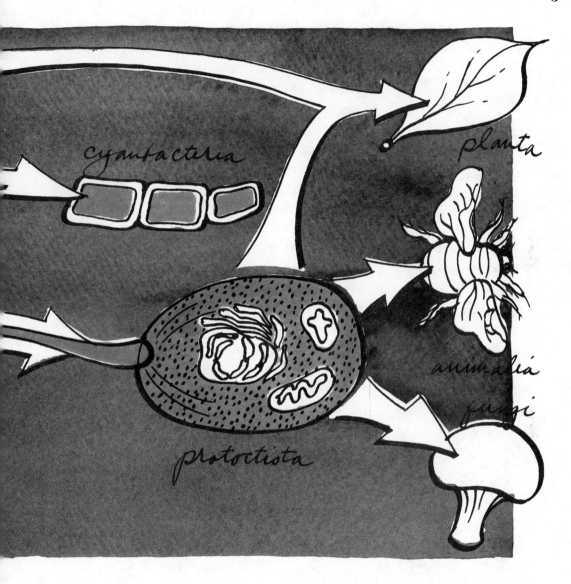

toward complete fusion [see page 302]. The nucleated eukaryote later acquired, by symbiotic association with an aerobic bacterium, the mitochondrion that conducts its respiration. From the diverse eukaryotic protoctista the animal and fungi kingdoms came directly, and from the association of a protoctista with a photosynthesizing bacterium came the plant kingdom. Cyanobacteria similarly arose from the symbiotic association of a photosynthesizing bacterium and a bacterium with aerobic metabolism.

The progressive fusion of partners is demonstrated in more than two dozen eukaryotes classified as archaeproctista. None of these ancient eukaryotes has mitochondria. The partner spirochetes have become the "mastigont" (Greek: "cell-whip"), the propulsive organelle of the chimera [left]. In many of these organisms, evidently the earlier, the mastigont is attached to the nucleus to which both partners have contributed their genes. At cell division, this "karyomastigont" replicates as a unit, with the spirochetal remnant playing its role as microtubular apparatus in separation of the replicated karyomastigont. In other, evidently later, organisms, the nucleus is disjoined from the mastigont, and the tubules inherited from the spirochete line have begun to serve in many other functions, including the symmetrical division of the chromosomes. Over perhaps millions of years, the mitotic spindle evolved from the mastigont system in the cells that gave rise to fungi, plants and animals.

The architecture of the cilia and undulipodia that protrude from the membranes of all eukaryotic cells that have them, whether protoctista or cells in specialized tissues in animals and plants, argues for the occurrence of such an event in the evolution of an early common ancestor. The connection of the mitotic spindle to the ancestral mastigont is established by the structure of the centrosome, from which the tubules of the spindle are seen to radiate radiate. In cross section it is identical to the kinetosome that drives the tubules of the propulsive system. The centrosome is seen in some cells doing double duty, serving as a kinetosome for the undulipodium upon completion of cell division.

For the displacement of the single hooped nucleoid of the prokaryote by the bundle of linear chromosomes there is no obvious scenario. Natural selection, it is clear, would have favored orderly filing of accumulating genetic information. Maynard Smith suggests that time was of the essence in the divergence of the eukaryotes from the prokaryote way of life. The arraying of genes on free-ended chromosomes, in place of the closed hoop, makes for more efficient use of time. On all chromosomes, made as they are of the same DNA, the "reading" of the DNA sequences by the replicating enzymes proceeds at effectively the same rate. This takes about 20 minutes on the single prokaryotic hoop. In that same 20 minutes, as many readings as there are chromosomes can go forward in the eukaryotic nucleus. The eukaryote can replicate a much greater genetic capacity in a finite lifetime.

The presence of introns in eukaryotic chromosomes suggests that the multiple-chromosome arrangement affords capacity to waste. It suggests still another advantage of the arrangement. There is some evidence as well as speculation that virus infections in the distant past of the eukaryote line installed the introns. RNA viruses thus transcribe their identity in DNA bases hijacked from their host's inventory. In eukaryotes that survived infection, the virus vestige persists in scrambled identity. Such intrusion in the single strand of a prokaryote would have been fatal and left no trace. It begins to appear, in any case, that the introns may serve a function in the management of the genetic information encoded in the exons. Coiling and uncoiling, they may store away exons not in operation and surface them for service when called for.

The bundle of chromosomes in place of the prokaryote hoop is open, topologically at least, to addition and subtraction of chromosomes. Such arrangement could facilitate the fusion of the genes of two organisms in the course of their symbiotic association. Evolution is driven, however, not by outcome but by immediate selective advantage. Closer acquaintance with the Leeuwenhoek animalcules may yet tell that story. The still largely unexplored kingdom of the protoctista is sure to hold clues as well as surprises.

Mitochondria are absent in many of the ancestral karyomastigont lines. Some have, however, acquired these energy-generating organelles.

The scenario of this next step in serial endosymbiosis has many living models in the symbiotic association of eukaryotes with aerobic bacterial heterotrophs. For the installation of the chloroplast in cells destined to create the algal protoctista and from them the plant kingdom, living models suggest their scenarios in turn.

Mitosis and meiosis

The evolution of sexual reproduction, it is agreed, must have proceeded according to the plan first proposed by L. R. Cleveland of Harvard University. First, there were cells with one set of chromosomes. In lean environments, then as now, cells cannibalize. From those that survive, offspring often carry two sets of chromosomes. Two sets of chromosomes may result also from failure of mitotic division after the doubling of chromosomes, an accident that proves advantageous on occasion. In some species, fusion of two cells, each with a single set of chromosomes, is the first step in the alternation of sexual with asexual reproduction. Chromosome doubling and mitotic division next produces two cells, each with two sets of chromosomes. Thereafter, by meiotic division these give rise to four offspring, each with one set of chromosomes. The four sets of chromosomes resulting from such division of the two second-generation cells will carry different assortments of the genes conferred by the first two grandparent cells [see illustration, opposite].

In the sexual reproduction of plants and animals—including human beings—the egg and the sperm cell are the product of meiotic division; each has one set of chromosomes. The union of egg and sperm yields a "zygote" endowed with two sets of chromosomes, one from each parent. The expression of the parental genes thereafter, in accordance with Mendelian law, ensures that no two individuals (except identical twins that arise from a splitting of the zygote upon its first division toward multicellularity) will have the same set of genes. The stock of genetic traits in a population is accordingly exposed in a diversity of expression to natural selection. From this source of variation, evolution speedily—in the last fifth of the 3.5-billion-year history of life—generated the planetary ecosystem in its present high diversity.

Prokaryotes do not enjoy the advantages of sexual inheritance.

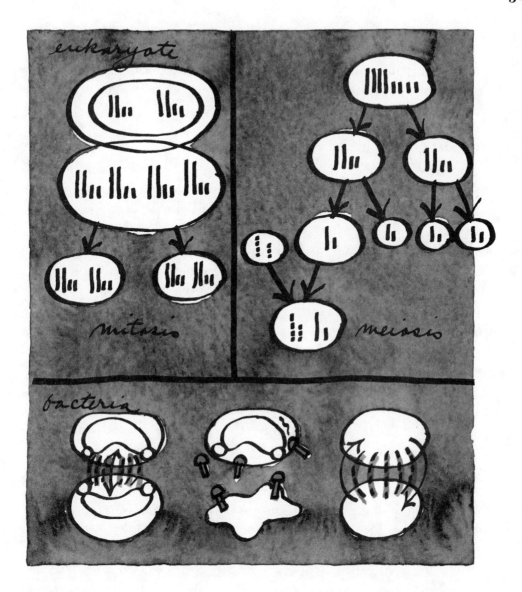

Eukaryote genes are conveyed *in mitotic division by the doubling [top left] of the two sets of chromosomes and the conveyance then of a two sets to each off-spring cell. Mitosis governs reproduction of tissue cells and protoctista. Meiotic division serves sexual reproduction by conveying one set of chromosomes from each parent to the egg or sperm; their union combines chromosomes from both parents [page opposite]. Bacteria exchange genes by conjugation [bottom left], indirectly by viral transfer [middle] and randomly via fluid environment.*

They do share and exchange genes, however, in various ways. Genetic exchange proceeds by contact, "conjugation," between two bacteria; by transport in bacteriophage, viruses that infect bacteria and carry genes from one to the next; and continuously by leakage of genetic molecules into the communal fluid environment.

Bacterial gene exchange

At Wisconsin in the early 1950s, Joshua Lederberg demonstrated the virtue of diversity in the genetic resources of a population. He sampled colonies of bacteria from a culture plate by pressing a sterile velvet pad on the plate. From the sampling, thereafter pressed upon a culture medium containing an antibiotic, a few of the colonies survived. Isolated from the original plate, bacteria from the same colonies produced successor generations on cultures poisoned with the antibiotic, confirming the presence of the gene or set of genes in the original population before any exposure to this powerful selective agent. The trait goes unexpressed until the antibiotic-resistant strain selected by the antibiotic becomes the population to which the human population, for example, is exposed by random prescription of antibiotics. Conversely, it has been shown, the frequency of a trait diminishes over many generations in a population not exposed to a selective agent. The gene remains nonetheless dormant in the genetic resources of the population. Earlier, working with Tatum at Yale, Lederberg showed that bacterial genes must be strung in the bacterial plasmid in the same organized sequential fashion as in the chromosomes of the eukaryotes, permitting the assortment and recombination of traits secured by protosexual exchange of genes among bacteria, however accomplished.

Sorin Sonea at the Université de Montréal argues that bacteria, in their prevailing genetic promiscuity, do not exist in differentiated species but rather form a continuously graded, single species. The best definition of a species bacteriologists can offer is the sharing of 85 percent of traits in common. This is not a distinction that endures and surely does not define a "species" in the eukaryotic superkingdom. Sonea calls for consideration of bacteria as a kind of worldwide superorganism. A bacterium, he observes, is "an incomplete organism," depending for its existence on the mutual sustenance afforded by the bacterial "chimera" assembled in the local ecosystem [see page 343].

Physics limits the size of single cells. Geometry shows that volume increases as the cube, and surface as the square of increase in the radius. The interior of the cell quickly recedes beyond the reach of diffusion from the outside, of oxygen to begin with. Organized in multicellular organisms, cells of optimal but still microscopic size may concert in action on the macroscopic scale. Increase in size then requires differentiation of cells in specialized tissues. Differentiation of tissue permits, in turn, differentiation of whole-animal life-forms. The peak of diversity of life-forms sharing contemporary existence has been reached only in the present era, along with arrival at peak of size in the blue whale.

Size by multicellularity

In gigantic organisms—among which, relatively speaking, may be counted the human—the surface of cells functionally related to the world outside keeps pace with their volume. The lungs spread a vast surface for gas exchange in the alveoli bubbles; they are so tiny that it is the force of surface tension that contracts them in the exhaling of the breath. The convoluted folds of the gut and the villi that line them multiply by many times the surface exposed to absorb nutrition, molecule by molecule. The hardworking cells in these tissues, sustaining the rest of the organism, have comparably the same ratio of volume to surface exposed to the world outside as cells in solo existence.

Contemporary natural history offers numerous models of the onset of multicellularity. From single-celled amoeba come descendants organized in multicelled slime molds, the masses of cells observed as apparent slugs or naked snails that creep on wet surfaces in damp forests. In their concerted existence they move together as a kind of super-organism. When occasion in their life cycle demands, they cohere in a stem that mounts upward from the main colony. At the top of the stem, some of the cells form a fruiting body from which spores disperse to establish new colonies. Any one cell in the colony can give rise to a colony, from which some members will differentiate to form the fruiting body and spores. Certain Myxobacteria enact the same life cycle, growing in wet soil, moving as an individual organism and developing fruiting bodies from which thousands of offspring bacteria disperse. In stromatolite mats, filamented Cyanobacteria have, from long ago, presented what must be regarded as macroscopic multicellular organisms with cell differentiation and programmed cell death.

Margulis cites the volvocine group of colonial algae as displaying a succession of steps in the development of multicellularity, from hollow spheres of identical cells to organisms in which cells have differentiated to serve specialized functions. The various species of *Gonium,* the most rudimentary in this group, form disks of 4 to 32 clonal cells glued together by a kind of gelatin. Their cilia, beating in unison, propel the communal body around. Any one of the cells can form a new colony.

The Volvox proper is a hollow sphere of 500 to 600,000 cells, depending upon the species. In these species an identifying feature of true multicellularity comes incipiently into force. Only some of the cells remain fitted for independent existence and reproduction of the organism. The others have surrendered the omnicompetence of the genes they carry to specialize in the expression of some feature of their genetic capacity that contributes to the life of the whole organism. In some species the reproductive cells in an individual Volvox specialize as egg and sperm. Their fusion effects exchange of characteristics shared by all the cells within the spherical organism. These have arrived at Mendelian crossing of the characteristics of the whole organism, which has become the unit of reproduction.

It was the eukaryotic cell line that went on to realize a wealth of possibility in these evolutionary inventions. That is to be seen in the four kingdoms of the diverse eukaryotic empire. From the founding protoctists, the eukaryote went on to generate the fungi, plants and animals that assert their presence in the visible world. Their existence remains nonetheless anchored in the life processes of the ancestral prokaryotes, which outweigh them in the planetary biomass.

6

Earth History and the Evolution of Life

The old 6,000 years of chronology becomes a kitchen clock.
Ralph Waldo Emerson

E arth in photographs taken by Apollo astronauts in the 1960s appears, as Archibald MacLeish exclaimed, "small and blue and beautiful" in the black void of space. By that time, the planet had harbored life for more than 3.5 billion of the 4.5 billion years of its history. Living organisms had fashioned the planet to suit their existence. The photographs show their creation: the atmosphere, blue and beautiful, of the "biosphere" in which they have enveloped the Earth.

On Earth, a chance combination of circumstances—the size and composition of the planet and its distance from a star of the right radiance—allowed atoms to assemble in highly organized ways. Most atoms in the universe are stripped of electrons and gathered by the gravitational force in violent mutual repulsion in the hot plasmas of stars. On Earth, 150 million kilometers from the Sun and cooled down from the heat of its gravitational collapse, electron shells are intact. The light elements—hydrogen, oxygen and nitrogen, all gaseous above 75 degrees K—are held by gravity inside their escape velocity. Hydrogen and oxygen, in mutual attraction, self-organize as water. In water, early in Earth history, the three light elements combined with carbon in large self-replicating molecules. Not long after, these molecules had organized in self-actuating, self-replicating living cells.

Within a billion years, the Earth was mantled in its first biosphere. By then, living cells were incorporating in their substance, otherwise constituted of the four elements from air and water, as many as 24 elements from the crust of the Earth. Life had become a geological force. Vladimir I. Vernadsky, the Russian "geo-biologist" who coined the term "biosphere" early in the 20th century, called life "the most powerful geological force." Today, 4.5 billion years into Earth history, much of most of the rock in sight on the continents has been processed through living cells in the course of Vernadsky's "bygone biospheres."

Life could not thus declare its presence were the Earth itself not geologically alive. Life evolved through 3.5 billion years of tumultuous Earth history. The door to that history opened only in the last half century. Not until then did people recognize on the Earth's surface a topographic feature as grand in scale as the continents and the oceans.

This is a mountainous ridge 75,000 kilometers long that runs nearly twice around the planet. It had gone undiscovered because it rises from the bottom of the ocean. The ridge is rifted by steep-walled canyons and marked by frequent earthquakes. Here and there volcanoes have erupted into the ocean, and hot springs eject smoky sulfurous plumes of superheated water.

The submarine rifted ridge manifests at the surface of the planet the action of a newly recognized geologic force. This force derives from heat-driven convection—a kind of slow boiling—of the plastic rock of the deep mantle below the brittle crust. Acting horizontally in concert with the vertical force of gravity, this force has made and remade the map of the world. Through the long past, convection cells turning over in the mantle rock have dragged continents over the surface through all latitudes, assembling them in supercontinents and then disassembling them and, in the course of this cycle, closing and opening oceans.

Through nearly 3 billion years of this history, bacterial cells exerted their increasingly powerful presence. The chain reaction of their replication, in cycles of less than an hour, pervaded the world ocean with the geological force of life. The sealing of 1 million peptide bonds per second in each cell, at the rate of 3.2 eV per bond, sped up significantly the energy turnover of the planet. Vernadsky and his colleagues likened the free energy of the communities of life to that a flow of molten of lava, dispersed at lower temperature in larger spatial volume. Selectively incorporating elements from the atmosphere, water and rock, the diversely specializing and mutually supporting bacterial ecosystems spread the reconstituted planetary substance in sediments on the ocean floor. The recently recognized cyclic turnover and regeneration of the Earth's crust have shown that those ecosystems hastened the granitization of the rock and the building of continents. Capturing the energy of the Sun and steering the chemical avidity of oxygen, the by then planetary oceanic ecosystem of bacterial cells began to reconstitute the atmosphere, wrapping the planet in its first biosphere. Perhaps 1.5 billion years ago, the symbiotic way of prokaryotic life generated the eukaryotic cell. The stage was set for the elaboration of new kingdoms of life—plants, animals and fungi—their conquest of the continents and, in the last half century, the reconstruction of this history.

Plate tectonics

The submarine ridge and distribution of earthquakes on the world map delineate the boundaries of the convection cells presently turning over in the Earth's mantle. Along those boundaries the Earth's crust is fractured into eight "tectonic plates." Where convection cells are active, as down the mid-Atlantic, their concerted upward roll has lifted and rifted the ocean floor. Molten rock floods upward from the mantle into the rift. As the new rock freezes into the crust on either side of the ridge, the horizontal return motion of the cells drags the tectonic plates away from the ridge.

Like great rafts, these tectonic plates float in the plastic mantle. In six of them, granitic continental crust rides along with attached plates of denser, thin, basaltic ocean crust. Two plates of oceanic crust, with no continents aboard, underly the vast Pacific Ocean. Where continental crust on one plate collides with oceanic crust on another, as along the Pacific coasts of North and South America, continental crust overrides the oceanic. In deep trenches at these boundaries, the oceanic crust plunges into the mantle and, melting there, offsets and balances addition of rock at the mid-ocean ridge [see illustration, next four pages].

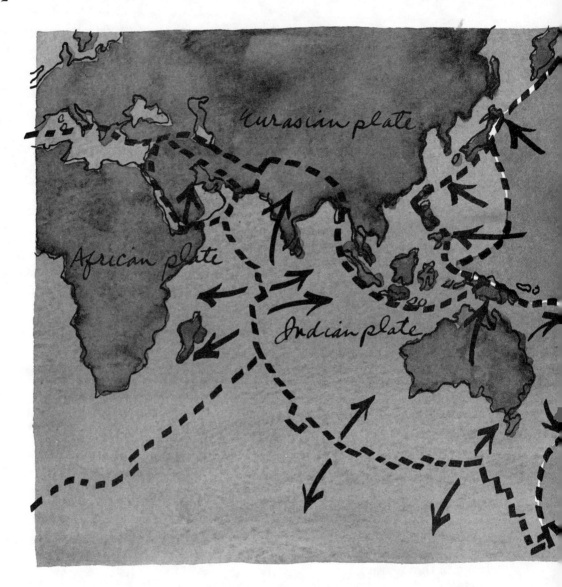

Submarine mountainous ridge and tectonic plates *manifest the turnover of convection cells in the plastic mantle below the brittle crust that have constantly assembled and broken up continents and opened and closed oceans over the past 3 billion years. Worldwide, the movement of the plates is away from the ridge. In major plates, such as the North American, thick continental crust moves with trailing edge of thin oceanic crust [see page 311]. On their western side the North and South American plates override oceanic plates; from the sub-*

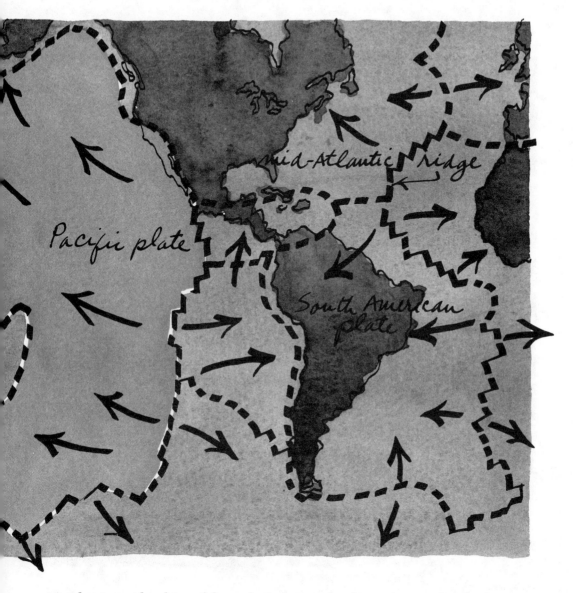

Pacific plate

mid-Atlantic ridge

South American plate

duction and melting of those plates deep under the continents rises the mountain chain reaching from Tierra del Fuego to Alaska. The eruption of Mount St. Helens, Washington, in 1980 was evidence of this activity. Around the west side of the Pacific Ocean, the other half of the "ring of fire" of earthquakes and vulcanism marks the subduction of the big Pacific plate under the Eurasian plate and under other oceanic plates from which rise the island arcs and archipelagoes reaching from New Zealand to the Aleutians [see page 316].

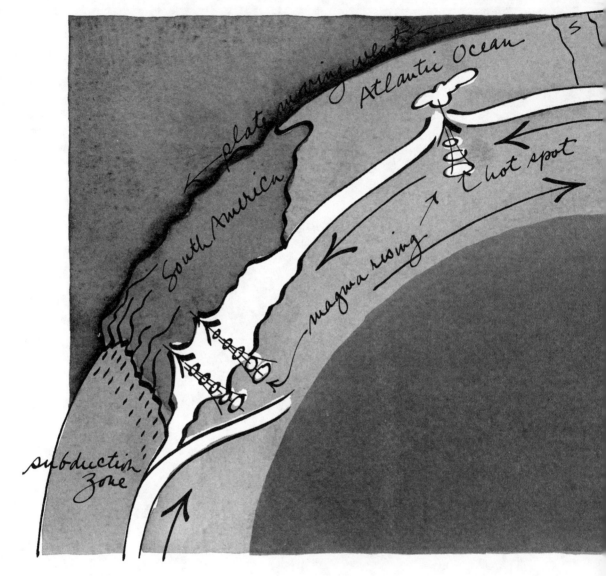

Upwelling of convection cells at mid-Atlantic ridge *is adding new rock to the crust, carrying away the continent-bearing tectonic plates and widening the ocean [see page 311]. The heat gradient in the asthenosphere between that under thick continental crust and under thin oceanic crust causes white-hot mantle rock at the bottom to creep toward mid-ocean and roll upward, lifting and breaking the crust. The horizontal return motion in the upper region of the cells carries the tectonic plates away from the ridge on either side. The addition*

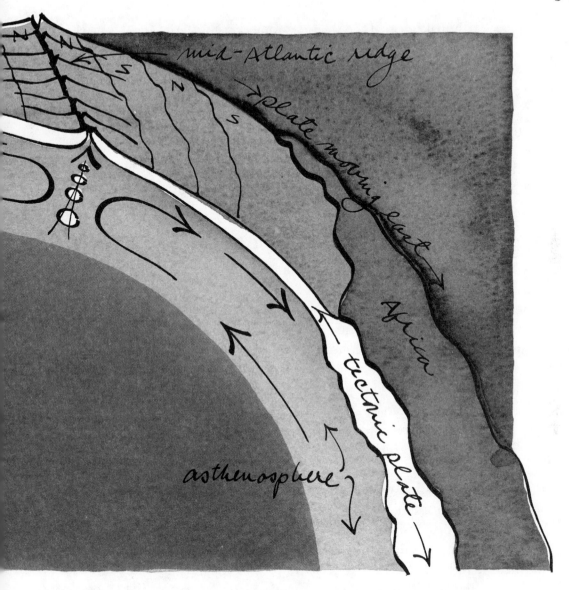

of new rock at the ridge is compensated by subduction of the Pacific plate, plunging deep in the mantle under the continents. In the melting there of the oceanic crust and its overburden of sediments leavened with light elements from the biosphere, life processes contribute to the "granitization" of the rock that returns to the surface in the cordilleran mountain chain that runs the length of the Pacific coasts of North and South America [see page 316]. The lithosphere thus undergoes a cyclic turnover comparable to that of atmosphere and hydrosphere.

Earthquakes signal the steep-slanting descent of the oceanic crust deep and deeper in the interior. Upon reaching depths between 600 and 700 kilometers, from which the deepest earthquakes sound, the subducting plate loses its solidity. Under heat and pressure there, the crust and ocean-floor sediments melt in a magma that is hotter than the surrounding mantle. The magma, enriched with lighter elements from the detritus of the biosphere, is lighter as well. It rises through the mantle and disrupts the continental crust. Volcanoes and "plutonic" intrusions of molten granite are building the great cordilleran mountain chain that reaches almost uninterrupted from Alaska to Tierra del Fuego.

The Earth's crust—its lithosphere—thus rolls over in a cycle analogous and comparable to the cycles of the planetary atmosphere and hydrosphere. Passage of the atmosphere through living tissue maintains the atmosphere's oxygen content at 21 percent and carbon dioxide near the greenhouse 0.03 percent that modulates the in- and outflow of heat from the Sun. The hydrologic cycle lifts water from the ocean onto the continents and washes the continents into the ocean. The crustal cycle carries the outwash from the continents and the debris of the biosphere accumulated in ocean sediments deep into the mantle. Reconstituted in the magma, this material returns to the continental crust in volcanoes and cordilleras that rise inland from oceanic trenches.

For the past 200 million years the breakup of the last supercontinent—Pangaea, in which Laurasia in the Northern Hemisphere was joined to Gondwanaland—has been widening the Atlantic and Indian Oceans [see illustration, pages 312–313]. From the ridge in the mid-Atlantic, the Laurentian (North American) and the Eurasian plates, once joined in Laurasia, are drifting apart about 2 centimeters per year. In the South Atlantic, the spreading ocean floor similarly parted the coastlines of Africa and South America, once joined in Gondwanaland. The opening of the Indian Ocean pushes the Indian, African and Australian fragments of Gondwanaland northward from Antarctica.

These movements of the continent-bearing plates have been closing the Pacific Ocean from all sides. All around the Pacific, active volcanoes and recurring earthquakes—the "ring of fire"—signal the encroachment of the continental plates and the submarine collision of oceanic plates. On the western side of the Pacific, this immense activity is mapped by the chain of island arcs and volcanic islands that reaches from Antarctica to Alaska: the islands of New Zealand, the Polynesian and Micronesian island arcs, the Indonesian, the Philippine and the Japanese archipelagoes, the Kamchatka peninsula and the Aleutian island arc.

The Himalayas, the Caucasus range and, farther west, the Alps and the Pyrenees thrust upward from the collision of continental crust with continental crust. The spreading floor of the Indian Ocean has crushed the Indian subcontinent against the Tibetan plateau. The sub-duction of the smaller plate from the south under the giant Eurasian plate heaves the Himalayas skyward. The shallow Mediterranean and Black Seas are what remain of an ancient ocean closed by the northward

movement of the African plate. The Pyrenees, Alps and the Caucasian massif mark the suture of the African and Eurasian plates.

The Appalachians and the Urals testify to an earlier collision of continental plates in the assembly of Pangaea. Over the long history since, glaciation and the more continuous processes of erosion have planed the skyline of once alpine mountain ranges.

Another kind of mountain buidling has its model in the Hawaiian Island seamounts. Such seamounts rise above sea level from a "hot spot" in the mantle. A plume of molten mantle presses the seafloor upward in recurrent volcanic eruption. Pointing southeast to northwest—the direction in which the Pacific plate is drifting—the islands trace the passage of the plate over the hot spot. Ahead of the chain, sunken seamounts bear witness to the past. Trailing the chain, Hawaii Island, the largest, with volcanoes still active, is the youngest.

The map of the continents and oceans and their topography is thus wonderfully explained. Plate tectonics has endowed the local, descriptive science of geology with a planetary, predictive general theory. It resolves not only global questions but also local puzzles that have troubled the science. Gravity, by itself, never satisfactorily explained the formation of geosynclines. Such horizontally compressed folds in deep sedimentary rock are today exposed to view by the deep cuts that take the interstate highway system through the eastern U.S. landscape. These formations, it is now understood, are offshore continental sediments that were heaved up and folded back on the continent upon its collision with the Eurasian and African continents in the assembly of Pangaea 250 million years ago. Here incorporated in the rock and exposed to view is one of Vernadsky's bygone biospheres.

The new understanding of Earth history is bringing Earth history and the evolution of life together in a single grand chronicle. Sunlight warms the Earth, but shallowly. The planet's own heat sustains life and the rolling of the convection cells. It is the heat of the planet's gravitational collapse [see pages 209 and 335]. The low heat conductivity of the rock has brought it upward from the deep interior, slowly and steadily over all the eons. The decay of long-lived radioactive elements segregated in the upper mantle, and more so in the continental crust, makes its contribution. That contribution was stronger through the first combined half-life of these elements, in the first billion years when the sunlight was fainter. Heat from the Earth's interior will be enough to sustain life on Earth until the Sun turns onto the Red Giant track [see page 173].

The oldest Earth rock so far identified, found at Isua in southern Greenland, has been dated to 3.8 billion years. The Moon rocks establish the beginning at 4.6 billion years ago. The rate at which the Sun consumes its hydrogen fuel affirms that the life-bearing Earth has still half of its history ahead.

By convention, we divide the newly discovered long past into three stretches of time. The most recent, the crowded foreground called the Phanerozoic (visible life), reaches back only 570 million years. The preceding 4 billion years are divided evenly and arbitrarily between the Proterozoic (earlier life) and the Archaean (primeval). Until the last half century, those billions of years were empty. Radiometric dating of the rock had not even yet established their passage. The scant fossil record has only since been opened to inspection. The reading of the biogeo-chemical record in the rock has just begun.

The first signs of life

The tectonic turnover long ago effaced or deeply buried the Earth's first crust. The rock at Isua already bears signs of life. This rock is an extensive sedimentary formation. Almost uniquely for rock of such age, it did not undergo complete metamorphosis. The sediments establish that an atmosphere and a hydrosphere had already then been engaged in the weathering of continental crust. Strata of fine-grained reddish rock rich in iron oxide alternate with more purely silicate grains of quartz. This "banded iron formation" indicates something more. Such formations from later history elsewhere in the world reflect the seasonal waxing and waning of local communities of life sustained by photosynthesis. Confirmation of the biogenesis in the Isua formation would indicate that life began within the first billion years of Earth history.

Evidence of life from elsewhere sets a later beginning. At the North Pole site in Australia, signs of life date back to 3.5 billion years; in the Fig Tree sediments in Africa, to 3.3 billion years ago [see page 293]. Not long after, colonies of prokaryotes were establishing local ecosystems. In rock 2.5 billion years old all around the world, life left unmistakable evidence of its presence in fossil cells and macrofossils of Cyanobacteria communities, called stromatolites.

Sediments from the late Archaean record, moreover, the not yet fully catalogued diversity and virtuosity of prokaryote biochemistry. As Preston Cloud of the University of California at Santa Barbara, an early investigator of the evolution of the biosphere, observed: "once life appeared on Earth, the geochemical record could never be the same again." Bacteria were dissolving from the rock or capturing from solution in the water such elements as sulfur and phosphorus and sodium, potassium, iron, manganese and magnesium. They were already bringing the lithosphere, along with the atmosphere and hydrosphere, into the creation of the biosphere.

The world's present biosphere continues to depend upon

prokaryotes to fix essential elements in organic compounds. Eukaryotic organisms can engage only three—hydrogen, oxygen and carbon—from the air and water and none from the rock. They must even depend upon prokaryotes to fix nitrogen from the air.

Sedimentary rock from the Archaean shows prokaryote ecosystems already occupying the onshore waters of all the world's landmasses. They must have proliferated throughout the sunlit top 200 meters of the ocean, where bacteria today are counted in concentrations of 500,000 per milliliter, and in many times that concentration in the sediments accumulating on the ocean floor. At the hot sulfur springs along fissures in the oceanic crust, they were almost surely setting the foundations for the diverse communities of life recently discovered there. Those ecosystems testify to the prokaryote's biochemical capacity to sustain life out of reach of sunlight. Bacteria may even have colonized the land. In deserts today these organisms, awakened by occasional rainfall, bind the surface grains in a crust that holds against erosion by wind.

Where prokaryotes specializing in particular metabolisms congregated, they concentrated one or another of the lithospheric elements: iron, manganese, magnesium, aluminum, sulfur and phosphorus. Those mineral deposits serve human convenience today.

Discovery of Earth history

Against the brevity of human life, 5,000 years once seemed a long time. Through the 18th century, catastrophes, whether from flood, earthquake or volcanic eruption, sufficiently explained the tumult apparent in the landscape. Before the end of the century, James Hutton culminated a lifetime of study of the landscape of his native Scotland with the publication of *Theory of the Earth*. In this two-volume masterwork, he catalogued a natural succession he had seen in rock. First came the igneous, frozen from the molten state deep underground. The breakdown of rock yielded sand, which then consolidated in sedimentary rock. Under sufficient pressure and heat, but not enough to melt, sedimentary rock consolidated further to metamorphic rock. Hutton gave those names to the three kinds of rock and proposed that the succession of rock from state to state must have proceeded "uniformly" through all time by freezing and thawing, erosion, accumulation and consolidation of sediment, and the occasional volcanic intrusion. Those processes shaped the landscape. Rain and frost and rain wore down the mountains and spread them on the floodplains; the rivers carried the continents to the ocean. Needed only was time. Of time, Hutton could see "no vestige of a beginning, no prospect of an end."

Time for uniformitarian processes

Evidence in Hutton's favor soon accumulated. Collectors found whole communities of fossil creatures especially in the Cambrian rock, so called after the Roman name for Wales. In the logic that placed more recent rock on top of older, it was seen that the communities had succeeded one another over what must have been long periods of time.

Early in the 19th century, on the immense authority of Georges Cuvier, the founder of paleontology, some thinkers were dissuaded of Hutton's great insight. Cuvier saw discontinuities in the fossil record as catastrophes that had intervened in the succession of fossil communities. Catastrophism, admitting the possibility of divine intervention, persisted late into the 19th century.

The fieldwork, scholarship and tireless advocacy of Charles Lyell ultimately established the time necessary for uniformitarian processes to do the work of creation. Lyell calculated, from the recorded change in the horseshoe configuration of Niagara Falls, how long the Niagara River had pounded rocks to wear the falls back upstream into that configuration and, from the record of the reverse process, the time needed for the deposition of the Mississippi delta. Lyell established the order of succession of the most common fossils, the calcium carbonate and calcium phosphate shells of the mollusks and brachiopods. By "index" fossils as well as mineral composition, he identified sedimentary formations that came to the surface at distantly separated locations. He showed that the strata of the Val de Noto in Sicily were younger than the English Crags, and that the Crags were younger than strata in Bordeaux.

Lyell assigned the names to the geologic periods and epochs by which they are still known, and suggested time frames for them that are close to measurements derived today by techniques he could not have anticipated. The "Recent," about 2 million years in duration, is preceded by the "Tertiary," a whole order of magnitude longer. Before the Tertiary, time stretched without end into the past behind the dying-away of the fossil record in the Cambrian rock.

The geologists had discovered an uninhabited Earth. For almost all of the newly established unimaginable time, there had been no human beings. There had scarcely been animals with backbones. For ages the ocean had nurtured invertebrates in abundance and in inchoate variety. Human existence could be seen as a recent episode in an un-heeding Earth history. Troubled contemplation of that idea apparently prolonged the years that Charles Darwin gave to the writing of *The Origin of Species*. He was born to an upper-middle-class English family at a time when such families, if not unanimously Darwin's own, lived secure in the presence of a Divinity Who had a purpose for humanity in Mind.

Voyage of the Beagle

Free to pursue a native interest in natural history, young Darwin in 1831 signed on as naturalist for a five-year shoreline-surveying expedition that took him around the world. He was able to make his own forays ashore along the way.

At Tierra del Fuego, he recoiled from the naked natives who welcomed the expedition. He had not conceived "how wide was the difference between savage and civilized man."

On the Galapagos Islands he was enchanted to find himself "surrounded by new birds, new reptiles, new shells, new insects, new plants." Then, across the Pacific, he went ashore collecting on Tahiti, New Zealand, Australia, Tasmania, the Maldives and, on the way home, St. Helena, Ascension and the Cape Verde Islands—all buoyantly recounted in *The Voyage of the Beagle*.

Darwin spent the rest of his life contending with what he learned on that voyage. He had gathered evidence that convinced him of the hypothesis of natural selection. Especially in the Galapagos, he had seen evolution at work. What must have been a single species of finch had apparently given rise to a half dozen species, occupying different niches, breeding in separate populations and diverging in behavior and appearance, especially the appearance of their beaks. A species of land tortoise, in populations isolated from one another on the several islands, was undergoing similar divergence in conformation and behavior.

Natural selection

Evolution had thus, over the ages, without any discernible plan, design or purpose, given rise to all those invertebrates, to vertebrates in much lesser variety and indifferently, among them, to the human species. Evidence established by others sustained this appalling proposition. Darwin gathered it all in his study at his country place at Downs. There, for two decades, he paced back and forth on the same garden path until the threat to his priority laid by the young naturalist Alfred Russel Wallace compelled him to publish in 1859.

The hypothesis of natural selection was already abroad in the last decades of the 18th century. Darwin was not even the first Darwin to address the subject. His grandfather, Erasmus Darwin, founder of the family fortune, polymath, *bon vivant,* and eminence in his own right, had written a series of works culminating in the publication of *Zoönomia, or the Laws of Organic Life,* before the end of the century. He proposed that competition among living things selected the survivors and, therefore, the"fittest" to beget the next generation. Fecundity and population growth, overrunning the means of subsistence, drove the competition. That idea attracted the interest of his contemporary, Thomas Robert Malthus. In his speculations, the elder Darwin allowed also the possibility that use and disuse of faculties in one generation might affect the inheritance of them by the next.

In England by the time the younger Darwin made his voyage, these ideas were already called Darwinian. In France, Chevalier de Lamarck embraced the elder Darwin's second thought. As each generation improved upon the characteristics acquired by the last, life-forms progressed from less to more perfect. The giraffe's neck stretched as the consequence of the animal's reaching for higher branches. Lamarck thereby reduced the inheritance of acquired characteristics to the absurdity that was to inspirit the *Just So* stories of Rudyard Kipling. Despite this embarrassment to the hypothesis of evolution, other English authors continued to argue for the idea of natural selection embedded, in those very words, in Erasmus Darwin's competition for existence.

In *The Origin of Species,* Charles Darwin established the concept of natural selection on the compelling evidence of his own observations and his scholarship in natural history. Darwin is rightly regarded as the founder of modern biology. His great work gave this descriptive science its first comprehending theoretical statement. Today, evolution is an observed fact of at least 3.5 billion years of Earth history. The theory of evolution sets out the best understanding of the fact at the time, as the accumulation of evidence affirms it.

Even in the first edition of his great work, Darwin acceded to the possibility of evolution by inheritance of acquired characteristics. In later editions, he admitted the possibility alongside natural selection. The aging Darwin apparently yielded to the public controversy and to his own hankering for design and purpose in the history of creation. With the controversy settled in the Victorian compromise—which, to this day, assigns to science the establishing of useful knowledge and reserves to religion the framing of purpose and value—his remains could be buried in Westminster Abbey.

Natural selection stood in any case as an incomplete theory of evolution. It offered no explanation for the origin of the traits that are subjected to natural selection. In the hypothetical mode of heredity then prevailing—the blending of parental traits in offspring—it could be shown that dilution would wash out a new trait in a few generations. The completion of the theory of evolution in the "modern synthesis" had to wait for rediscovery of Mendel's gene.

The Earth's interior

Until the mid-20th century, geology, too, worked with an incomplete governing theory. The vertical force of gravity could be seen in isostatic equilibrium of the continental crust with the ocean and the crust beneath it. The geological record showed the height of continents above sea level depressed by the weight of glaciation. Such change in equilibrium, it was thought, would be attended by block faulting and tilting of continental slabs. This went to explain the occurrence of

mountains, such as the Appalachians, not generated by volcanic eruption. Such events do play their part in the shaping of local landscapes
but, as now understood, in consequence of larger-scale proceedings.

The jigsaw-puzzle fit of the facing Atlantic coastlines of Africa
and South America, first plainly disclosed in 17th-century maps, had put
the extraordinary notion of continental drift in the heads of some of the
first people who inspected them. For such evidence of the action of a
horizontal force there was no ready explanation.

Gravity accounted well enough for the internal structure of the
Earth. Sorting the elements by weight, it concentrated the heaviest abundant element, iron, in the center and floated the lighter elements upward,
ultimately into the crust [see page 327]. For studies of earthquakes,
physicists had developed the seismometer early in the 20th century.
Earthquakes set seismic waves—compression waves analogous to waves of
sound—in motion in all directions through the Earth. Perfection of the
seismometer put sound waves at the service of the geologist, much as the
spectroscope put light waves to work for the astronomer and physicist.
Geologists had found a way to look inside the Earth.

The velocity of seismic waves [above] changes with changes in
temperature, pressure, density and elemental composition encountered
on their passage through the Earth. As at the surface, pressure squeezes
atoms together and heat pushes them apart. Analysis of seismic waves,
taken together with other considerations, shows that the temperature

rises most steeply down through the first 700 kilometers of the mantle. It reaches 1,000 degrees C within the first 100 kilometers, 1,500 degrees C at around 200 kilometers and 2,000 degrees C at 700 kilometers. From there to the core, 3,000 kilometers down from the crust, the temperature rises, at less than a degree per kilometer, to the estimated ultimate 5,000 degrees C.

Under maximum pressure overriding the force of heat, the innermost core of the planet, about 1,500 kilometers in radius and composed principally of iron, is solid. An outer surrounding core, about 2,200 kilometers deep, is of molten iron. Here heat overrides pressure. The slow convective turnover of the vast tonnage of iron in this outer core carries the enormous flow of electric charge that generates the Earth's magnetic field. An abrupt slowdown in the travel of the seismic waves at the boundary between the inner and outer core signals the change in phase there from the solid to the liquid state.

An abrupt increase in velocity marks the transition from the liquid to solid at the boundary between the outer core and the mantle. The velocity of the waves through the 3,000-kilometer depth of mantle rock changes relatively smoothly with pressure up to 700 kilometers below the surface. Variations in velocity in this upper region, not nearly as abrupt as at the boundaries of the core, give evidence of phase changes there. They are blurred by the mixture of elements in the mantle rock, as compared to the relative purity of the iron in the core. In this zone of steeply rising temperature, gravity has concentrated the radioactive elements that remain in the mantle. Their heavier weight is offset by bonding to lighter elements.

At the temperatures and under the enormous gravitational pressures that characterize the planet's interior, changes in phase reflect behavior of matter not seen in the phase transitions at Earth's surface. Inside the planet, matter exists in many unfamiliar states. A change in phase may crush outer electron shells, as Percy Bridgman first showed. The outer core, for all its liquidity, has high solidity. Even at the pressure prevailing there, however, heat at the estimated 5,000 degrees C dissociates the crystal structure of iron. The iron flows as a liquid, though imperceptibly on the human scale of time and distance.

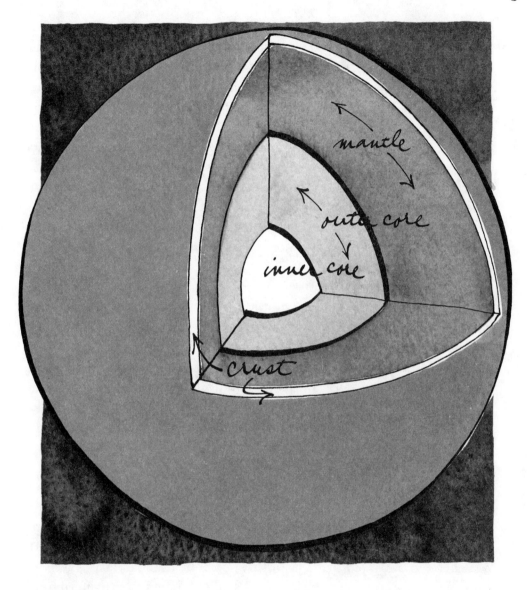

Earth's structured interior *reflects the gravitational sorting of its constituent elements and phase transitions from solid to liquid to solid under changing equilibria of heat and pressure. Iron, the heaviest abundant element, settled in the central core, solid to diameter of nearly 3,000 kilometers, and liquid through a depth of 2,100 kilometers. Oxides of manganese, iron and other metals constitute the "mafic" plastic mantle 3,000 kilometers deep; oxides of silicon, aluminum and other lighter metals, the brittle "sialic" crust [see page 333].*

The emerging picture of geology was disturbed in the 1920s by the serious advocacy by serious geologists of the "absurd" idea of continental drift. By this time geologists at work on the continents of the Southern Hemisphere recognized that they had been finding the same distinctive fossil flora in the coal seams of South Africa, India, Australia and South America. Called Glossopteris, after the tongue-shaped leaf of the most conspicuous genus, a variety of seed fern, it was found in the coal seams of Antarctica as well by the first geologists who ventured there. The Swiss geologist Eduard Suess offered a cautious explanation of these findings: the southern continents had once been united in a single continent by land bridges. He called this hypothetical reconstructed continent "Gondwana-Land" after the mining district in India where Glossopteris abounded in the coal.

The principle of isostasy, however, forbade the sinking of the lighter continental rock of the hypothetical land bridges into the sea bottom. In 1908, the American geologist Henry B. Taylor proposed that the southern continents had once been fused in the single great continent of Gondwanaland. Upon the breakup of that continent by subterranean forces, its daughter continents had drifted into their present locations on the planet. As to the mechanism of continental drift, he made a prescient suggestion: close examination should be made of the submarine ridge that had been encountered at mid-Atlantic in the laying of the first transoceanic telegraph and telephone cables.

Continental drift

A meteorologist framed the question of continental drift for its final debate. In writing *The Origin of Continents and Oceans,* Alfred Wegener demonstrated that geology can be done in the library as well as in the field. He consulted the literature from his base in a provincial university in Germany. The evidence he amassed showed that not only the southern but also the northern continents must once have been joined in a single supercontinent, to which he gave the name Pangaea. He pointed to similarities among living as well as fossil flora and fauna, especially the tropical vegetation in the coal seams of presently temperate and polar landmasses; to the transoceanic continuity of geologic formations and to scars of common glaciations on now separate continental land-masses—all arguing for the union of those continents.

Wegener made another fundamental contribution to understanding the grand-scale topography of the planet. Averaging elevations of the landmasses and known elevations in the ocean depths, he distinguished two levels in the Earth's crust. These levels correspond to the distinctly different crusts of the continent and seafloor. He pictured the continents as mesas standing 5 kilometers high on the ocean floor.

Published in 1915, Wegener's work did not get much attention until the end of the First World War. The controversy then ran through the 1950s. Southern Hemisphere geologists supplied new evidence on the Wegener side. The South African geologist Alexander du Toit found the same genus of living earthworm in the far southern ends of both Africa and South America, closely abutting regions in Wegener's map of Gondwanaland. Fossils of *Mesosaurus* were found in both South Africa and South America. Alfred Romer, at the American Museum of Natural History, showed *Mesosaurus* to be a freshwater reptile that could not have swum the Atlantic.

Evidence from the Southern Hemisphere did not carry great weight with the majority of geologists whose interests centered on the Northern Hemisphere. In the absence of any plausible mechanism to account for it, continental drift was lunacy if not heresy. The evidence for the necessary horizontal force lay concealed on the ocean floor.

That *terra incognita*—on the maps, 70 percent of the Earth's surface—came under systematic exploration for the first time at the end of the Second World War. From the Lamont Observatory of Columbia University, Maurice Ewing and Bruce E. Heezen went to sea to map the mid-Atlantic ridge. From the Scripps Institution and the University of California at San Diego, Roger Revelle and Henry W. Menard put the Pacific floor, with its trenches, seamounts and portentous East Pacific rise, on the world map.

Geologists are indebted to the U.S. military for the new technology of sonar that had played a decisive role in naval warfare. They had adapted return echoes from surface explosions to sound and discriminate rock strata deep underground. Sonar emits pulses of sound and times their echoed round-trip for target location, sounding and ranging. With generous support from the Office of Naval Research, a half dozen research vessels went to sea equipped with sonar, along with gravimeters, magnetometers and other remote-sensing devices.

For two decades, on monotonous voyages back and forth along systematically spaced gridlines, sonar literally sounded the bottom of all the planet's oceans. Along with the map of the ocean floor, sonar unexpectedly showed very small variation in the 4.7-kilometer thickness of the oceanic crust. The echo measured as well the thickness of the sediments spread on the ocean floor.

On familiar maps of land and ocean, the continents occupy 29 percent of the Earth's surface. From their shores, it was known, shelfs extend some distance under the ocean. The new map, displaying the ocean bottom, shows continental crust occupying 40 percent of the planet's surface. Continental shelfs slope out under the ocean as much as 300 kilometers to the 1,000-meter depth line and then fall steeply to the oceanic abyss. The continental mesas appear as Wegener imagined, rising 5 kilometers above the seafloor.

Most prominently, the new complete world map displays the rifted ridge over its full 75,000-kilometer reach. Along the 25,000 kilometers of its presently most active stretches, the ridge averages 1,300 kilometers in width and rises to heights of 2.5 kilometers. Other instrumentation working in concert with sonar revealed intense activity along those stretches of the ridge. Heat-sensing devices measured strong heat flow on either side. There, gravimeters found weakened gravitational force, signifying upward flow of material. Seismometers recorded numerous shallow earthquakes. Slower travel of the seismic waves indicated the presence of molten rock under sides of the rifted ridge, in confirmation of the readings by the sonar and heat-sensors.

In the North Atlantic, the ridge comes above sea level as Iceland. The rift appears as the graben, the deep cleft that crosses the rocky island from north to south. As if to confirm what the oceanographers were then deducing, the new volcanic island of Surtsey came suddenly above sea level in 1963, just south of Iceland.

At the trenches on both sides of the Pacific Ocean, the instruments recorded complementary readings. Heat flow is weak; the gravitational force stronger, and seismic waves from earthquakes originate at increasing depth down the steep incline of the subducting oceanic plate.

Clock and compass

Core samples of rock drilled from the ocean floor established a startling new difference between the continental and oceanic crust. Nowhere was the oceanic basalt found to be more than 200 million years old, compared to the eons that measured the age of continental granite and basalt. On both sides of the ridge, the age of the oceanic crust was found, moreover, to increase in constant ratio to distance away from the ridge. The overlying sediments, thinner at the ridge, proved to be younger as well; they increase in age with depth toward the foot of the continental shelfs, where they lie intermixed with the runoff of sediments from the continents.

Radioactivity ticks the clock that reveals the ages of rock with such assurance. Physicists had shown that each element decays at its own fixed rate and yields characteristic daughter and granddaughter elements [see page 124]. At its cooling, basalt rock in the ocean floor—or igneous granitic rock intruding in the continental crust—holds some minute sampling of these radioactive elements. The ratio in the rock of one or another of these elements to its daughter elements fixes the time of its cooling and crystallization.

Geologists were equally indebted to physicists for the discovery of a compass in the rocks. In the molten state, atoms of magnetic elements, such as iron, line up with the geomagnetic field. Like compass needles, they not only point to magnetic north; they also dip from the horizontal at an angle that depends upon their distance from the magnetic pole. Cooling freezes the orientation of these compass needles.

Early in the 1950s, the remanent magnetism fixed in the rock attracted the attention of P. M. S. Blackett and S. K. Runcorn at University College, London, and Patrick M. Hurley at MIT. They perfected instrumentation to read the weak magnetism. Gathering rock from around the world, they took care to see that the location and orientation of each specimen were recorded at each site. From age to age on the same continent, they discovered, the fossil compass readings pointed north in different directions from different latitudes. Sure that the magnetic pole had done no such wandering, Blackett, Runcorn and Hurley were satisfied that the changes in azimuth and dip in the rock samples had to be attributed to the drifting of the continents.

In the end, it was the compass in the rocks that settled the question of continental drift. It was known from the remanent magnetism in rocks ashore that—for reasons not known—the polarity of the Earth's magnetic field undergoes reversal at long, irregular intervals. In the seafloor of the Atlantic and then the Indian Ocean, remote-sensing magnetometers found these reversals recorded symmetrically in wide lanes

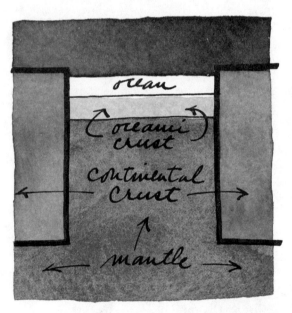

running north and south on either side of the rifted ridge. From lane to lane, the remanent magnetism pointed alternately north and south [see illustration, pages 314–315]. From reversal to reversal, as on magnetic tape, the new oceanic crust had registered the prevailing polarity. There could be no doubt that the Atlantic and Indian Oceans were widening.

A few geologists—Egon Orowan at MIT, Walter M. Elsasser at the University of Maryland, John Verhoogen at the University of California, Berkeley, and F. A. Vening Meinesz at the University of Leiden—had been working on a mechanism to explain continental drift now so convincingly established by plain and by subtle physical evidence. From the seismographic picture of the Earth's interior, empirical knowledge of petrology and geochemistry and the physics of high pressure and the solid state, they managed to frame a plausible hypothesis. In 1952, Vening Meinesz summarized their work in a paper that described what is now agreed to be the driving mechanism of plate tectonics.

Gravimeters and seismographs had long since established that the granitic crust of the continents is, on average, 34 kilometers thick and so much thicker than the basaltic ocean floor, a uniform 4.7 kilometers thick. Continental crust may be likened to the slag that floats up from molten iron ore. Composed principally of silicates—silicon and aluminum in compound with oxygen—it is referred to as "sialic." The dense oceanic crust, by contrast, is composed principally of manganese and iron in compound with oxygen and is called "mafic." That is close to the composition of the mantle, which is more malleable than the crust owing to incorporation of water in its crystal structure. The outgassing of water in the crystallization of the crust makes it brittle.

Isostatic equilibrium weighs a figurative column of continental crust against a corresponding column of ocean water 5 kilometers deep, plus the 1.3-kilometer average depth of the sediments, plus the 4.7-kilometer thickness of oceanic crust and plus a depth of mantle rock sufficient to balance the two masses [opposite]. The sialic rock of the continental crust has about 0.8 the density of the mafic rock of the mantle and oceanic crust. Like the tip of the iceberg (which has 0.9 the density of water), just the top of the continental column floats above the mantle. Above the 5-kilometer depth of the ocean, the continents rise an average 300 meters. In this equilibrium, an elevation of the continental surface must be complemented by corresponding descent of the bottom of the crust into the mantle; the elevation would settle otherwise at lower height, as does a melting iceberg.

The bottom of both crusts is signaled at the Mohorovich discontinuity ("the Moho") by sudden acceleration in the velocity of seismic waves. Under the thin oceanic crust, a relatively solid top layer of mantle overlies the discontinuity and smoothes the bottom of the tectonic plate where the oceanic crust joins the thicker continental crust, as along the eastern edge of the American continents.

Horizontal boiling

The Vening Meinesz paper explains the horizontal force beginning with the observation that the thick continental crust lays not only an insulating but also a heating blanket on top of the mantle. Deep in the crust, the decay of radioactive elements adds to the heat generated by those elements in the mantle. Heat builds up accordingly under the continents. Through the thin oceanic crust, conversely, heat escapes. A heat gradient thus runs in the crust: from the warm middle of a continent and to the cool middle of the adjacent ocean.

On such a gradient, heat flows by conduction from the hot body to the cold. Heat moves also, however, by mechanical motion in and of its medium. Water boils because convection carries off heat faster from water than can the conductivity of water. Meinesz and his colleagues saw that the same must be true of the transport of heat in the mantle. Rock does not, of course, flow like water. Like other solids, however, rock may nonetheless be made to creep. Hot and cold rolling, pressing and forging exploit this property in metals. At the white-hot temperature of 1,000 degrees C, mantle rock becomes malleable and freer to creep. Heat moves faster in the mantle by creep than by conduction and sets the rock rolling in convection cells.

The convection cells turn over in the upper 700 kilometers of the mantle, with its anomalous phase transitions. Given the 40,000-kilometer circumference of the Earth, this is a shallow zone. From mid-continent to mid-ocean, a typical cell reaches several thousand kilometers. The cell and the convection turning over in it to a depth of perhaps a few hundred kilometers must be seen as horizontal; the boiling goes sideways, not vertically as in a kettle. The rock in the bottom third or so of the cell creeps laterally toward the mid-ocean ridge, where it meets the convection cell from the other side of the ocean. Moving upward—if molten, streaming upward—the two cells yield heat and deliver new rock to the crust. Away from the ridge the rolling convection cells drag the tectonic plates and the solid layer of mantle underlying the oceanic crust. Softening of the upper mantle, in this so-called asthenosphere, apparently facilitates horizontal motion in the cells. At around 70 kilometers depth, the increase in temperature overtakes the rise in pressure. Seismic waves do not recover velocity until 130 kilometers farther down.

The convection cells roll over, of course, at geologic speed—at the rate of a few centimeters a year. Instruments aboard Earth satellites show the North American continent moving northwestward at that rate.

The protoplanet that was to be Earth grew by accretion of plan-etesimals, swept up by the lengthening reach of its gravitational field. Upon arriving at a critical mass, it went into gravitational collapse. To the immense heat generated by the collapse, the long- and short-lived radioactive elements in the rock added the heat of their decay through their first combined half-life. The temperature reached the melting point. In the symmetry of its gravitational field the molten planet fell into a sphere. Over time, millions of years, gravity brought the molten iron and other heavy elements trickling through the melt to settle in the core. The lighter elements migrated correspondingly toward the sur-face. As the surface cooled, a first crust of basaltic mantle rock, leavened with lighter elements, crystallized. In crystallizing, it outgassed the first atmosphere and a shallow ocean that covered the entire surface. Convection cells, smaller and more numerous and churning deeper in the softer mantle, bulged the crust outward and pulled it inward.

The Hadean period

The parting of the waters and the emergence of the continents were now mediated by interplanetary forces. In this Hadean period, the planet was still sweeping up planetesimals within range of its gravita-tional field. The larger meteoritic impacts punctured the thin primordial crust. From the molten interior the first continental crust, more mafic than sialic in composition, welled upward in volcanic eruption. The "cratons," nuclei of the continental shields, thus formed brought about 20 percent of the Earth's present land area above sea level. Around them, in the course of the next 2.5 billion years, the generation of con-tinental crust from below built the continents.

Dating of the Moon rocks assigns most of the cratering of its sur-face to the first 500 million years of solar-system history. The persistence of craters on the Moon and on Mercury, Venus and Mars testifies that their interiors are not stirred to the turnover that shapes and reshapes the face of the Earth. In a few extinct volcanoes, Mars shows signs of ancient geologic activity. Moreover, a Mars orbiter has recently detect-ed remanent magnetism in the surface of the planet, evidence that Mars once had a magnetic field. If so, the mass of the smaller planet did not sustain temperature to keep its core molten beyond its first billion years. Satellites and robot landers will no doubt continue to search for signs of the life that may have started up in those years—and for life that may persist in waters in the Martian crust.

The 65 meteoritic craters—or "astroblemes"—catalogued on the Earth's present surface date from recent times. As will be seen, impacts of extraterrestrial objects have played a role in the evolution of life.

The clock of radioactivity sets an approximate date for the formation of the Earth's continental crust. In the sedimentary rock at Isua in Greenland, 3.8 eons old, were found grains of zircon. This semi-precious stone is notable for hardness and resistance to abrasion. It is lower in iron and manganese and higher in light metals and silicate and so a distinctly continental mineral. The ratio of uranium to its daughter lead in the Isuan zircon dates its crystallization to 4.4 billion years ago. That places the freezing of the earliest continental crust within the first 200 million years of Earth history.

The oldest rock

Zircon grains dated to 4 billion years ago have turned up in ancient sediments and conglomerate rock on the continents of North America, Australia, Antarctica and Africa. This is evidence that cratons of continental crust from earliest times have nucleated the growth of those continents. Still other continental landmasses must have had their beginning at the same early time, but little of the ancient rock has been found. Uniformitarian processes, weathering, sedimentation and conglomeration, have reworked the earliest rock and buried it kilometers deep under later overlying continental crust.

The archetype of cratonal rock is the so-called greenstone of the Laurentian shield, exposed over large regions in Arctic and sub-Arctic Canada by the last glaciation. It is a recycled conglomerate of material younger than the primeval rock, but the most widespread of the oldest. It is similarly exposed, owing to ancient glaciation, on the Australian continent. More basaltic than granitic, more mafic than sialic, these pavements look from the air like the beginning of time.

The most thoroughly investigated of the ancient cratons is the

Kaapvaal craton underlying the Witwatersrand district of South Africa. This region is well studied because its rocks have yielded half of all the gold—40,000 metric tons—in human hands. Gold is "siderophile"— iron-loving—and its deposits are associated with the iron and manganese of mafic rock. Since the Witwatersrand gold rush of the 1880s, the ransacking of a cumulative—not all in one place—12-kilometer depth of rock laid down on this craton over the first 3 billion years has yielded priceless insight into the history of the Earth and the evolution of life.

Zircon in the metamorphosed sediments certifies the Kaapvaal craton as primeval. In later Kaapvaal sediments, at the Fig Tree site, Elso Barghoorn identified the oldest fossil cells, reliably dated to 3.2 billion years ago. The Kaapvaal sediments have since yielded evidence of life (in the skewed ratio of carbon-12 to carbon-13) dated 200 million years earlier, to 3.4 billion years ago. Preston Cloud took this to be the signature of the first in the succession of ecosystems that he traced through the Kaapvaal sediments laid down over the next billion years.

To a conscious observer, 3.4 billion years ago the Earth would have presented a mournfully empty seascape. The shallow endless ocean was scarcely ruffled by the winds of the thin atmosphere. It was stirred principally by the rise and fall of the tides in the course of the eight-hour day of the then-faster-rotating Earth. Low-lying barren cratons broke the surface at lonely removes from one another. It was a murky ocean, holding elements vaporized and dissolved from the crust, especially manganese, iron and sulfur. Inflow of juvenile water from undersea hot sulfur springs and volcanoes at the boundaries of tectonic plates kept the mineral content constant against its settling out in the sediments. Gaseous exhalations of volcanoes under the ocean and ashore renewed the atmosphere of carbon monoxide, carbon dioxide and nitrogen. Downwind from volcanoes, the atmosphere carried hydrogen chloride and nitric and sulfuric acid, all borne in steamy clouds of water vapor.

Rarest of all elements in the atmosphere was oxygen. The Sun's unshielded high-energy radiation, continuously dissociating atmospheric molecules of water, did free atoms of oxygen. Those atoms instantly reacted with hydrogen again or with one or another of the unoxidized, naked or reduced elements everywhere in the atmosphere and in solution in the ocean. Oxygen vanished so instantaneously that its presence in the atmosphere could not have exceeded 10^{-14} of its now prevailing atmospheric level. Living tissue generated the present atmosphere, 21 percent of it oxygen, in its evolution over the next 3 billion years.

Where on Earth life began will likely never be known. When life began will remain uncertain. The Isuan and Kaapvaal sediments are the oldest known to hold rudimentary signs of life. The same sources of energy acting on the same atmosphere—and infalling meteorites— would have tainted the ocean everywhere with abiotic organic compounds. In shallow and impounded water at the shorelines of the cratons, evaporation would have increased the concentration of those compounds. Proto-organisms antecedent to the common ancestor could have been under development at many sites.

Ecosystems flourishing today in the environment of hot sulfur "black smoker" springs along the mid-Atlantic ridge and the East Pacific rise suggest another site where life may have had a start. These communities teem with eukaryotic organisms—krill, clams and tubeworms. Far out of reach of sunlight, they are sustained in part by the detritus of life and oxygen dissolved in the water from the world above. They receive sustenance also from chemoautotrophic bacteria, serving in the role of plants. If ecosystems of such bacteria had developed on the ocean floor at the dawn of life, the turnover of the oceanic crust would have erased all trace. The argument for their existence then is the communities they support on the ocean floor today.

Manna from heaven

When and wherever life started, it must have been confined at first to local ecosystems. Kaapvaal sediments dated from more to less ancient times show increasing presence of life in the changing ratios of carbon isotopes. In accordance with then-prevailing ideas about the origin of life, Preston Cloud conjectured a first ecosystem of heterotrophs, nourished by the same ready-made organic compounds in which life must have originated. The life of this ecosystem would have been extended—if it did not begin—with the arrival of autotrophs on the model of the anaerobic Archaebacteria. From hydrogen, naked or in compound, with sulfur in HS_2, for example, these chemoautrophs secure electrons to push the energy-requiring uphill phosphorylation of ADP to ATP. The energy in that third phosphate bond went to fix carbon in the primary organic compound CH_2O to sustain their anaerobic fermentative metabolism. On downstream products of this inefficient process the community of heterotrophs presumably thrived.

In the further skewing of the carbon-12 isotope ratio, in sediments dated to 3.35×10^9 ago, Cloud found the onset of the second Kaapvaal ecosystem. By this time, some chemoautotrophs had begun to utilize the energy of sunlight in manufacture of their nutrients. From their living fossils, it is deduced that they would have been equipped with chlorophyll*a*. With this molecule they enlisted the energy of sunlight to race the chemoautrophic reduction of carbon with hydrogen donated by HS_2. Through the downstream products of their speeded-up metabolism, they brought sunlight to the support of the ecosystem.

Oxygen crisis

For the arrival of ecosystem 3 in the Kaapvaal craton, Cloud's benchmark was the appearance of the first banded-iron formation in the sediments. The free oxygen captured in oxidation of the iron in these formations can have no origin other than as by-product of the aerobic noncyclic Photosystem II [see page 281]. As their descendants now can switch to HS_2 when and where oxygen is short, these first true photoautotrophs could capture hydrogen from H_2O instead of H_2S to reduce carbon from carbon dioxide in the air. On water and in air, they could live anywhere in the world. From the local ecosystem they required only the essential trace elements that they could not themselves fix in organic compounds from the environment.

Cloud called BIF (banded-iron formation) the "geologist's DNA." BIF begins to appear all around the world in rocks laid down before the end of the Archaean, 2.5 billion years ago. In Kaapvaal sediments, they make their appearance around 3.3 billion years ago. This momentous development followed by no more than 100 million years the first appearance of life on the Kaapvaal craton. Here, barely discernible in the sediments, was a beginning of the self-sustaining, mutually supportive community of life that was to remake the surface of the entire planet. The only earlier BIF is known from the Isuan sediments.

By liberating free oxygen into the environment, the first bacterial photosynthesizers menaced every living thing around them. Against this poisonous element, all familiar living organisms are now protected by internal enzymatic defenses. Built into every cell, these defenses instantly bind stray atoms of free oxygen.

At the onset of aerobic photosynthesis, primitive life had no built-in protection against free oxygen. Protection was provided, however, by the finely divided iron particles vaporized from mantle rock that had accumulated in abundance in ocean waters and sediments. Capture of oxygen in iron oxide took the poison out of circulation.

Perhaps it was with periodic exhaustion of the supply of iron in local water that a population of oxygen-releasing photosynthesizers died back; with seasonal upwelling of bottom water to restore the supply of iron, the population would revive. In the interim, oxygen releasers switched to anaerobic photosynthesis, as their descendants managed to do today [see page 295]. Such periodic oscillation explains the lamination in banded-iron formations.

Encountered all around the world today, banded-iron formations testify to the success, all at once, of aerobic photosynthesis, to the abundance of iron in the waters and to the time required for life to secure its defenses against oxygen. Industrial civilization has scraped off only the top layers of BIF laid down on all continents beginning 3.3 billion years ago. In deposits exceeding 10^{14} tons, iron is stored in the Kaapvaal craton, in the Hammersley and Nabberu basins in the Australian outback, in the São Paulo state of Brazil, in eastern Canada and in Ukraine. The industrial revolution has thus far dug its iron from mere 10^9-ton deposits more conveniently located in the Lake Superior region of the U.S. and Canada and in western Europe.

Against the menace of oxygen, primitive bacterial algae developed the now universal enzymatic defenses. Soon, one or more lines improved on that strategy. They engaged the by-product oxygen in aerobic metabolism of the product of their aerobic photosynthesis. The first such metabolic cycles fell short, undoubtedly, of the efficiency of the

Krebs cycle [see page 285]. The improving efficiency of aerobic metabolism is nonetheless evident in signs of life everywhere in the last 100 million years of the Archaean. Colonies of filamentous Cyanobacteria appear in stromatolite macrofossils on all continents. On the Kaapvaal craton, Cyanobacteria may have appeared as early as 3.2 billion years ago, if the Elso Barghoorn dating of the Fig Tree fossil cells stands. This would date the beginning of Cloud's fourth ecosystem.

By the end of the Archaean, that ecosystem was enveloping the planet in its first biosphere. The prokaryotes, in the judgment of Lynn Margulis, had by then evolved all of the diversity of metabolisms they now practice. Random mutation, in the right places at the right times, brought new resources in reach of their metabolism. Bacteria fix in organic compounds all of the elements from the lithosphere that presently contribute to the life processes of eukaryotic organisms, such as calcium in bone; copper in enzymes; iron in hemoglobin; magnesium in chlorophyll; potassium, sodium and lithium in the electrolytes; and, in trace amounts and other services, still other elements. Teeming in the floor of the oceans and its waters and making the fresh waters of the land agreeable to their habitation, they were cycling the elements of water, air and soil as they continue to do today.

First planetary ecosystem

With aerobic metabolism of the product of aerobic photosynthesis, the prokaryotes had crowned their biochemical innovation. By that innovation they began the transformation of the atmosphere. From 10^{-14} they raised the concentration of oxygen by 11 orders of magnitude to 10^{-3} of the present prevailing level by the end of the Archaean. In the Cyanobacteria, they had given rise to the class of dominant organisms that was to hold sway for the next billion years, longer by many 100 millions of years than any other organism can be said to have dominated the Earth. If there was an Age of the Dinosaurs, the Proterozoic was the Age of the Cyanobacteria.

The arbitrary boundary that divides the 4 billion years of the pre-Phanerozoic or Precambrian between the Archaean and the Proterozoic does mark a turning point in Earth history. In the course of the Archaean, the cooling of the Earth by the churning convection cells had been carrying the lighter elements distilled from the deep mantle rock into the upper mantle. In exothermic reaction with one another, these elements melted out in great pools of magma. Preferentially, the pools formed under the heat-conserving insulation of the cratons. In the closing hundreds of millions of years of the Archaean, volcanic eruption and massive plutonic upwelling flooded "modern" sialic granitic rock into and over the primeval mafic basalt. The Proterozoic opened with 60 percent of the present continental crust in place. There followed an intermission of perhaps 200 million years in the generation of new continental crust. Then, another spasm of plutonic upwelling set in and, over the next 250 million years, brought a near-50 percent increase in the volume of continental crust, to 85 percent of its present volume.

The start of plate tectonics

The extrusion of continental crust in the Archaean brought the cratonal landmasses to something like continental size. They were large enough to blanket the escape of heat and bring the upwelling of the mantle convection cells underneath them. Then began the cycle of plate tectonics that ever since has been redrawing the map of the planet. Mantle convection had been moving the cratons about earlier in the Archaean. The radiometric clock and compass aboard the Kaapvaal craton show it in random motion between 72 degrees north latitude and the equator as its first ecosytems succeeded one another, wandering north again to latitude 30 around 3 billion years ago and back to the equator as the Proterozoic opened.

Early in the development of the theory of plate tectonics, J. Tuzo Wilson of the University of Toronto made a pertinent observation. He saw that heat must accumulate under the middle of a large continent and, especially, a supercontinent. The middle of the nearest ocean would be too distant to establish the outward heat-flow gradient at the bottom of an ocean-rifting convection cell. The convection cells set in motion by the flow of heat under the middle of the supercontinent would roll upward, therefore, rather than downward. Rolling in opposite directions, they would eventually rift the crust and, in time, tear the continent apart. The opening of new oceans between fragments would drive them into collision with other continental plates. In the fullness of time, Wilson predicted, the fragments would reassemble in a new super-continent. The Wilson cycle now stands confirmed by evidence of the assembly and breakup of a number of large continents over the ages and of at least two supercontinents that incorporated virtually all of the planet's landmass during the most recent billion years.

Around the landmasses, at whatever latitude, the vast increase in extent and volume of shallower onshore ocean waters correspondingly enlarged the environments in which life throve. The stromatolite macrofossils of Cyanobacteria turn up in increasing abundance in the rock of all the continents.

Evolution proceeded slowly through these ages. In the Archaean the prokaryotes had established their versatile biochemistry. The eons of the Proterozoic saw no increase in their size or elaboration of their form. They ranged from round coccoid to linear spicule, with variety supplied by the undulating spirochete. Prokaryotes in each local community con-certed for their mutual support their respective genetic capabilities.

"A bacterium," Sorin Sonea has observed, "is not a unicellular organism; it is an incomplete cell." The "complete cell," in this under-standing, is the community of bacteria, each exploiting its niche in the local environment, contributing the downstream products of its metabo-lism to the support of the community and reciprocally enjoying what the rest of the community had to contribute to its support. Taken together, the community is a chimera. For the local chimera, the decomposers of the lithosphere extracted such essentials as calcium, potassium, sodium and silicon, as well as elements required in micro and trace amounts. The photosynthesizers supplied organic molecules. Where essentials were scarce, members of the community concentrated them by huge multiples of their local availability. Prokaryotes have been shown thus to concen-trate manganese 1.2 million times and iron 650,000 times their respec-tive concentrations in the immediate environment.

The prokaryote communities raced to the maximum the turnover of the elements taken up in their substance and metabolism. Everywhere, the density of life would be limited at last by the element in shortest local supply. The saturated environment exerted in return the pressure of natural selection. That pressure the prokaryotes eluded by the transfer of genes that proceeds so easily among them. Speciation among the prokaryotes is not only uncertain but mutable [see page 306]. The variable, adaptable, chimeric prokaryote superorganism made the most of whatever was at hand around the planet as it proliferated throughout the ocean.

The ubiquitous Cyanobacteria, with command of energy from the Sun, secured the supply of carbon compounds to local ecosystems wherever they found sunlight, air and water. They in turn acquired by genetic transfer the biochemical capacities of the indigenous local specialists. The range of capacities qualifies this group of prokaryotes down to present times for the title of "unexcelled ecological generalists." They are so designated by J. W. Schopf and colleagues in their comprehensive *Earth's Earliest Biosphere:*

> Cyanobacteria are extremely difficult to exterminate.... *Nostoc commune* has been revived after 107 years of storage as a dried herbarium specimen. *Oscillatoria princeps, O. subtillissima,* and *O. minima* have been reported to survive immersion in liquid helium at −269 degrees C for 7.5 hours; several cyanobacterial species have been shown to survive nuclear test-site explosions within a distance of about 1 kilometer from ground zero, the radius from which surface soil was not completely stripped away by the nuclear blasts; among these nuclear survivors was *Microcoleus vaginitus,* a species which also survived experimental cobalt exposure ... two orders of magnitude greater than that lethal to eukaryotic microalgae.

Oxygen in the atmosphere

By 2 billion years ago, the cycling of air and water by the worldwide population of Cyanobacteria raised the concentration of oxygen in the atmosphere another order of magnitude, to 10^{-2} of the prevailing level. The presence of that much oxygen reflected, as well, progress in the development of internal enzymatic defenses among heterotrophs dependent on photoautotrophs. Free oxygen now swept the oceans clean of iron. Deposition in banded-iron formation gave way to deposition of iron in red beds and the secondary oxidation of the BIF. Anaerobic prokaryotes retreated to refuge in niches they now occupy, excluding, of course, those provided later on by eukaryotic hosts.

Enough oxygen was moving into the upper atmosphere to begin the veiling of the planet in its ozone layer. The ozone layer reduced the intensity of high-energy radiation reaching the Earth's surface and opened habitats to an increasing variety of life in the sunlit "photic" zones. At the same time, the rising concentration in the atmosphere of carbon dioxide, the by-product of oxidative metabolism, began to block the reradiation of sunlight from the Earth's surface. That and increase in conservation of the flow of energy from the Sun offset the decline of heat flowing from within the planet.

In the sediments laid down 1.5 billion years ago are found cells distinctly larger than those of the prokaryotes of the 2 billion years preceding. Over the next 300 million years, these big cells became more common. With the appearance of so-called acritarchs they diversify in form, developing spiny exoskeletons. By 1.2 billion years ago, there is no doubt that eukaryotes have put in their appearance. Evolution has begun to accelerate from the pace set by the prokaryotes over the preceding 3 billion years.

In the genesis of the eukaryote, evolution had found a shortcut, as related in Chapter 5. The eukaryote is a chimera, the outcome of symbiotic prokaryote existence that went on to the fusion of the mutually dependent cells. The heterotrophic eukaryotes—the cells of animals and fungi—fuse in their anatomy two different prokaryotes. By the same reckoning, the autotrophic eukaryotes—the plant and algal cells—incorporate three. The mitochondrion, one of the ancestral symbionts transformed to a eukaryotic organelle, conducts the intricate cycle of aerobic respiration. Installed in a eukaryotic cell in numbers proportioned to its energy requirement, mitochondria supply it with the energy output of much the same number of aerobic prokaryotes. The eukaryote embodies in a single cell the intercellular symbiotic association of the prokaryotes that had so well prepared the Earth for its arrival.

The size alone of those first giant—as compared to the prokary-

Cyclic turnover of atmosphere and hydrosphere is significantly promoted by the biosphere. Oxygen in water, H_2O, delivers hydrogen to photosynthesis of glucose, CH_2O, in plants and goes to the atmosphere [see page 280]. From respiration of animals and plants, it returns to the atmosphere in CO_2 [see page 285]. In high-altitude ozone layer, absorption of high-energy radiation in the cycling of O_2 through O and O_3 (ozone) protects life in the biosphere below. The oxygen in the atmosphere is maintained there by life processes and would speed-

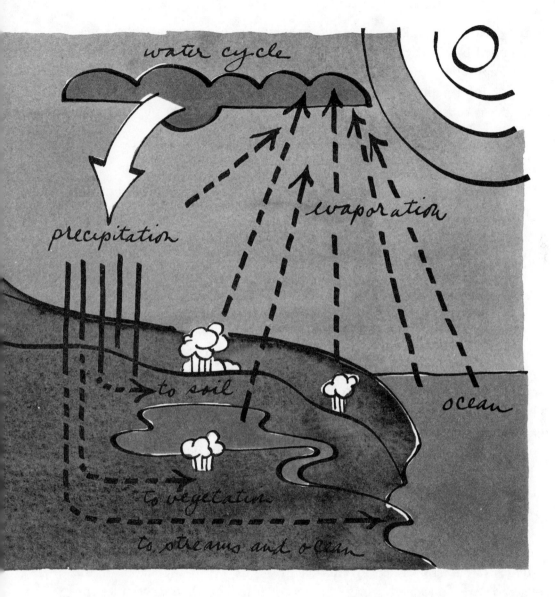

water cycle

precipitation

evaporation

to soil

ocean

to vegetation

to streams and ocean

ily vanish in their absence. Carbon cycles in CO_2 from the atmosphere through photosynthesis to CH_2O and through respiration to CO_2. Decay of plant and animal tissue in soil returns much CO_2 to the atmosphere. Human activity amplifies carbon turnover by 25 percent. Oceans hold 97 percent of Earth's 1.5 billion cubic kilometers of water; 75 percent of fresh water is locked in polar ice and mountain glaciers. Through the .03 percent in atmospheric vapor, solar energy drives the hydrologic turnover between ocean and land.

otes—cells of 1.5 billion years ago suggests their identity as eukaryotes. Equipped by a nucleus and a mitotic spindle to manage the complexity of its replication, these cells possessed an open-ended capacity to acquire genetic information. From these organisms stemmed four whole new kingdoms of life to complement the ancestral prokaryotes in enlarging and diversifying the biosphere.

Four new kingdoms of life

The eukaryotic kingdoms arose abruptly. The first paleontologists were in awe of the sudden appearance of fossils in the Cambrian rocks, now dated to 570 million years ago. They consigned all older strata to a single "Precambrian" era of mystery and unknown duration. The efflorescence of the kingdoms has now been traced in fossils laid down in the 100 million years or so preceding the Cambrian.

In the five-kingdom classification scheme, the prokaryotes are designated by their more familiar name, Bacteria. Not-elsewhere-classified eukaryotes—the amoeba and the plasmodium of malaria, for example—are distinguished as the Protoctista. In direct line from the first eukaryotes come the multicellular Animalia and Fungi. Neither of these heterotrophic kingdoms could exist, of course, in the absence of the fifth kingdom, that of the autotrophic Plantae. Not even the ubiquitous, protean Cyanobacteria could satisfy their enormous demand for oxygen. The plants also supplied in abundance the nutrient glucose, the fuel burned by cellular respiration.

The presence of Animalia, such as the amoeba, as well as Plantae among the Protoctista betrays this kingdom as a catchall. It includes also analogues of the Fungi. Improving acquaintance with its membership suggests that still other inventions from the symbiotic habit of the Bacteria are concealed there. Some are fitted by their ancestral symbionts to share niches with all but the hardiest prokaryotes. Their chimeric endowment suggests multiplicities of ancestral collaborators. Some Protoctista, such as the kelp and red algae "seaweeds," are multicellular and well known to the beachgoer.

Toward diversification of form—as different as the mite from the whale—the evolution of the early eukaryotes made a number of critical transitions. One was sexual reproduction [see pages 304]. Genetic exchange, until then, had proceeded largely at random. In random exchange, as the relatively featureless planet of the prokaryotes testifies, the trend is to homogeneity. Sexual reproduction channels the flow of genes through individual organisms. The shuffling of parental genes in the fertilization of the egg ensures that no two offspring (with the exception of twins from a divided fertilized egg) come exactly alike. The trend, in consequence, is to diversity. To populations sex secures the dissemination in the gene pool of traits that are favored by selection in a given and then in a changing environment.

With chromosomes to contain growing libraries of genetic information, the eukaryotes opened a new era of innovation in biochemistry. They fashioned molecules unknown in the prokaryotes' repertories. One and then another line of cells began synthesizing large polymers. An early innovation was chitin: a tough cuticular material composed of polysaccharide chains bound in a matrix of protein. Chitin supplies structural material to the most distantly related invertebrates. It constitutes the exoskeletons of the arthropods—that is, all of the insects, spiders and crustaceans. It is the substance of the bristles, hinges and shell covers of snails and clams and other mollusks and of scallops and other brachiopods. The appearance of chitin in the cell walls of fungi and even green algae indicates this polymer was most likely first synthesized by a common ancestor. The spiny exoskeleton implicates the acritarch. For the vertebrates, keratin does much the same service as chitin. A protein polymer, it is the material of hair and fingernails, hooves and horns, turtle shells and baleen whale "bones."

Some forerunner(s) of the plants began the manufacture of cellulose, a polysaccharide. Later on, emerging plants synthesized lignin, the glue that binds cellulose fibers in wood. These two substances hold the green leaves up to sunlight. They now constitute 50 percent or more of the total, planetwide eukaryotic biomass.

An early innovation in eukaryote biochemistry was the incorporation of calcium carbonates and calcium phosphates in organic molecules. Coral reefs, the White Cliffs of Dover and chalk deposits around the world testify to the scale of this activity among the protoctists.

"Biomineralization" produced the first hard fossils, the housing and exoskeletons of the mollusks and other marine invertebrates. Later, it went to build the internal skeletons of the vertebrates.

Much of this innovation in biochemistry must have been occasioned by the most crucial innovation in eukaryotic cellular anatomy after the nucleus: the development of the cell membrane [see page 274]. No mere container, the membrane is an active organ that relates the cell to the world around it.

First predators

To the planet of the prokaryotes, where evolution was stalled in homogeneity, the eukaryotic membrane brought a revolution. The flexibility of the membrane allows intake of nutrition by phagocytosis. This is the engulfment of particles, as contrasted to the osmotic absorption of molecules to which most prokaryotes are confined in their stiff and waxy capsules. The membrane in contact with the particle or prey invaginates, drawing prey or particle into the cell's interior. By phagocytosis, a eukaryote could swallow up a whole live prokaryote. The eukaryotic cell membrane brought predation into evolution.

Predation is a force for diversity. It is not, especially when mediated by a membrane, "Nature red in tooth and claw." Nature must be seen rather as a concert of mutually fructifying populations of organisms, difficult as suppression of anthropomorphic impulses may be.

Paleontologists have called the sudden appearance of multicelled organisms in the Cambrian period "the foremost unresolved problem of paleontology." The eukaryotic heterotrophs, it is clear now, solved that problem. By feeding on the blue-green algae, they opened the way for the multicelled kingdoms of life. That is the persuasive explanation of this turning point in the history of life advanced by Steven M. Stanley of Johns Hopkins University.

It is a principle of the new science of ecology that "cropping"— a more neutral term than predation—brings diversity to an ecosystem. Observation shows a close relationship between the diversity of herbivores in tropical rainforests and the high diversity of vegetation there. On the Gondwanaland continents, last glaciated 200 million years ago, these ancient forests have had time to arrive at high diversification of plant species and their predators. Trees tend to grow near parents. Herbivores specialize, however, in host plants. In tropical rainforests, each herbivore compels the observed dispersion of the tree of its specialization, hence the remarked-upon diversity of tree species per hectare. Herbivore insects and parasite fungi of the *Hevea* rubber tree interdicted its plantation in its native Amazonia. Again, tourists learn that the Serengeti Plain owes the diversity of its ungulates to the diversity of the cats—lions, leopards, cheetahs—that prey on them.

Experiment confirms such observation. The removal of the principal carnivore in a tidal basin results in immediate reduction of diversity among the herbivores there. Safe from the predator, the most successful herbivore(s) crowd out the others. Similarly, removal of grazing sea urchins brings the takeover of the territory by the dominant algal species. In random encounters, the carnivore and the herbivore consume more of their more numerous prey, opening ecospace for its competitors. Such turnover in ecosystems proceeds so reliably as to find generalization in nonlinear equations.

The Cyanobacteria took over the oceans when there was no predator capable of gulping them whole. Significant evolutionary change in the world of the bacteria transpired over billions of years. The arrival of the heterotrophic eukaryotes and their multicelled progeny set evolution on a course of accelerating diversification to the peak of diversity that not long ago welcomed the arrival of the human species.

First multicelled organisms

Multicelled organisms make their now earliest known appearance in sediments deposited perhaps 700 million years ago—in the last 130 million years of the Precambrian—along the western shores of the Laurentian supercontinent. Those sediments are upheaved in the McKenzie Mountains in the Northwest Territories of Canada just below the Arctic Circle. The fossils of the organisms, discovered and brought under study there by Canadian geologists in the late 1980s, occur in strata through a depth of 2.5 kilometers. They richly set out the evolution of this fauna over the next 100 million years.

352 THE AGE OF SCIENCE

McKenzie life-forms, especially later ones, have been found at 25 different sites around the world, from Spitzbergen, in the Arctic Ocean, to Australia. After the site at which they were first discovered in Australia in 1949, they are known as the Ediacara fauna.

Other evidence in the rocks compels a finding almost as astonishing as the appearance of multicelled life. It shows that this surge in evolution transpired in an epoch when the Earth had turned more hostile to life than at any time since life began. Everywhere, the debris of a worldwide glaciation underlies the fossil-bearing sediments. After its Spitzbergen site, it has been called the Varanger glacial epoch.

Earth's longest winter

Recent study of its aftermath in the Congo craton has established this glaciation as the Earth's longest winter. Pack ice floated in the oceans in the middle latitudes. At the edges of continents in those latitudes, glaciers were grounded on the seafloor, as the East Antarctic glacier is grounded today. The Earth was a "snowball."

At the McKenzie site, this glaciation *overlies* the sedimentary rock that bears the earliest fossils of 700 million years ago. Underneath that stratum lies the debris of a still earlier worldwide glaciation.

As the evidence of these glaciations was accumulating in the 1950s, paleomagnetism began to plot the migration of the continental landmasses during the late Proterozoic. The clock and compass in the rocks showed that all the major landmasses had been assembled a billion years ago in a single supercontinent. With Laurentia at its center, this so-called Rodian supercontinent lay athwart the equator. The Antarctic craton, then north of the equator, abutted Laurentia on its western flank. The western flank became the western shore of Laurentia upon the drifting off of the Antarctic craton and the opening of an ocean between them. On that shore the fossil-holding sediments, now high and far inland in the McKenzie stretch of the Rocky Mountains, were laid down [see illustration, pages 354–355].

With the breakup of the supercontinent there began a prolonged ice age. Millions of years of glaciation alternated with millions of years of thaw, terminating in the severe and prolonged glacial episodes that coincided with the first appearance of multicelled life. Through all of that time, the clock and compass in the rocks show, the principal land-masses drifted in the middle latitudes. Geologists had now to answer the question: how could glaciers have taken over continents lying within 30 degrees north and south of the equator? The answer comes from biology as well as Earth physics. It shows again how mortally life on the planet hangs upon the presence of life on the planet. W. Brian Harland of the University of Cambridge proposed the answer in 1964, early in the accumulation of evidence. In 1992, Joseph L. Kirschivink extended it, in the form now confirmed by the evidence.

The Sun's radiation warms the Earth principally by its absorption in ocean water of the middle latitudes. Through the late Proterozoic ice age, sunlight-reflecting and -rejecting landmasses displaced a critical area of heat-accumulating water in those latitudes. That reduced global temperature. Concurrently and paradoxically, the flourishing of life reduced the global temperature some more. The breakup of the supercontinent brought continual increase in shoreline. That meant an enlarging habitat for life. The intake of carbon dioxide from the atmosphere by photosynthesis—and the ultimate burial of the carbon in the ocean floor after passage through the ecosystem—overtook the injection of carbon dioxide into the atmosphere by volcanoes. The collapse of the greenhouse brought the temperature, even at the equator, below freezing. The first snows increased the planet's reflectivity, its albedo. In runaway positive feedback, increase in albedo further decreased the planetary temperature, spreading glaciers worldwide.

With the near shutdown of life, active volcanism attending the breakup of the supercontinent restored the carbon-dioxide greenhouse. The icehouse melted and the cycle started over again.

This scenario found spectacular confirmation in the record of the terminal glaciations exposed in the Congo craton. Paul F. Hoffman of Harvard University and his colleagues reconstructed the story, especially of the second glaciation, in 1998. The carbon isotope ratios in sedimentary rock held the principal clues.

In the periods preceding the glaciations, the ratio reflects the vigor of life. Carbon fixed in organic compounds constituted as much

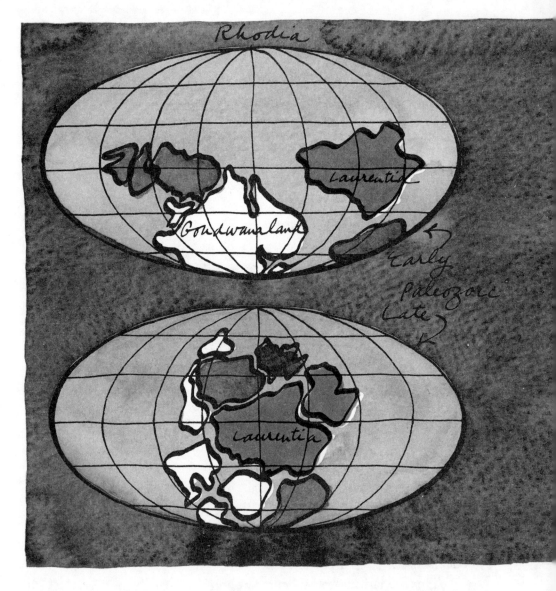

Supercontinents of 800 and 200 million years ago *were sites of crucial developments in the evolution of life. In the more ancient, called by some the Rodian supercontinent, the cratons, primordial continental nuclei, of South America, Africa, India and Australia assembled for the first time in Gondwanaland. The planet's first ice age brought glaciation from south polar region but not from the craton of Antarctica, which then crowded the northwest coast of Laurentia. On that coast, now in the McKenzie region of the Canadian*

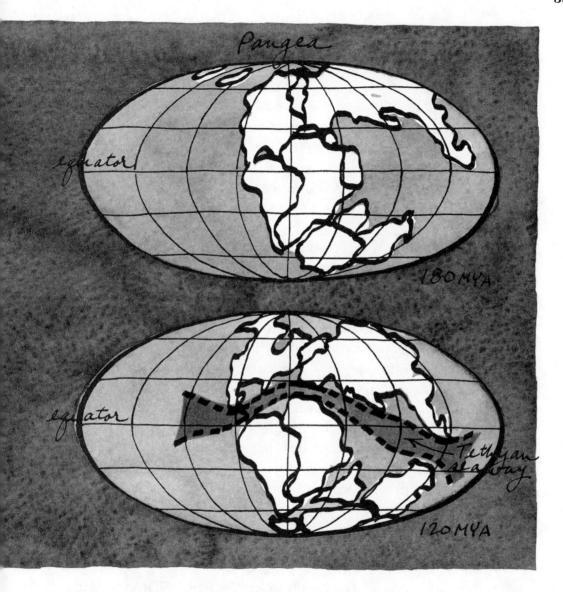

Rockies, the first known fossils of multicellular organisms that were to give rise to plants, animals and fungi were laid down 700 million years ago [see page 357]. In Pangaea, present continents can be recognized. The therapsids, ancestral to reptiles and mammals, dominated the intact continent in a great variety of species that penetrated every ecosystem from pole to pole and both shores of the world ocean, Panthalassa. Dinosaurs flourished in the tropical Tethyan seaway that opened with the breakup of Pangaea [see page 367].

as half the carbon in the seafloor. Just before the glaciation, the ratio registers abrupt decline in that contribution until, locally at least, life processes must have shut down. No sediments younger than the earliest Proterozoic have shown ratios establishing such exclusive deposition of inorganic carbon. The decline, in parallel, of sedimentation from continental erosion points to shutdown of the hydrologic cycle, of the weather itself. Water vapor had frozen out of the atmosphere.

The McKenzie sediments bear special evidence of the return to early Proterozoic times. Banded-iron formations were deposited once again, after nearly 2 billion years, in the sediments there.

Back in the Congo craton, the next overlying strata bear dropstones and other debris scoured up by the glaciation and deposited in its retreat. The second postglacial deposition is blanketed to great depth by inorganic carbon. The carbonate "capstone" proclaims the abrupt return of the greenhouse. With life still at low ebb, carbon moved from the mantle through the volcanoes directly to burial in the seafloor. Its deposition was accelerated by renewal of the hydrologic cycle and continental erosion. Worldwide, local measurement indicates, the mantle yielded enough carbon to bury the planet in carbonate to a depth of 5 meters. Up through this deposition, the isotope ratio moves slowly in favor again of the burial of carbon processed through living cells.

The best estimates indicate that these glaciations lasted 9 million years. It is, perhaps, not difficult to account for the persistence of prokaryotes through this period. The eukaryotes must have found warm refuge somewhere, perhaps along continental shores heated by volcanic and tectonic activity.

Prokaryotes in retreat

The entire record would seem to lend weight to Baron Cuvier's scientific catastrophism [see page 321]. It does advance another ecological explanation—to go with cropping—for the sudden surge in the diversity of living forms. The 9-million-year freeze-up cleared the oceans of the monopolizing Cyanobacteria. To the surviving diversity of organisms—including the Cyanobacteria, of course—the thawing planet presented something like a level playing field. Eukaryotic heterotrophic predators were there to keep the monopolists at bay. Formerly pre-empted niches opened up to the new eukaryotic autotrophs, the green algae forerunners of the plant kingdom.

The fossils discovered in the McKenzie sediments were there by the rare geological accident that saved the imprint of soft tissue. Biomineralization had yet to make the shells of the marine invertebrates that were to mark the bottom strata of the Cambrian rock 130 million years later. The earliest McKenzie fossils are scarcely recognizable as organisms. Their imprints in the shale and fine-grained sandstone appear as rings, 1 to 3 centimeters in diameter and .2 millimeter in cross section, and disks of about the same diameter with a ring around the circumference and a bump in the middle. A few have radial ridges. They could be taken for the accidental outcome of some mechanical deformation of the clay and sand in which they formed. But they are too numerous, distinguishable and repetitive in form to be thus accounted for. Similarly primitive forms appear in later Ediacara assemblages of more recognizable fossils found in Russia, China, Namibia, Australia and elsewhere. By 650 to 620 million years ago, multicellular life had spread worldwide.

The earliest McKenzie fossils, dated to 700 million years ago, are assigned to three taxa or types rather than three different stages in a life history. These annular and disk organisms surely had their simpler antecedents. The least complex of living multicelled animals, *Trichoplax adhaerens,* barely visible to the naked eye, would leave an even more featureless fossil imprint if it left any at all. The McKenzie preglacial organisms had elaborated past *Trichoplax* to radial symmetry. Like *Trichoplax,* they were doubtless constituted of undifferentiated tissue, perhaps two layers of cells arrayed for absorption of nutrients and oxygen. The McKenzie sample represents an infinitesimal fraction of what may have been a much wider diversity of organisms, although hardly more distinguishable from one another than those in the McKenzie deposit.

Among the postglacial McKenzie fossils are forms found at other Ediacara sites. The brief interval of 50 million years—brief enough compared to the 3 billion years preceding—brought considerable elaboration of form. There are fronds that resemble the sea pens familiar to snorklers today. There are imprints of medusoid, jellyfish-like, organisms. Still more elaborate are the fossils of segmented annelid worms. From their descendants or analogous living forms, it can be deduced that evolution has passed significant milestones. Features of the basic body plan of animals commonly recognized as animals are in place.

Behavior in evolution

The plan starts, as in the sponges, from a cavity enclosed by a tissue of undifferentiated cells. In the "blastula" stage of embryonic development, all animals are momentarily such a cavity. The cavity, in the medusoid, is contained within another cavity, the coelum, and each is of a different tissue. The radial symmetry of the medusoids yields, in the annelids, to the start toward bilateral symmetry that characterizes the animals most familiar to people today. Those worms also appear to display differentiation of tissue into the three types—ectoderm, mesoderm and endoderm, corresponding roughly to the skin, muscle and digestive tract—found in most animals. The worms, moreover, had a skeleton: a hydraulic skeleton of fluid in the coelum surrounding the gut cavity. With their segmented musculature to squeeze the fluid back and forth in the coelum cavity, the worms could burrow in the silt of the seafloor. Behavior now began to play a role in evolution.

Of great importance to future evolution are the Ediacara "trace" fossils. They appear in the sediments more often without than with the fossils of the worms that traced them. Marine worms leave such traces in the ocean bottom today. They and other ooze-eaters conduct a critical phase of the carbon cycle. Ingesting the buried carbon, they return carbon dioxide to the atmosphere. Rendering the same service, their late Precambrian forerunners must have helped to restore the greenhouse at the end of the last glaciation.

Over the years that began with Lyell's collecting, the rocks of the Cambrian period—570 to 510 million years ago—have yielded tons of fossils. All this time, paleontologists have been discriminating among and classifying the life-forms saved in them. Their labor, over generations, now compels a momentous conclusion. The Cambrian congregation of life-forms holds representatives of all but one of the 30 to 40 phyla of animals alive in the world today. All but a few of the invertebrate phyla have stayed where they started, on the sea bottom. Anchored there, many are taken to be plants. The one post-Cambrian phylum, another sessile marine animal, the Bryozoa, is so called for its resemblance to the mossy Bryophyta.

In taxonomy the phylum is the first rank of classification under kingdom. Phyla are distinguished from one another by such significant differences in body plan as can be seen between the vertebrates and all the invertebrates and, among the invertebrates, by the differences among sea pen, mollusk and Bryozoa. Whether the number of phyla is 30 or 40 depends on whether the taxonomist is a "lumper" or "splitter."

It is the new understanding from the Precambrian rocks that makes the consensus of the paleontologists momentous. The evidence establishes that the animal phyla diversified from one another in less than 200 million years from the featureless assemblage of life-forms saved in the earlier McKenzie sediments.

Another treasure trove of fossils suggests that the Cambrian period saw the tryout of a still wider diversity of life-forms than that here to enchant human existence today. This is the Burgess Shale, discovered in the Canadian Rockies near Banff, Alberta, in 1909 by Charles D. Walcott, paleontologist and secretary of the Smithsonian Institution. These fossils were laid down in the sediments on the western shores of Laurentia, some 1,500 kilometers south of the McKenzie deposits and 180 million years later. By another lucky accident, the soft parts of animals now possessing hard parts were preserved. Quarried out of the mountain and taken to the Smithsonian, the Burgess Shale have been under study there and in other institutions ever since. The diverse anatomies, sorted into 120 different kinds, make apparent use of all of the polymers devised by their single- and multicelled eukaryotic forerunners. They challenge observers to identify them with living phyla.

At the University of Cambridge in the 1970s, H. B. Whittington and his students sorted the 120 anatomies into 18 sufficiently distinct and different phyla. Of the 18, they managed to tie 8 to phyla that survive to the present day. All but two of the living phyla are confined to the sea and the seafloor: medusoids, sea pens and annelids persisting from Ediacara times and creatures that can be matched to brachiopods, mollusks, echinoderms, cnidarians and other familiar and unfamiliar aquatic phyla. Present besides are forerunners of the arthropod, the most successful of all phyla on the scales of differentiation and abundance of progeny in and on the sea, the land and the air.

Of anthropomorphic interest are 60 specimens assigned to the phylum Chordata. From that phylum stemmed the eight classes of vertebrates, the 35 orders of mammals, the seven families of primates, the four genera of Hominoidea and the genus Homo.

More than whimsy prompted Stephen Jay Gould at Harvard University to ask what if the Chordata had been one of the 10 Burgess Shale phyla [opposite] not found today. Had the dice rolled another way, he concluded, evolution might not have generated a human or humanlike organism with conscious intelligence. In the Burgess Shale is frozen the action of the blind force from which Darwin recoiled.

Evolution in the "modern synthesis"

Contributors to the understanding of evolution in its modern synthesis have had the advantage of seeing the story, thus, from its beginning. Darwin was constrained to consideration of its outcome. In the new perspective, evolution exhibits the protean force of the Mendelian genetic generation of living forms.

The fossil record is, of course, incomplete. Most animals lived and died in watersheds, where geological turnover may be witnessed in a human lifetime. With bones and other hard parts to bear witness, chance kept the record. The statistical sample discloses nonetheless plainly the central theme. The chapters of evolution that follow the Cambrian reenact, again and again, the original scenario of the differentiation and elaboration of the phyla. Each phylum realized a form that was somehow potential in the nearly featureless fossils of the McKenzie sediments. The phyla radiated new forms from potential more apparent in the Burgess Shale. Thereafter, in successive cascades of enlarging and realized potential, evolution reached the diversity of life-forms now discerned by the latest to arrive.

Given the long history of the planet, the 500 million years of post-Cambrian evolution may be likened to the pyrotechnician's grand finale. Each spark radiates a shower of sparks. The imagery suggested is, in any case, happier than the Malthusian pruning hook in neo-Darwinian popularization of the principle of natural selection.

George Gaylord Simpson was a principal author of the modern synthesis. During the 1930s and 1940s at the American Museum of Natural History, he made the definitive study of the evolution of the horse. It began, late in post-Cambrian history, with the tiny four-toed *Eohippus.* From that beginning, evolution to the single-toed modern, large and useful *Equus* might be taken as linear "progress." In fact, the evolutionary transformation traced by Simpson shows no trend at all. In a succession of "adaptive radiations," numerous genera of three-toed, then two-toed and one-toed, animals occupied different niches "opportunistically" in the environments of North and South America and Eurasia. The radiations came quickly, each in a few million years. They show no trend but adaptation to local environments. The species that persisted stemmed from those that were there to make the adaptation from browsing to grazing as grasslands spread over the northern continents. The extinction of all but *Equus* and the all-but-extinct *Przewalski* horse transpired over much longer times than the initial radiations.

What he found in the evolution of the horse Simpson proposed as a general law. Evolution proceeds at different paces. From the Greek, Simpson drew a vocabulary to distinguish, as *bradytelic* and even *hypobradytelic,* the tempo of evolution in the Proterozoic from its *horotelic* rate in the Phanerozoic. In general, with the onset of a new form and radiation of its genera and species into new ecospace, evolution accelerates to the *tachytelic.* Upon saturation of the ecospace, it slows again to the prevailing *horotelic* rate. Simpson's law is better known today under the title "punctuated" evolution popularized by Gould.

On the Simpson timetable, the arrival of all extant animal phyla before the end of the Cambrian must be reckoned the first of the *tachytelic* radiations. Considering just the vertebrate phylum, the next round gives rise in 175 million years to five of the eight vertebrate classes: the jawless fishes; the jawed, armored and extinct fishes called placoderms; the cartilage fishes; the bony fishes and the amphibians.

Milestones under water

In the jawless fish, the essential tissues and internal organs of the vertebrate were realized. Representatives of two immortal classes from this phylum, the lamprey and the hagfish, survive to be dissected. The three cell layers first seen in the annelid worm had completed their differentiation. As recapitulated in the embryonic development of all vertebrates, the ectoderm had given rise to the epidermal tissue of the skin and to the central nervous system and along with the keratin of tooth enamel, hair and fingernails. From the mesoderm had come the skeleton, musculature and cardiovascular system; from the endoderm, the gastrointestinal system and accessory organs. That wondrous invention, the vertebrate eye, was there in working order. With the arrival next of the jawed fish, the tetrapod plan of the vertebrate was complete. It can be seen that the vertebrates passed all these milestones under water.

Between water and land, the distance was closed by the adaptation of the fish's flotation bladder to service as lung. Lungfish are counted in roadkill on highways through the Everglades today. From them stemmed the amphibians. Starting life still as legless and gilled aquatic animals, amphibians complete their metamorphosis to air-breathing, land-dwelling quadrupeds under water.

Amphibians came on land to stay around 380 million years ago. They gave rise to the three other vertebrate classes. First came the reptiles and, from them, the mammals and the birds.

Between amphibian and reptile the critical transition is in mode of reproduction. Mating in the amphibian, as in the fishes, is external. Egg and sperm fuse at random in the ambient water. In that hostile environment, the gel-coated egg goes through embryonic development. Fertilization in the reptiles is internal. The advanced behavior of mate selection here enters the natural selection process. The shell-enclosed egg holds the embryo and its first amniote nourishment. It incubates in a variety of niches and may or may not hatch under parental attention.

Radiation proceeds faster at the next level, that of the order. From the reptiles came 6 orders in their first 50 million years and 10 more in the next 50 million years. Two orders went to sea, and one took to the air. The terrestrial orders found their way in all avatars—reptile analogues of the tree-climbing squirrel, the ground-burrowing shrew, the swamp-grazing hippopotamus, the browsing giraffe—into every eco-space. To carnivores of appropriate dimension the herbivores supplied a superimposed ecospace. The celebrated dinosaurs were 2 of those 16 orders, and one of them, the *Ornithischia* ("bird-hipped"), gave rise to birds. Three orders—*Chelonia*, the turtles; *Squamata*, the snakes and lizards; and *Crocodilia*, the crocodiles and alligators—survive from the 200 million years of the Age of Reptiles.

The class of mammals meanwhile stemmed from one of the first six reptilian orders. If evolution had a plan, there might not have been an Age of Reptiles. The *Therapsidia* were more "advanced" than the dinosaurs. That is to say, they were more mammal than reptile in all but one respect: they were egg-layers. They were mammal-like in skeletal structure, and they evidently also had the mammalian advantage of endothermy, the metabolism that maintains body temperature independent of surroundings. These vigorous animals underwent adaptive radiation into 20 families of herbivores, insectivores and carnivores in less than 10 million years. They were the "dominant," the most numerous, large animals for 50 million years, inhabiting all ecosystems on the supercontinent of Pangaea.

All but one family of the therapsids perished, however, in the unexplained extinction that overtook the world 300 million years ago. That extinction left the world to the dinosaurs. From the surviving *Dicynodon* came the mammals. Gould might count this as the second lucky roll of the dice for Homo *sapiens*.

A principal selective advantage of the mammals is, of course, their mode of reproduction. Around 250 million years ago, descendants of the *Dicynodon* completed the transition from the amniote egg to the embryo brought to term in the maternal womb and the infant at the nipple of the mammary gland. Transitional mammals survive. The duckbilled platypus lays eggs but nurses its hatchlings. The marsupials—the opossum on the American continents, the panda in the bamboo forests of China and the kangaroo and numerous other marsupials in Australia—deliver immature embryos to a kind of external womb, the pouch, where they suckle to completion of their embryonic growth.

Even as the dinosaurs bestrode the Earth, the mammals gave rise to three orders, all extinguished in the course of the dinosaur regime. A meteorite, it is agreed, relieved the mammals of the presence of dinosaurs 65 million years ago. Within 10 million years, 18 orders of mammals radiated into the vacated ecosystem and 8 more in the next 20 million years. Two orders of mammals returned to the sea and one took to the air. In all, over 65 million years, 32 orders have risen in the mammal class; 20 survive, as different as whales from elephants, elands from lions, bats from shrews.

The nearly 20 orders of mammal count 4,500 species in their numerous families. Families may be as diverse as the fissiped bears, raccoons, dogs, cats, hyenas and primates; genera, as different as lions and tigers; and species, as different as Kodiak and grizzly bears. Turnover, or radiation and extinction, at these levels proceeds in shorter average times and varies, of course, over wide ranges. The fossil record shows mammal species coming and going in a few million years.

The plants came first

There could have been no ecospace on land for this 400-million-year exuberation of vertebrate life had not plants first occupied the continents. Ahead of plants came, inevitably, the Cyanobacteria. They changed the landscape first. The runoff of water from the uplands in shallow, braided streams began to give way to flow in rivers confined in banks consolidated by their infiltration.

On land thus secured for occupation, the first plants washed ashore around 430 million years ago. They were little more than disks of undifferentiated green algae cells. Against desiccation they developed the protection of a coating of cutin, a polymer synthesized "opportunistically" in the exposure of these organisms to the open air. Soon, the cells of these protoplants differentiated into root, stem and leaf tissue. Thereafter, they radiated swiftly into the new land niches. In 50 million years, they forested the continental lowlands with trees of considerable variety. On the way, they differentiated the phloem and xylem tubular tissues that transport nourishment and water in vascular plants.

During those 50 million years, these tree-fern forests wrought environmental revolution. Their photosynthesis raised the concentration of oxygen in the atmosphere to its present 21 percent. At that level, the respiration of land-dwelling arthropods, vertebrates and fungi, along with the respiration of the plants themselves, has held it ever since. Oxygen escaping into the far upper atmosphere completed the establishment of the ozone layer. The absorption there of high-energy solar radiation in the constant making and breaking of the bonds of the unstable ozone molecule, O_3, has protected terrestrial life from that radiation. The turnover of carbon dioxide soared with the increase in terrestrial life. Nonetheless, its concentration in the atmosphere—set by feedback loops interlocking its output from volcanoes to its turnover in photosynthesis and respiration to its sink in the inorganic chemistry of the oceans—held steady in the greenhouse range favorable to life.

The presence of these forests is remembered in the world's poorer coal seams, laid down between 360 and 290 million years ago. The alternation of the seams with strata of sediments records the success of the forests in withstanding the recurrent draining and flooding of the swamps in which they grew. Over those years the sea level rose and fell 100 times with the advance and retreat of glaciers on Gondwanaland. In that huge continent the present southern continents were already assembled in the Southern Hemisphere, reaching from the South Pole to the equator. The polar ice cap flooded on some of those numerous occasions into tropic latitudes and melted back.

Need for water and humidity limited the range of the first vegetation. In the sexual cycle of the ferns and mosses, the sperm find its way to fertilization in water on the stem or underside of the leaf. In a transition corresponding to that from the gel-coated egg of the amphibian to the shell-enclosed egg of the reptile, the spore-bearing plants gave way to the Gymnosperms. Wind carries the "naked seed"—actually the embryo—of these plants, known today in the rare cycads and gingkos and the family of the familiar conifers. In a second swift radiation, starting 300 million years ago, the Gymnosperm forests replaced the fern forests in time to furnish the landscape of the dinosaurs.

The flowering plants

Again, as in the transition from reptiles to mammals—and around the same time—the naked seed gave way to the protected amniote seed—again, the embryo—of the Angiosperms. These, the flowering plants, engage animals in their reproduction as well as the wind. Symbiosis has set plants and insects especially on countless paths of coevolution. Plant hosts and their parasites as well as symbionts have sustained their mutual existence over millions of years. The adaptive function of the flower is the attraction of insects, birds—such as the hummingbird—and mammals—bats. Each flower attracts its own pollinator(s), ensuring fruitful distribution of its pollen to other flowers of its species. By various strategies, plants also enlist insects, birds and mammals in the distribution and planting of their seeds. The amniote nourishment then supports the germination of the seed.

Beginning 65 million years ago, the radiation of the flowering plants speedily dressed the world in its present vegetation. The deciduous hardwoods of the Angiosperms share forestland with the conifers in the temperate zones and tend to displace them. It is the herbs—nonwoody flowering plants of everyday discourse—and grasses that brought the greatest innovation in the landscape. Their local ecosystems hold vast territories formerly prone to erosion. The grasses developed their continuously growing habit as recently as 30 million years ago, in coevolution with the grazing horses and ungulates that so outnumber the browsing species. During the last 10,000 years, the herbs and grasses have engaged the human species in mutually sustaining existence.

When the first protoplants washed ashore 430 million years ago, the fragments of the Proterozoic supercontinent were reassembling. The Southern Hemisphere continents of the present world map were already joined in Gondwanaland; they took turns under the south-polar ice cap as they drifted over the South Pole. Over the next 100 million years, as the amphibians gave rise to the reptiles, the collision of Laurentia and the landmass that was to be northern Europe began the assembly of Laurasia. At 250 million years ago the joining of Laurasia to Gondwanaland gave the therapsids the single supercontinent of Pangaea to roam from pole to pole. The first dinosaurs, succeeding the therapsids, shared some of their ground-covering capacity. They found their way around the central deserts of the continent to habitations on both sides of the supercontinent; on each side, to a shore of the world enveloping Panthalassa [see illustration, page 355].

In time for the arrival of the big dinosaurs, around 200 million years ago, Pangaea began to break up. The Tethyan seaway transected the continent along the equator between Laurasia and Gondwanaland. For the largest of all land animals, only the swamps and marshlands that came to flourish along this tropical seaway could have supplied sufficient forage. Flotation of their huge mass in the shallow waters lifted some of the burden from their musculoskeletal systems. Dinosaurs found forage also along an inland seaway that joined what are now the valleys of the southward flowing Mississippi and northward flowing McKenzie river and divided the North American continent for 20 million years.

During this period, the subduction of oceanic plate under the northwest coast of Laurentia brought island arcs and cratonal fragments into collision with the continent. The sediments holding the first signs of multicelled life were heaved high in the McKenzie stretch of the Rocky Mountains and the Burgess Shale farther south.

Cretaceous catastrophe

A measure of the lush vegetation of these years is its presence today in more than half of the world's coal deposits. In the United States and Canada, these are the seams, 100 meters thick, laid down along the continental seaway at the foot of the Rocky Mountains. In the same years, bacteria processed the anoxic depths of the Tethyan seaway and laid down more than half the world's petroleum. It is trapped now under fossil water of the seaway underlying the Eurasian continent from the Arabian peninsula to Kazakhstan behind the Himalayas.

The dinosaur paradise apparently came to an abrupt end 65 million years ago. The fossil record, a statistically unreliable sample of the life in the time from which it dates, holds no certainties. Some dinosaur types, perhaps not adapted to paradise, had been fading out. There seems to be no doubt, however, that the major dinosaur populations at that time vanished in a geological instant. The consensus is hardening on the possibility that the agent of this extinction was an asteroid. The physicist Luis Alvarez and his father, the physician Walter Alvarez, called attention in 1979 to the evidence represented by a thin stratum of iridium-rich rock dated to 65 million years ago. A class of meteorites is rich in iridium. The stratum has now been found in rocks around the Earth. Core drilling has pinpointed the Caribbean bay enclosed by the Yucatán peninsula as the site of the asteroid's impact. Dust and volcanic gases would have veiled the Sun for many years. In the long winter, it is thought, the dinosaurs and all but the three surviving orders of reptiles froze and starved to death. The shutdown of photosynthesis terminated the dominance of the Gymnosperm vegetation with equal finality. The way was open to the contemporary world of mammals and flowering plants.

Catastrophe and extinction have thus had their role in evolution. Ice ages and warm spells, cold oceans and anoxic oceans have, again and again, occasioned the mass extinction of marine invertebrates. Those animals live in the intimate embrace of their environment. Water conducts heat 1,000 times faster than air at atmospheric pressure. The advance and retreat of ice ages have worked extinction on land as well. Animals and plants that could not translocate their habitat ahead of the advancing glacier disappeared under it. The collision of the last glaciation with the Alps left European forests and fields impoverished in diversity of species compared to North America. There, retreat of species close to those of Europe was not foreclosed.

Natural selection: an outcome

However large their evolutionary consequences, these instances of extinction cannot be regarded as outcomes of natural selection. They are calamities from which the survivors escaped more by luck than by the adaptation that natural selection secures.

Natural selection, as George Gaylord Simpson insisted, is not an operation performed on a phylum or a species from the outside. Nor, as Simpson further insisted, is selection significantly a process of elimination. Living form and the environment do not present a dichotomy, one arrayed against the other. Natural selection is an outcome that common parlance reifies to standing as a force. Change in living forms proceeds with change in the environment. Organisms respond from inside to stimuli and pressures originating in the environment. As constituents of the environment, they themselves change the environment.

The unit of selection, favorable or unfavorable, is not a mutation or even, most of the time, a single gene, as in the Beadle-Tatum Neurospora experiments [see page 257–258]. It is a trait, binocular vision perhaps, that more or less significantly relates the organism to the environment. It most often expresses a constellation of genes. Selection is not for the trait in the individual, but for its presence in the gene pool of a breeding population. It is in the population that Mendelian assortment and recombination of genes keep the trait in dynamic interaction with the environment. The random circumstance favors or disfavors this or that aspect of the trait, there by random chance. It is a question of the match of the trait to the circumstance, the heritability of the trait and the nature and "intensity" of the environmental stimulus or pressure.

It is perhaps easier to see selection as a positive process in the invention of the diverse metabolisms by the prokaryotes during the Archaean. The accumulation, then, of accidents of mutation, occurring at random, made new connections to the environment and brought new elements and compounds from the lithosphere into service. Metaphorically, such genes were evoked—"selected"—by the environment.

Over times long enough, evolution proceeded even to the fashioning of the vertebrate eye. This outcome of evolution, it has been noted, disconcerted Darwin [see page 289]. The argument from "intelligent design"—that no feature of the eye can serve a function without all the rest—summons the Watchmaker. With its evolution traced from the beginning, however, the eye proves to be no less wondrous and no more than natural [see illustration, page 372].

Along with gravity, light pervades the planetary environment. A

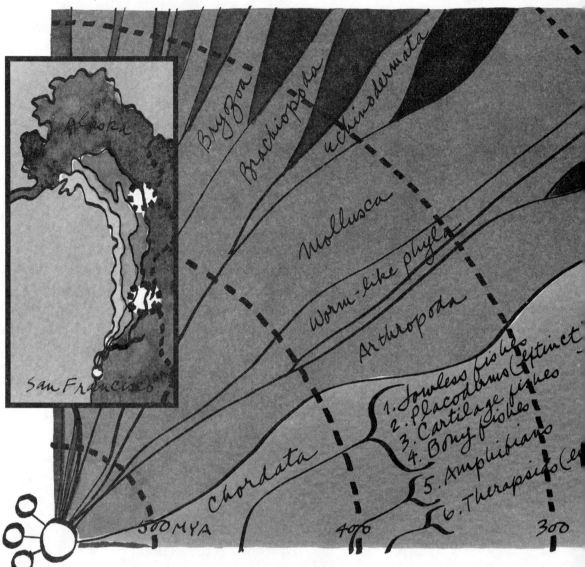

Protoctista

Alaska

San Francisco

Bryozoa

Brachiopoda

Echinodermata

Mollusca

Worm-like phyla

Arthropoda

Chordata

1. Jawless fishes
2. Placoderms (extinct)
3. Cartilage fishes
4. Bony fishes
5. Amphibians
6. Therapsids (ex

530 MYA   400   300

Ever-increasing diversity of life *evolved from the several almost indistinguishable, first-known ancestral, multicelled organisms of 700 million years ago,
fossilized in hyperfine oceanic sediments now upraised in the McKenzie region
of the Canadian Rockies. The first "radiation" generated all of the 30 to 40 now-
living phyla by 530 million years ago. From each phylum successive radiations
brought geometrical increase in diversity—in the order Arthropoda (insects,
crustaceans) more than 80,000 species today. Emphasis here is on the chordate*

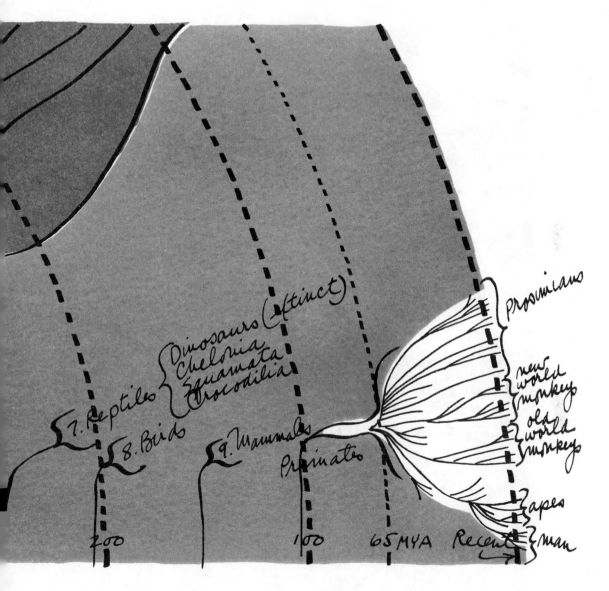

phylum and its vertebrate subphylum. By 350 million years ago, it had given rise to four classes of fish, of which three survive, and to the amphibians that brought the vertebrate plan—basically, four appendages and a jaw—ashore. From them, in 100 million years, came thousands of species of reptiles, birds and mammals. Some 65 million years ago, the primate order budded from the mammalian line, giving rise to the genus Homo about 2.5 million years ago and to Homo sapiens in the last 200,000 years [see pages 391 and 395].

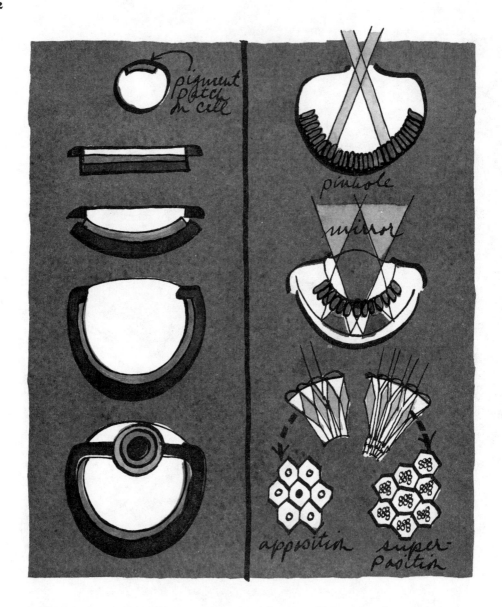

Evolution in stages of increasing complexity, *all found in living organisms, explains the intricacy of the vertebrate eye that confounded Darwin [see page 289]. A patch of pigment on its membrane orients a cell to the light; a patch of pigmented cells does same for a multicelled animal; depression of patch in a rounded pit improves orientation, and so on. Eyes as intricate as that of the vertebrate are seen in the crustacean concave mirror and the arthropod compound eye that produces compound or superimposed images [see page opposite].*

localized patch of light-sensitive pigment in the external membrane orients the motion of certain Bacteria and Protoctista to their adaptive advantage. A patch of pigmented cells, along with gravity, orients the growth of most sessile invertebrates. The same is true of plants. In some organisms, concave depression of the patch improves orientation to the source of light. The pit viper, already equipped with the vertebrate eye, developed another pair of eyes in those pits, sensitive to infrared radiation from its target prey, on each side of its nose. A spherical fluid-filled pit with a pinhole opening describes the image-forming eye of an abalone and the nautilus, a primitive mollusk. In a clam and several crustaceans, layers of a polysaccharide make a concave mirror at the back of their eyes that focuses the image on light-sensitive cells arrayed in front of the mirror. Numbers of mollusks, some sea worms, spiders and some other arthropods have more conventional image-forming eyes with a lens in the light aperture. The eye of the octopus more closely approximates that of the vertebrates, with the lens behind a cornea and an iris. Its retina does not, however, serve as an extension of the brain as it does in the vertebrate eye, where the retina conducts preliminary information processing before delivering impulses to the optic nerve.

Most arthropods, insects and crustaceans have compound eyes with many lenses. Investigators have identified five different designs. In the most common eye—found in most daylight insects, some crabs and all lower crustaceans, including the horseshoe crab—the lenses project their images next to one another, making a compound image. The lenses of the four other eyes—found in nocturnal insects and in crabs and lobsters and other crustaceans—superimpose their images. Superimposition, in effect, amplifies what light the night or the sea depth affords.

Dan E. Nilsson, zoologist at Lund University, has matched the design of the many different kinds of animal eyes to their analogues in just about all of the optical devices ever made by man. Organisms under natural selection "found" the same uses for light. Animal eyes feature even the only recently devised optical fiber and the corner reflector, placed on the Moon for monitoring, by laser beam from Earth, the play of the gravitational force in the solar system.

Nilsson and his colleague Susanne Pelger, a geneticist, arrived at a "pessimistic estimate of the time required for an eye to evolve." They assumed a generation time of one year, common for small aquatic animals, and appropriate constants for mutation and genetic exchange in a breeding population. From light-sensitive patch to camera eye, their model showed, the course of evolution would run some 346,000 years.

Evolution, in the metaphor of the molecular biologist Francois Jacob, is a *bricoleur*. In the Larousse dictionary, a *bricoleur* is a householder who fixes things with what's at hand, without help from the *métiers*, the Plumber or the Carpenter or the Watchmaker.

The Earth's gravitational field bathes the organism in a force more pervasive and constant than light. In response, from the time of the jawless fish, the inner ear of the vertebrate has presented an instrument suited to its function as elegantly as the eye. It is the three bone-enclosed, looped channels of the inner ear, the semicircular canals of the labyrinth. The loops are arrayed at right angles to one another in the three dimensions of space [opposite]. Inside the loops, a heavy fluid bathes whiskers of the universal multipurpose cilia, each one protruding from a nerve cell. The inertia of the fluid lags movement of the head. The bending thereby of the cilia excites nerve impulses in the three dimensions.

Evolution is a geometer. The fashioning of this Euclidian instrument by evolution certifies the objective truth of geometry and the truth as well of the mathematical reach of the human imagination into the universe beyond sensory experience.

The hankering for design

Under environmental stimuli as pervasive as light and gravity, it is not surprising that widely diverse phyla should have evolved similar sensory organs. Such convergence of organic design excites those who hanker for design and direction in evolution, if not for the purpose and meaning that escaped Darwin. Simon Conway Morris, a student at Cambridge with Whittington of the Burgess Shale, cites examples of the convergent evolution of whole animals. As a "classic example," he cites the dolphins, "which evolved from dog-like animals [and] are shaped like fish because there exists an optimal shape for moving through water." He points also to "another example: both placental mammals and marsupials produced a large saber-toothed carnivore on separate continents [North and South America, respectively]."

Morris goes further. Responding to Gould, he argues that convergent evolution would have produced human intelligence in some other evolutionary line had the Burgess Shale chordate lines perished. Elliptically, he declares, "If such a quality as intelligence can arise both in human beings and in the octopus ... then perhaps there is a course and direction to evolution."

Here, it is evident, Morris has in mind, though he does not mention, the similarity of the octopus and vertebrate eyes. Nilsson and Pelger were careful to note that no eye ever evolved in 346,000 years. "An eye makes little sense on its own," they said. A worm, served by light-sensitive skin patches, would have little use for the eyes of a fish

without a fish's central nervous system: "the worm would need to become a fish." To have an eye like that of Homo *sapiens*, Morris argues implicitly and elliptically—overlooking all the other sighted vertebrates—the octopus must also be endowed with intelligence.

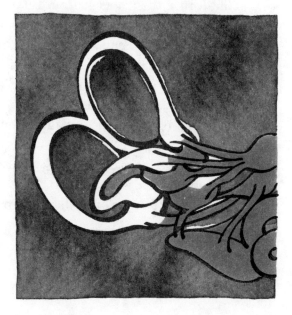

The annelid worm in the McKenzie sediments did become a fish with a vertebrate eye. Gould would argue, however, that the growing constellation of genes on the way to vertebrate form evolved in opportunistic interaction with random environmental

stimuli and pressures encountered in random order over some millions of years. Equipped with eyes and other organs—including semicircular canals in the inner ear, the vertebrate class of the chordate phylum radiated into ecospaces not defined until they occupied them.

Responding to Morris, Gould observes that the dolphin and the fish and the two saber-toothed tigers are all vertebrate descendants of the Burgess Shale chordate. "What odds," he asks, "would anyone place on the evolution of [the vertebrate tetrapods] if no ancestral creature had the precursors for these structures?"

No phylum has gone extinct since the Cambrian. Immortal representatives of ancient orders persist to document the transformations of life-forms worked in the course of evolution. Extinction of species is, however, as natural as the death of individuals. Fauna and flora both disappear from the land with the alteration of their habitats beyond the range of their capacity to adapt. Changes in habitat occur rapidly with the onset of ice ages. Presently, no more than 10 percent of the species that have ever lived remain alive. Yet, in consequence of the persistence of the earliest—including survivors from the first 100 million years after the Burgess Shale were laid down—as well as the addition of the latest life-forms, the planet has never known such diversity of life.

Nor has the planet ever carried such abundance of life. The total mass of living tissue, it is estimated, comes to 1.2×10^{12} to 1.4×10^{12} tons. Of the total, the tissue of the Plantae accounts for nearly half; a large percentage of that is extracellular cellulose and lignin. The tissue of all the Animalia and Fungi adds up to no more than 2 or 3 percent of the total biomass. If tonnage is the test, then the supreme invention of evolution is the earliest. Bacteria weigh in at half, or more, of the tonnage of the biomass. In the soil and waters, in the ocean-floor ooze, in unconsolidated sediments on and deep in the continents, the prokaryotes go on living as they have from near the beginning of Earth time.

7

Tools and Human Evolution

A civilization cannot ... put on science like a suit of clothes
—a workday suit which is not good enough for Sundays.
Jacob Bronowski

Upon publication in 1859, *The Origin of Species* excited the public to controversy over the origin of the *human* species. Thomas Huxley made the case for natural selection in *Evidences as to Man's Place in Nature* in 1863. He had had his famous exchange about choice of ancestor—whether ape or bishop—with Bishop Wilberforce, when Darwin at last addressed the subject. *The Descent of Man,* published in 1871, set out the features of anatomy that so apparently relate the human species to the primates and directed attention to the crucial role that the prolongation of infancy must have played in the departure of human beings from the primate line.

In 1853, five years before he submitted the précis of his thoughts about natural selection to Darwin, Alfred Russel Wallace had posed his question about the origin of humankind. In *Travels on the Amazon and the Rio Negro,* reflecting on his encounters with the indigenous people who had assisted his collecting there, he wrote: "Natural selection could have endowed savage man with a brain only a few degrees superior to that of an ape, whereas he actually possesses one very little inferior to that of a philosopher. With our advent there had come into existence a being in whom that subtle force we term 'mind' became of far more importance than bodily structure."

The importance—the power—of Wallace's "subtle force" has its measure in the value that mind has come to place on life. A healthy consciousness of individual existence holds that value absolute. People have been known to give their lives for other values they held absolute. Most

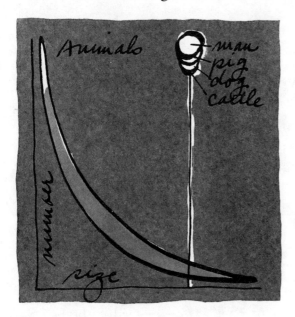

would resist, however, their negative selection in favor of the fitness of the species. As consciousness dawned in the line of descent—or ascent—to human, individuals contrived to foil, in increasing number, the diverse, random agencies of natural selection. With successful stratagems for survival conveyed from generation to generation by the transgenetic modes of teaching and learning, human evolution has raced ahead of the biological to the social rate of change. In consequence, the human population now exceeds, by at least 10,000 and by as much as 100,000 times, the population of any animal of comparable size, the human among the largest. That is, except for the populations of domesticated animals, raised also to the billions in the custody of their keepers [above].

Individual life expectancy in half the world population runs well beyond the reproductive years, the only years that matter in natural selection. The materially best-off 20 percent, especially, enjoy years of engagement in activities other than the struggle for subsistence. They have time to explore and to realize the unique endowment that each brings to the onetime experience of human existence.

Humankind now possesses and the Earth affords the means to extend to the rest of the world population the opportunity of that full human lifetime. The means are in hand to end the poverty that blights the biological development of 8 in every 10 children.

This prospect has improved materially in the course of the 20th century. The last half of the century brought increase in the means of subsistence well ahead of population growth. Living longer, as people do today on every continent, they are correspondingly restraining their fertility. The procreative drive has proved, in human biology, to be subject to control by the mind. Able to future-dwell, people make the rational calculation that for fewer there can be more. In that best-off 20 percent, families are smaller, and fertility has declined to the replacement rate and below. Fertility in the next best-off (and next longer-lived) 20 percent is approaching the same low levels. The rate of growth of the world population, having peaked at just under 2 percent around 1970, is in decline. In the last decade, fertility began to follow the death rate downward everywhere except in the countries of the desperately poor, as in sub-Sahara Africa, where the death rate has stopped falling, and, in some countries, turned upward again.

The demographic transition

The entire population, it appears, is otherwise entrained in the demographic transition—that transition from near-zero growth at high death rates and high birth rates, through population explosion, to near-zero growth again at low death rates and birth rates [see page 41]. Vital statistics, kept by all nations and assembled by the United Nations, promise that the present doubling of the world population will be the last. It goes without saying that a smaller population will exert smaller stress on the Earth's resources. The struggle for subsistence at the margin might exceed human capacity for the organization of social order. Established trends show the population settling at a size that will allow people to live in dignity and liberty. In sight, as early as the end of this century, is the possibility of humankind at home on Earth.

The familiar historical contingencies, it must be conceded, discount that possibility. The same 20th century that saw the Age of Science saw atrocity on unprecedented scale and capacity for still larger-scale atrocity drawn from the work of science.

The future is jeopardized as well by fumbling first employment of the means to realize it. The most recent doubling of the world population is the first to be sustained by increase in the output of food per hectare rather than by increase in the number of hectares under cultivation. Agriculture is losing topsoil to erosion, however, by an average depth of a centimeter per year. From the Gondwanaland continents, where the world's largest and poorest populations live, the rivers are carrying off what remains of the stock of biologically vital, soluble mineral elements that were ploughed from the lithosphere by the last glaciation there, 200 million years ago.

Environmental impact

For mechanical energy, the liberator of human energy, industrial civilization continues to depend upon fossil fuels. It does so even with solar and other alternative energy technologies—modes of solar-energy conversion, ocean-temperature conversion, deep dry-earth heat and not excluding nuclear fission—at the ready. The four-fold increase in combustion of fossil fuels during the past 50 years injects carbon dioxide into the atmosphere in a volume exceeding 25 percent of the natural turnover of the gas. Global warming, in consequence, threatens a worldwide shift in climate prospectively catastrophic to agriculture. The projected rise in sea level endangers the habitations of a third of humankind. If the present doubling of the world population is to be the last, a human existence must be extended to all. That will require another four-fold increase in energy. Such increase cannot come from fossil fuels; it must come from alternative technologies.

Human activity has raised the cycles of other gases in the atmosphere—nitrogen, for example, to twice its natural turnover, and nitrogen and methane well in excess of that. Now it is injecting into the atmosphere compounds new to nature—the notorious fluorocarbons, for example. About one untoward consequence of these unnatural emissions there is no debate. Observation has established and now monitors the ominous thinning of the high-atmosphere ozone layer over the Southern Hemisphere. Increase in high-energy solar radiation at the ground has begun to disrupt photosynthesis, starting with the phytoplankton in the Antarctic Ocean. This is a considerably more serious matter than the rising incidence of skin cancer in that hemisphere and more immediately menacing than global warming.

The Montreal Convention of 1989 and its stiffening after-amendments may yet curb release of the gases that most insidiously disrupt the ozone layer. This international convention was adopted within a decade of the publication of the scientific paper establishing the peril. Here is ground for hope that the human species will not lose the future open to the means at its command.

"Everything has changed but our way of thinking." So Albert Einstein observed in the aftermath of Hiroshima and Nagasaki. A half century later, that way of thinking persists. Nuclear armament continues to proliferate, vertically as well as horizontally. First- and second-generation weapons in the possession of unstable rogue nations present the familiar danger. Of unknown dread are the diabolical third- and fourth-generation weapons still under development.

The economic problem

"Our way of thinking" answers the primary anxiety besetting individuals and societies. This is the economic problem; as John Maynard Keynes defined it, "the struggle for subsistence, always hitherto the primary, most pressing problem of the human race—not only of the human race, but of the whole biological kingdom from the beginnings of life in its most primitive forms."

Return from labor on "the problem" has always been uncertain and often insufficient. From the earliest days of recorded history, accordingly, the values and institutions that organize human behavior have secured inequitable distribution of the return. Under this arrangement, a few have lived better and longer. They got on with the adventure of history and the creation of high civilization. In support of their enterprise, under one mode of coercion or another, the great mass of humankind rendered up what remained from the little they scratched up to sustain their existence. They lived just long enough to reproduce themselves. Until most recent times, the world population increased by the small difference between the high death rate and the resolutely maintained birth rate of the great mass of humankind.

Now conquest of the material world portends the end of scarcity. This development is already inducing some revision of prevailing values.

With its problem approaching solution, economics turns from the production of goods to their distribution. Wassily Leontief stated the new economic problem in a fable:

> Adam and Eve enjoyed, before they were expelled from Paradise, a high standard of living without working. After their expulsion they and their descendants were condemned to eke out a miserable existence, working from dawn to dusk. The history of technological progress over the past 200 years is essentially the story of the human species working its way slowly and steadily back into Paradise.
>
> What would happen ... if we suddenly found ourselves in it? With all goods and services provided without work, no one would be gainfully employed. Being unemployed means receiving no wages. As a result, until new income policies were formulated to fit the changed technological conditions, everyone would starve in Paradise.... The setting aside of the Puritan "work ethic" ... is bound to prove difficult and long drawn-out.... In popular and political discourse on employment and unemployment, with its emphasis on incomes rather than the production of goods, it can be seen that the revision of values has already begun.

Objective knowledge

Thinking need not change to a new way. It is the accumulation of objective knowledge that brings the different future in sight. That knowledge proceeds from a way of thinking that long antedates the agricultural revolution and the beginning then of recorded history. The corpus of objective knowledge comprehends the cumulative experience of the physical world by humankind from the outset. The first stone tools, discovered in East Africa in the 1950s, bear incontrovertible evidence of the command of objective knowledge by the earliest forerunners of Homo *sapiens* now known.

Those tools, from 2.5 million years ago, manifest the divergence of the genus Homo from the family Hominid. The stone tools, in their multiplication, diversification, specialization and refinement, have been called "the most common fossil of the Pleistocene," the last million years of Earth history. These fossils of behavior constitute the principal record of the evolution of the brain that is the locus of mind. Until the middle of the 20th century, it had been thought that H. *sapiens* was the first toolmaker. Toolmaking was the status symbol of humankind. Now it is established that a pre-*sapiens* hominid made the first tools.

The Hominids, or man-apes, departed 4 million years ago from the ape, or Hominoidea, line to radiate into the grassland habitat then spreading across southern Africa. The skulls of the first toolmaker found with the first tools are not otherwise readily distinguishable from contemporary skulls of man-ape cousins that long ago went extinct.

Toolmaking was the decisive adaptation in the line of natural selection that leads to humankind. This not to say that tools made man. By toolmaking, it may be said, man made himself.

The dawn of purpose

Other vertebrates are known to make, as well as to use, tools. Chimpanzees, in the wild and in captivity, have been seen to devise tools in original solution of problems presented by novel situations. One of Darwin's finches fills the woodpecker niche with a thorn held in its bill. The difference between human and other toolmakers is, perhaps, one of degree, not kind.

The earliest toolmakers left evidence of behavior not observed in other animals. It is compelling evidence that they made their tools for conscious purpose. They made them at, if not in, tool shops. They made them in quantity "now" for purposed use in a future "then." They used their stone tools, moreover, as machine tools, to make tools of other materials, such as tools from bone found with them. This was purpose framed to enable a second purpose, a logical cascade of purpose.

Like their man-ape cousins, the toolmakers were omnivores: herbivores primarily and carnivores opportunistically. Opportunity was presented in the kills left on the African grasslands by the big cat carnivores; they leave carcasses with plenty for other animals when they have had their fill. The toolmakers joined the jackals and vultures in scavenging. Evidence shows that they butchered desired parts from the carcass and carried them to the homesite. There animal bones have been found with principally two kinds of tools: blades that cut meat and gristle and hand axes that crushed bones to get at the marrow.

The purpose served by toolmaking can be understood, therefore, to have been a social one: food-sharing. This behavior too is observed in other animals—in all the land-dwelling vertebrates except the amphibians. The toolmakers would have found the purpose of their toolmaking reinforced in the social setting. The onset of consciousness of self in relation to others would carry purpose across the delay in time between the making of the tool, the securing of the food and the carrying of the food to the homesite.

In the rhythm of biological time—over tens and hundreds of thousands of years—progressive mastery of the environment and elaboration of social behavior secured the prolongation of helpless human infancy. The embryonic development of the human brain in social interaction outside the womb, prolonged in mutual feedback with increasing sophistication in toolmaking, moved evolution of the species onward. Over a million or more years, the adaptive advantages of the heavy skeleton, thick skull, strong jaws and large canine teeth of the formidable first Hominids yielded to the lighter skeleton, thinner skull and enlarging brain of the emerging species of the genus Homo. Another million years brought the arrival at last of the Homo *sapiens* in whom Alfred Russel Wallace recognized the mind of a philosopher.

Arrival of the primates

Before toolmaking and the elaboration of the philosopher's brain could begin, a still longer period of slower biological evolution had tautly wired the laws of physics in the circuitry of the brain. This phase in human evolution began in earnest with the appearance of the primates around 65 million years ago in Gondwanaland. That southern half of the supercontinent Pangaea was then separating from the northern Laurasian half. A planetary calamity had extinguished the dinosaurs and cleared the dominant gymnosperms from the landscape. The primates were among the mammals that came to thrive in the angiosperm vegetation. They followed its spread throughout Pangaea; their fossils are known in North America.

The first primates were mouse-sized herbivores and insectivores. They secured their niche in the leafy canopy of the new forest where they were safe from most mammalian predators. Primate genera and species came and went in large number through the first 20 million years. The early primates shared the canopy, as warblers are observed to do, finding micro-ecosystems within it, distinguishable by tree species and by micro-climate and insect population at different levels in the canopy.

Life in that aerial environment selected for acute tuning of the semicircular canals. The Galilean parabolic trajectory—in which vertical acceleration quickly overtakes horizontal motion—commanded the reflexes. The critical judgment of distance from here to there selected for binocular vision. Roundish skulls holding brains large relative to body size and eyes looking forward identify members of the primate order. The organization of the brain, with strong linkage of motor centers to visual cortex, is their own.

Over the years that organized the primate brain, travel in the treetops brought reorganization of the skeleton in some lines of descent. The first primates were clingers and creepers. Access to nutrition selected for jumping and grabbing. The spine, a horizontal bridge in grounded quadrupeds, became a derrick pivoted at the pelvis. The forelegs became forearms, and the forefeet, grasping hands. In the course of such reorganization, these primates perfected a mode of locomotion—brachiation: swinging from hand to hand, to tree limb from tree limb—that is all their own.

Success in these developments brought early radiation of the order in diversities of specialization. Representative genera and species survive from those days. The now-squirrel-sized Prosimians—tarsirs, lemurs, lorises—carry on the primate way of life in its earlier phase of 65 to 45 million years ago. They survive in the forest canopies on the Gondwanaland continents. Until recently, they thronged in isolated refuge on the island of Mozambique. Deforestation and shrinkage of that habitat now lay the final threat to their existence there.

Around 45 million years ago, the monkeys, growing larger in size and mastering travel by brachiation, diverged from the ancestral line. In formal acknowledgment of the resemblance noted by Huxley and

Darwin, they are called Anthropoidea. The organ grinder's, hang-by-the-tail monkeys diverged from the main Anthropoidea line around 40 million years ago. Isolated on the South American fragment of Gondwanaland by the opening of the Atlantic Ocean, they come down to modern times as the New World monkeys.

The Old World monkeys went on to a wider future on the African fragment of Gondwanaland. They gave rise to the rhesus, macaque, mandrill and other species that abound in Eurasia as well as in Africa. From these Anthropoidea, the still more human-resembling Hominoidea, the apes, diverged around 16 million years ago.

The Hominoidea

In the change of climate then to cooler and dryer, forests on the African continent were giving way to grasslands. Herbs and grasses began to support populations of ungulates and their carnivores. Growing large and formidable, the apes came down from the trees to this inviting new habitat as ground-living herbivores and then omnivores at the forest margins. The ancestral adaptation of their skeletons to tree-top travel made them no ordinary quadrupeds. On the ground, they could be nearly bipedal, the spine at a 45-degree or higher angle to the ground, stabilized by long-armed knuckle-walking. Freed from dedication to locomotion, their hands served other useful functions. As can be seen in zoos today, they grasp food, to begin with, and keep onlookers at a respectful distance. Proconsul—the most remarkable representative of these early apes, named after a celebrated gorilla in the London Zoo—was flourishing in Africa when the northward drift of the continent first brought it into collision with Eurasia. The fossil remains of Proconsul have turned up all over the Eurasian continent. That is surely witness to the mobility afforded by knuckle-walking.

The modern gorilla diverged from the Hominoidea main line around 10 million years ago. A more gracile ape, *Ramapithecus,* appears in the fossil record about the same time in India. Celebrated for a while as the "missing link," *Ramapithecus* is now recognized as the forerunner of the orangutan and other Asiatic apes. The earliest Pan, the chimpanzee, appears in the fossil record around 7 million years ago.

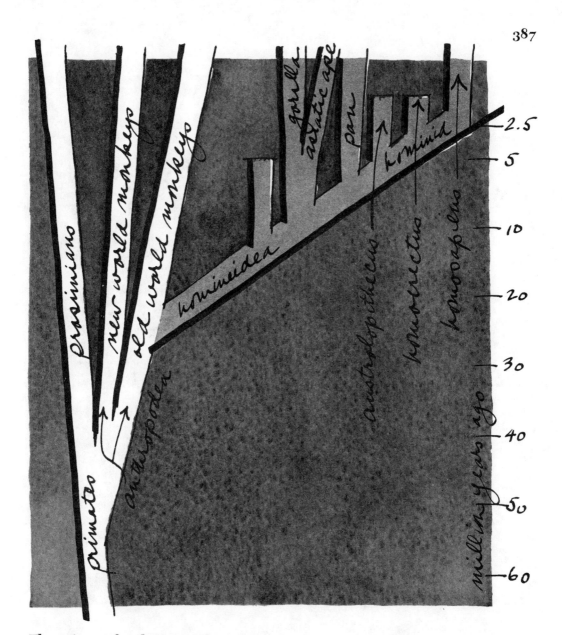

The primate family tree *took root 65 million years ago in the forest canopies of Gondwanaland; the prosimian tarsirs, lemurs and lorises carry on the arboreal way of life today in Gondwanaland continents [see page 328]. Anthropoidea began their radiation 45 million years ago; the Gondwanaland breakup isolated New World monkeys in South America. From Old World monkeys stemmed the Hominids about 20 million years ago. Man-ape toolmakers started the Homo genus 2.5 million years ago; sapiens arrived 200,000 years ago.*

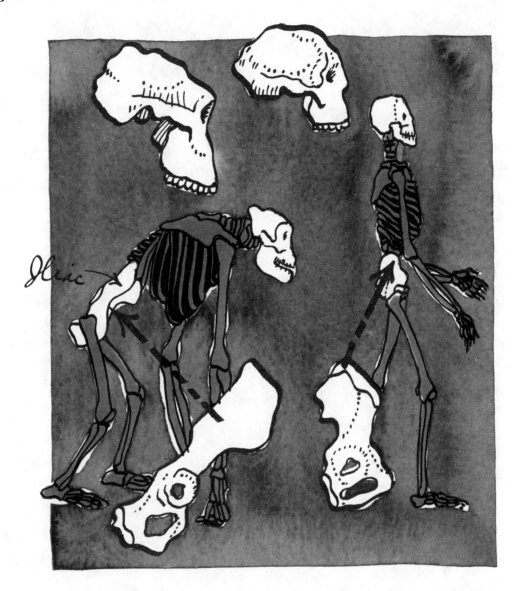

Evolution of pelvis and skull most radically distinguishes the departure of the genus Homo from the Hominid line. The iliac in the pelvis of the knuckle-walking great apes is extended, accommodating the suturing of the muscles of their carriage at the 45-degree angle. The iliac in the pelvis of Homo erectus was already shorter and the musculature reorganized to conduct fully bipedal walking [see opposite page]. The great-ape skull has a smaller brain case and heavy facial bones to carry a big-toothed jaw adapted to chewing raw vegetation.

The sequencing of DNA bases in living primates now confirms what the fossil record has thus disclosed about the primate family tree. DNA evidence serves equally to mark the distance of each of them from the human line. The more recent the appearance of each in the fossil record, the closer the relationship. Pan proves closest of all.

Bishop Wilberforce might have been mollified could he have known what the fossil record now shows. Neither the gorilla nor the chimpanzee, much less any monkey, appears in the ancestry of humans. On the family tree all diverge as branches from the main line, defined as that which leads to humans.

The man-apes

The man-ape, *Australopithecus,* appear on the main line 4 million years ago. The fossil record exhibits a decisive innovation wrought by natural selection. The iliac crest at the back of the pelvis rises high in the knuckle-walkers. In the man-ape, the iliac has shortened. This and the location of the regions on the pelvic bones to which the leg and back muscles anchored establish the man-apes as comfortably adapted bipeds. That is why they are called man-apes [see illustration, opposite].

Walking erect, *Australopithecus* had hands completely freed for other uses. With fully opposable thumbs, their hands could pick up, hold and carry things. Hand and brain, in interactive feedback, opened a new pathway in evolution.

Walking gave *Australopithecus* extended access to the grassland habitat; they left the forest margins to the apes. Massive molars, fit to grind tough fiber, identify them as herbivores. With small game abundant on the grasslands and the kill of the big cats to scavenge, they could be as carnivorous as they needed to be. The males, at 5 feet and 90 pounds—and the females, as well, at 4 feet and 60 to 75 pounds—well muscled and large-jawed, did not present attractive prey to predators.

How many species of *Australopithecus* arose in their long history—many times longer than that of H. *sapiens*—cannot be told from the fossil record. No complete skeleton and not more than a few incomplete skulls have survived at their living sites. Sexual dimorphism in any one species makes it difficult to discriminate among species. At least two different physiques persisted coevally over that long history. *Australo-pithecus robustus* was the more heavily boned and thick-skulled. In some specimens, cheekbones wide enough to have hidden the ears anchored the muscles of massive jaws. *Australopithecus gracilis* left bones that fit that adjective only by comparison. No trend from one to the other physique is discernible.

In all the known *Australopithecus* skulls, brain size varies between 450 and 650 cm^3. This is less than half the normal H. *sapiens* range of 1,200–1,800 cm^3. Of equal importance, however, is the ratio of brain to body size. At 90 pounds, *Australopithecus* differs from humans by that reckoning no more than humans differ from one another. Brain size proves not nearly so important as brain organization.

The cultural revolution

Louis and Mary Leakey found the bones of one of the first tool-makers in association with his (or her) tools in Olduvai Gorge, in Tanzania, in 1959. They dated their find at first to 500,000 years ago. With better physics, they later fixed this find at 1.5 million years ago. On the western shore of Lake Turkana in Kenya, the Leakeys and other pale-oanthropologists have since been working mines of evidence that take tool-making yet another million years deeper into the past.

From here on, the fossil record shows the biological evolution of the new genus. The time interval from turning point to turning point shortens from tens of millions and millions of years to hundreds of thousands of years. Biological and social evolution proceed together.

Looking for these first tools requires imagination tempered by circumspection. A hand ax—a wedge-shaped stone of a size conveniently grasped between fingers and palm—may show small evidence of the toolmaker's working [opposite]. Surer signs of the worker's hand often distinguish the blade tools. At Olduvai Gorge and Lake Turkana, the raw material preferred for blade tools was a fine-grained volcanic tuff. The Leakeys identified the "quarry" for this raw material, located kilometers away from the site where they found the tools. Such appreciation of the properties of stone must be seen as a considerable as well as early start on the accumulation of objective knowledge.

Some 2,683 fragments of worked stone at one Lake Turkana site testify as well to the attainment of considerable craftsmanship. By refitting blades and discarded fragments, the paleoarchaeologists managed to reconstruct several original cobbles. It can be seen that the toolmaker "knapped" with precision. His tool was one of the hard hammer-stones, also found at the site. Hammer-stone in hand, he struck the cobble,

held in the other hand, with anticipation of how the stone would fracture. He had done so repeatedly, and had secured a set of satisfactorily duplicate blades from the core stone. The Leakeys had well named their first toolmaker Homo *habilis,* "handy man." The site was quite evidently a tool shop. With connection between action and purpose extended in consciousness, the knappers were engaged in manufacturing.

At other Lake Turkana sites, the Leakeys with their son, Richard, exhumed stone tools along with jawbones and incomplete skulls and skull fragments of 100 individual Hominids in strata reaching across 900,000 years, from 2.4 million to 1.5 million years ago. They made confident reconstruction of the skulls of six individuals. Careful appraisal and much debate assigned three of them to the genus *Australopithecus,* species *robustus.* One item of evidence was the negative: no tools had been found with them. The other three skulls were at the *gracilis* end of the scale. Two of them, associated with tool finds, could be assigned to Homo *erectus,* as H. *habilis* is now designated. About the third skull, labeled ER-1470, there was no clear consensus. It was distinctly *gracilis*

and had the suggestively large volume of 775 cm³. No tools had been found in association with it, however.

A skull offers some evidence of the organization of the brain it once housed as well as its volume. A plaster cast shows the topography of the inner surface of the skull. The imprint of the bulging meningeal arteries locates the boundaries of the major structures of the brain. The occipital lobe, site of the visual cortex, is relatively larger in the ape. The temporal (locus of memory), the parietal (sensory integration) and the frontal (motor-adaptive and the locus of Broca's area, the speech center) are all relatively larger in man.

Homo *erectus*

An endocast of the ER-1470 skull, showed the vanished brain to have been "essentially human." Until displaced by another candidate, ER-1470 may be regarded as the skull of the first toolmaker, the Adam of the H. *erectus* line.

The head of H. *erectus* was poised on top of the spine, not hanging forward from it as in the *Australopithecus* line. H. *erectus* stood fully erect. Association of their remains with those of *Australopithecus* affirms that the two genera occupied the same ecospace on the grasslands for more than a million years. Over that time, signs of *Australopithecus* disappear. Nearly 1,000 encampment sites in sub-Sahara Africa document increase in the H. *erectus* population. Not until a million years ago did H. *erectus* follow Proconsul into Eurasia. It is not known whether *Australopithecus* ever did. The evidence for the early presence of H. *erectus* in Eurasia is the profusion of stone tools. *Australopithecus* would have left no such durable evidence.

Paleoarchaeologists in the 19th and early 20th centuries had identified what they called, after a site in southern France, the "Acheulian" culture as the earliest and most primitive of what they called the Stone Age cultures. The Acheulian is now called the Oldowan culture, after Olduvai Gorge. The stone tools are recognized today as the work of H. *erectus*. A half million years have been added to the record of human evolution on the Eurasian continent.

The distribution of tools of the Oldowan culture across the continent shows that H. *erectus* made themselves at home everywhere in Eurasia. Over the course of 800,000 or more years, the northern margin of their occupation of the continent must have retreated and advanced through the four Pleistocene glaciations that began perhaps 650,000 years years ago. As early as 700,000 years ago, they had ventured into Java, then attached to the Eurasian continent by the lowered sea level of the second glaciation. The discovery of Java Man in 1891 was occasion for another missing-link celebration. He is now firmly identified as H. *erectus.* Peking Man, found in 1929 and similarly celebrated, is now dated to 600,000 years ago and firmly placed in the H. *erectus* line.

From Chicken-Bone Hill Cave, 50 miles southwest of Beijing, paleoarchaeologists and anthropologists in the early 1980s dug out a continuous record of H. *erectus* life over the 200,000-year period from 400,000 to 200,000 years ago. The dig uncovered 13 layers of occupancy. It collected evidence of fire at all levels. At the bottom level, this stands as the earliest assured evidence of the use of fire by "man." Here it is associated with ample evidence of the fire-maker's presence; other, older sites offer little more than fossil charcoal.

Hand axes from the lower layers are large, 8 centimeters across and 50 grams in weight. The blades are made from sandstone. The tools in upper levels show the looked-for progressive trend to diversification and refinement. Diversification and specialization are evident in the ax heads, blades, pointed stones used as drills and small, flat, sharp-edged "burins" that were held between thumb and forefinger. The knappers have established the superiority of flint over sandstone.

Replicas of ancient tools have been made in the laboratory and put to the uses imputed to them. From the materials on which they worked, these tools were found to acquire characteristic chipping, wear and polish. Comparison under high magnification has established the uses to which the ancient tools were put.

From such evidence it has been shown that the assemblage of tools from Chicken-Bone Hill Cave were used to make articles of wood, bone and hide. Arrow and lance heads and the bones of 3,000 animals, principally deer, show that hunting had succeeded scavenging. Seeds found at all levels are durable evidence of food-gathering.

Evidence from caves does not make H. *erectus* a cave man. They occupied Chicken-Bone Hill and other caves seasonally and intermittently, taking refuge from weather, especially in periods of inclement glacial climate. Most of the time, H. *erectus* would have been on the move, following migratory game and seasonal vegetation cycles along accustomed trails. Campsites, where they have been found, indicate that the family economic unit counted no more than 25 members. On their seasonal migrations, they could travel unencumbered by provisions and possessions. They lived off the world around them by knowledge gained from trial-and-error experience with the useful and the noxious.

These bands must have encountered one another. Such encounters must have been welcome occasions for mating of daughters and sons outside the familial band. Whether in consequence of learning or natural selection, the incest taboo is universal. Shared experience brought the same innovations in toolmaking all across Eurasia over those tens of thousands of years. Vulnerable to drought and storm, predators, accident and disease, few H. *erectus* survived to leave offspring. Those few hairy, naked ancestors who survived from generation to generation, however, learned and taught their offspring their learning.

Between 200,000 and 100,000 years ago, populations of "archaic man" left considerably more evidence of their existence in the western regions of Eurasia. Neanderthal man, first found in a cave near Düsseldorf in 1856, left traces from western Europe to Central Asia of their Mousterian culture. Heavily boned and massively skulled, Neanderthal was physically H. *erectus*. The Mousterian kit of stone tools suggests the fashioning of shelters, skin tents perhaps, evidenced by at least one ring of mammoth bones around a hearth. The mammoth bones testify to prowess and, without doubt, collaborative enterprise in hunting. Rock drawings and burials with adornments and tools testify to symbolic communication. Speech and language must have begun to facilitate the teaching of what was learned. At the eastern end of Eurasia, Solo Man, coming after Java and Peking Man, was elaborating a corresponding culture during the same period.

By this time, by the estimate of Edward S. Deevey Jr., ecologist and director of the Geochronometric Laboratory at Yale University, some 35 billion toolmakers had lived their lives. The population on the African and Eurasian continents had reached a million. H. *erectus* could be reckoned a success.

Deevey called this period of evolution the "cultural revolution." In full analogy with the agricultural and industrial revolutions still to come, it had brought on a population explosion. The increase to a million from the tiny breeding population of 2.5 million years before, plotted on a horizontal and vertical logarithmic scale, describes the same curve as the much-deplored population explosion, plotted on the same scale, that has attended the industrial revolution.

The cultural revolution must be reckoned the most consequential of the three. From it issued H. *sapiens*. In the brief time since, the discernible evolution of the human species has been social. The intervals of evolution shorten to tens of thousands and thousands of years.

Homo *sapiens*

Between 100,000 and 50,000 years ago the large-brained, smooth-browed, light-skeletoned H. *sapiens* displaced the populations of archaic man all across Eurasia. Students of this phase of prehistory disagree on just how and when H. *sapiens* first appeared.

Some hold that archaic men and women evolved to modern men and women *in situ,* wherever they lived on the African and Eurasian continents. From the same genetic stock, encountering substantially the same selective pressures, in chains of contact that reached across the Eurasian continent, all had evolved in the same direction to H. *sapiens*. Isolation of the breeding populations from one another, according to this scheme, would account for the racial differences in the species.

Others contend that H. *sapiens* diverged from H. *erectus* in the homeland of Hominid evolution in Africa between 200,000 and 100,000 years ago. Toward the end of that period, H. *sapiens* came out of Africa into Eurasia in the third—from Proconsul—wave of Hominoid invasion. Genetics supports the belief that the common ancestor of the two species lived in Africa.

DNA sequencing has by now sampled the DNA of the mitochondria in the world's major population groups. Conveyed only in the mitochondria of the capacious egg, mitochondrial DNA (mtDNA) undergoes no sexual recombination. Differences in mtDNA sequences are owing, therefore, exclusively to mutation. Since mutation happens at a constant rate, the number of substitutions in the mtDNA sequences serve as a reliable measure of genetic differentiation over time.

Backward counting of nucleotide substitutions in population groups around the world traces the ancestry of all to "Eve," the females in a breeding population in Africa 200,000 years ago Counting from Eve forward, the substitutions establish separation of Eurasian populations from African populations at 100,000 years ago and separation of Asian from European populations at 50,000 years ago.

Racial differences among human populations, geneticists argue, are too trivial to have developed over the hundreds of thousands of years of genetic isolation implied by the *in situ* model. Not more than four genes—and recessive ones, for that matter—account for the black skin of the African population most distantly isolated in time from the other population groups. H. *sapiens* must have evolved in one place; the so-called races, by genetic drift, diverged afterward.

The scant fossil record has begun to show H. *sapiens* in the right places at the right time to fit African genesis. At 90,000 and 75,000 years ago H. *sapiens* appears in Africa; at 60,000 years ago, in West Asia; at 45,000 years ago, in East Asia. At 60,000 years ago, H. *sapiens* appears in southern Europe and then coeval with H. *erectus Neanderthaliensis* as late as 45,000 years ago. There is no sign of interbreeding between the two populations, affirming them as different species.

To the *in situ* hypothesis, there is the final genetic riposte: genes have descendants; not all fossils do.

H. *sapiens* ventured onto the Eurasian continent when the Wisconsin glaciation was nearing its peak. They carried the classical Old Stone Age culture across the entire continent as the glaciation waned between 45,000 and 10,000 years ago. Around 35,000 years ago, people crossed the Bering Strait land bridge exposed by the glaciation-lowered sea level. Signs of human occupation as early as 6,000 years ago have been found in Tierra del Fuego.

The Inuits must have been among this venturesome population;

they adapted to glacial life and stayed in the Arctic. Another venturesome population reached New Guinea and Australia 40,000 years ago across at least 70 kilometers of ocean lowered by the last Pleistocene glaciation. Surely the record ocean crossing at the time, it pales by comparison with the voyages of the stone toolmaking navigators who began the peopling of the Pacific islands around 3,500 years ago.

Homo *erectus* displaced

The displacement of H. *erectus* by H. *sapiens* would seem to present a classical case of the neo-Darwinian "struggle for existence." William W. Howells of Harvard University, a proponent of the *in situ* evolution to *sapiens,* took that idea for granted. As recently as 1979, in evidence against displacement of *erectus,* he observed that no cave had been found "containing the remains of killed Neanderthals in association with Upper Paleolithic tools"—tools, that is, of H. *sapiens.*

H. *sapiens,* it has to be agreed, is capable of violence. Evolution has been shown, however, to proceed more often by positive selection than by pruning. In the modern synthesis, "fitness" comes down to surviving progeny. George Gaylord Simpson set the matter straight:

> Advantage in differential reproduction is usually a peaceful process in which the concept of struggle is irrelevant. It more often involves such things as better integration into the ecological situation, maintenance of a balance of nature, more efficient utilization of available food, better care of the young, elimination of intragroup discords (struggles) that might hamper reproduction, exploitation of environmental possibilities that are not the objects of competition or are less effectively exploited by others.

The 1 million population of H. *erectus* occupied the African-Eurasian landmass at a density of one per 83 square kilometers. Allowing that two populations of 1 million would seek the same kinds of habitat, there would still have been room for everyone. *Australopithecus* and H. *erectus* shared the grasslands of Africa for a million years.

Displacement of H. *erectus* must have followed from greater mastery by H. *sapiens* in the several categories of behavior listed by Simpson.

398 THE AGE OF SCIENCE

All across the Eurasian continent, over tens of thousand years, the gradually outnumbered Neanderthals and their cousins would be displaced to marginal environments.

At times there may indeed have been violent collisions between individuals and bands of the two species. The statistically rarest site of such encounters would have been a cave. Elsewhere, time would have erased all trace of what might be regretted as deeds of Cain.

No primitive language

From the few fossil skulls of the first H. *sapiens* intruders on the Eurasian continent not much can be told about their capabilities and way of life. The best insights are offered by the Stone Age cultures that still survive for study by anthropologists. Best known are the denizens of Amazonia, the Pygmy bushmen of Africa, and the Inuits on the Arctic shores of Eurasia and North America.

As to capability, none of these people—and no indigenous people encountered on any continent—have been found to speak a primitive language. The capacity for speech is the conclusive phase in the biological evolution of H. *sapiens* in the social setting of family and clan. There is no such thing as a primitive language. Every language ever studied builds upon the same deep syntactical infrastructure. That infrastructure is plainly exposed in the Creole languages of the world.

Creole languages are the invention of the children of speakers of unstructured and rudimentary pidgin speech. Workers imported from elsewhere, for example, pick up a minimum repertory of essential words in the language of the new country. As has been observed by linguists in Hawaii, the children of the pidgin speakers learn to speak amidst the linguistic confusion of the parental tongue, pidgin, and the language of the country. With no consistent model to follow, they construct their own language. The bad English brought from poverty to school by many African-American children is a Creole spoken by their forebears in the first generations of children born in slavery. The Creoles are languages in their own right. Each solves, in its own way, the same set of problems, from the distinction between the singular and the plural to that between the declarative and the subjunctive voice.

The evidence strongly supports the thesis, advanced most notably by Noam Chomsky of Massachusetts Institute of Technology, that a universal grammar underlies all human language. It expresses the capability for symbolic communication wired and otherwise installed by natural selection in the brain. The event must have transpired throughout the biological evolution that eventuated in the arrival of H. *sapiens*. The earliest H. *sapiens* skulls housed the same mind that, over time, has accumulated the present stock of objective knowledge about the physical world. It invented, as well, all the imagined worlds that have filled human heads in the most primitive cultures and the highest civilizations.

Observers of surviving hunter-gatherers never fail to be instructed by their comprehension of the world they know. Ethnobotanists, following Wallace on collecting expeditions into Amazonia, have yet to exhaust the native lore. In the high diversity of tropical vegetation, the indigenous people distinguish and have use for an apparently endless number of plants. Fruits and tubers in their cuisine are coming on the world market. The typical shaman commands a pharmacopoeia of nerve poisons (for the points of arrows and darts), antibiotics, contraceptives, psychotropics and apparent cancer-inhibitors. The pharmaceutical industry is making a business of ethnobotany.

Living with Inuits on the Arctic coast of Canada early in the 20th. century, Vilhjalmur Stefansson found them comfortable and generously hospitable to visitors in—what visitors found to be—their barren and hostile environment. They made fires for cooking but not particularly for warmth. With casual competency, they conserved their body heat in the design of their clothing and dwellings. From the seal they secured the raw materials of their clothing, shelter, tools, hunting weapons and oceangoing transportation.

In the forests of the Congo watershed, the *bamiki nde ndura*—people of the forest, as the Pygmies call themselves—live as integral constituents of their ecosystem. Their way of life remains essentially unaffected by long-standing trading relationships with settled agricultural villages on the forest margins. Colin Turnbull of the American Museum of Natural History was the first to follow them into the forest, in the early 1970s. He estimated their population then at 40,000.

In bands numbering 100, of perhaps 30 families, they occupy peacefully bounded territories of around 100 square kilometers. The small size of the families bespeaks the high infant mortality and short life expectancy of the Stone Age. The machete, from trade with the villages, has replaced the primary stone tools, and some of their arrows carry metal tips. Otherwise their tools, weapons and implements are made from materials from the forest.

No discernible hierarchy governs the band. All join in the tasks of their communal existence, each contributing as qualified by age and sex. At a new encampment, the men supply the materials and the women erect the domed, leaf-shingled huts. Hunters, joined by women and the older children, beat the forest, driving forest antelope and smaller animals into nets strung across strategic forest alleys. There, other hunters dispatch the prey with spear and bow and arrow. Women, in this division of labor, know where to find fruit and edible roots. The encampments move periodically; the forest goes fallow. On the move, the band takes nothing but their metal tools. The forest fills with their voices in high-spirited talk and song; this has the adaptive advantage of flushing leopards and other predators out of their path.

No primitive art

The conquest of the Eurasian continent by H. *sapiens* has its contemporary record and bravest celebration in the Stone Age art that adorns the walls and ceilings of caves in France and Spain. The caves at Lascaux and Altamira are the best known of 100 caves. In these cathedrals of the Stone Age, painters were at work over the 20 millennia from 30,000 to 10,000 years ago. Their oldest works depict reindeer that went north with the last glacier. They depict the woolly mammoth, the European bison and the aurochs, a wild cattle, in the extinction of which the painters' contemporaries are implicated. The ceiling of one immense chamber in the Lascaux cave bears heroic paintings of aurochs bulls, some 5 meters from nose to tail. Present also from the earliest millennia are the rhinoceros and big cats that must have retreated to Africa with the advance of the last glacier. Through all the millennia, the red deer and the horse (still then a game animal) remain faithfully at hand.

Religious significance is inevitably imputed to these paintings. Such significance is indicated also by the presence of highly abstract as well as graphic depictions of human genitalia. Without doubt, concern for the sustenance and fertility of the species drove the artists. Whatever the prayer or magic they intended, their veneration of their subjects is plain. The bison's shoulders, the red deer's antlers, the mammoth's tusks, in just the right overstatement, bring the presence of these beasts across tens of millennia.

The Lascaux artists were masters of a considerable technology. Their pigments were not at hand there at the cave. For some pigments the artist traveled 20 kilometers. They ground the pigment minerals to fine powder with stones serving as mortar and pestle and mixed with cave water on flat stone palettes. Color saturation was adjusted with porcelain clay or finely ground quartz. Happily, the high calcium content of the cave water helped secure adhesion of the pigments to the damp cave wall and ceiling.

The discovery at Lascaux of a hank of rope—two thick strands, each twisted from many twisted thin strands—testifies to another vital, universal craft. In most ancient habitations, only the tools of the spinner and weaver remain. Junius Bird of the American Museum of Natural History found 3,000-year-old brocade in a unique state of preservation in the middens of the Moche culture of Peru. From this he concluded that there are no primitive textiles, just as there is no primitive language. People of the Stone Age had textiles, as well as hide, to wear. Some of the bounteous fertility effigies from this period, the Venus of Willendorf best known, are adorned in brocade [preceding page].

Out in the open on the Eurasian continent through those millennia, subarctic conditions generally prevailed, punctuated by occasional centuries of warming and renewed glacial advances. Of human experience at this time most is known about the region westward from the Urals to the far Atlantic shore of the British Isles, then attached to the continent by lowered sea level. The Baltic was an inland freshwater lake. At glacial maximums, the open treeless tundra reached south to the Pyrenees and the north coast of the Black Sea.

Following the glaciers

With bow and arrow hunter-gatherer bands followed the migrations of the reindeer. With stone-tipped lance they took the mammoth and the aurochs. In the tundra, they found berries and roots. Like H. *erectus* they used mammoth bones to anchor the hide-covered huts of the earliest few of their encampments found in Russia and Ukraine. Reindeer succeeded the hunted-down mammoth in supplying bones and antlers to later encampments.

In warming intervals, the tundra retreated northward before the advance of the taiga forest of evergreens and birch. Life was easier for the generations that lived in the centuries of the waxing and waning of this more diverse ecosystem. Deer and small game were to be had, and more immediate nourishment was found in berries, nuts, seeds and edible leaves and roots. In the short span of their lives, people made the best of the world as they found it, whether tundra or taiga.

Some 1,200 encampments, from between 14,000 and 10,000 years ago, have been located across northern European plains to the Urals. Evidence indicates that three or four families joined in a hunting band. The size of the economic unit may be taken as a measure of the resources afforded by the environment and accessible to the technology at the command of the unit.

The population at 10,000 years ago has been estimated by Deevey at 5 million. People occupied the land, now including the New World as well as the Old, at the density of one person per 25 square kilometers. By that time, 30 billion human beings had come and gone. If not always brutish and bloody, life was still short: 25 years. The intervals of social evolution now shorten to thousands and hundreds of years.

Preagricultural civilization

With the northward retreat of the continental ice, beginning around 10,000 years ago and proceeding rapidly thereafter, came transformation of the Eurasian landscape. West of the Urals forests of evergreens and then closed-canopy forests of evergreen (gymnosperm) and deciduous (angiosperm) trees succeeded the taiga. In a few millennia, the northern margin of the forest crossed the Baltic—once more, a sea—into Scandinavia. The biological diversity and wealth of the new ecosystem—the sheer increase in plant biomass—sustained corresponding increase in the animal—including human—population. In addition to deer, the forest teemed with small mammals, such as the hare and the squirrel as well as fowl, such as pigeon, quail and partridge. The lakes, streams and oceans had waterfowl, fish and shellfish.

Such resources are available in cycles of abundance and scarcity. There is, first of all, the cycle of the seasons and the migration of important prey. The population of small mammals, such as the hare, oscillates on another, longer-term, cycle with the population of their predators, terminating on occasion in the crash of both populations. Especially as the new ecosystem was making its way northward, the hunter-gatherer bands persisted in their migratory way of life.

Over time, with increase in the abundance of the environment and in yield secured from it, the migratory gave way to a more sessile way of life. Encampments were occupied for longer periods. The next encampment was closer to the last, a move merely from streamside or beach to forest upland. By what Robert J. Braidwood of the University of Chicago, called "living into the environment," people increased the yield of their habitation to their well-being.

By that expression, Braidwood summed up the deepening knowledge and sharpening ingenuity by which people far to the south in Asia Minor recognized and made use of resources in the world around them and went on to agricultural revolution. In the northern forests of Europe, around the Baltic Sea, such mastery established another way of life that endured for as long as two millennia. By 5,000 years ago, people came to live in permanent "forager" settlements. From forest and meadow, lakes, streams and estuaries within a radius of 10 or so kilometers, intensification and specialization in food-getting brought sustenance in security and sometimes abundance.

The migratory families now made a community of households. As households, they divided the communal tasks. Variously, they specialized in hunting or fishing, in gathering this plant food or that or in practice of one or another of the new supporting crafts. Pottery found its way to the Scandinavian peninsula from agricultural settlements in Asia Minor. This testifies that the forager settlements had food to store.

Specialization brought commerce between settlements. A flint mine discovered near Maastricht in the Netherlands produced an estimated 100,000 metric tons of nodules over a millennium or so beginning around 5,000 years ago, enough material to produce more than 150 million ax heads. Commerce had begun.

In the 1980s Marek Zvelebil of Sheffield University found evidence of class distinction associated with the division of labor in these settlements. The evidence comes from excavation of 20 substantial cemeteries. The dig at Oleneostrovski Mogilnik (Deer Island) north of Moscow tells a convincing story. The wealthy there were buried with necklaces of bear teeth, the most prized; people of middle rank were buried with necklaces of beaver and elk teeth; the poorest with no necklaces. Some men, held in special, probably religious, awe, were buried standing upright, much adorned, in their graves.

The legends of Nordic mythology may well have their origin in folk memory of this protocivilization. In that case the conquest of the Eurasian continent may be said to have its romantic celebration in the *Ring* operas of Richard Wagner. The Nordic legends, it is true, conflate earlier with later prehistory and equip their warriors with tools from the Iron Age. There, Wagner has his mortal and divine heroes beset and undone by a villainous clan of miners.

In those legends, given substance now by archaeology, the primary social-economic unit had grown to be the clan. Typically, a clan consisted of a few dozen families of sufficiently extended kinship to per-mit marriage within. The ordered anarchy of simpler times yielded to hierarchy headed by chiefs, who sometimes went by grander titles.

Agricultural civilization

Meanwhile, people living into other environments went on to agricultural revolution. This revolution ensured to the human species our daily bread. It must be reckoned as Robert Braidwood reckoned it: the first achievement of our species comparable to the initiation of tool-making, 2.5 million years earlier, by the man-apes who became our fore-bears. By agricultural revolution, people began the increase in yield from the planet to support their numbers that continues to this day. Favorable hunting grounds in North America sustained, it is estimated, five people per 25 square kilometers. The technology of the forager set-tlements of northern Europe sustained perhaps five people per square kilometer. By 5,000 years ago, agriculture was supporting 50 people per square kilometer of settled land in Asia Minor.

Around 15,000 years ago, on the southern piedmont of the mountains of what is now the border of Turkey on the north and Syria and Iraq on the south, people began to cultivate wheat. Across the Eurasian continent in the Yangtze valley, at about the same time, people began to grow rice. Later, by one or two millennia, people began to grow maize on the central highlands of Mexico. At all three times and places, people apparently arrived at the same solution to the problem of sustenance in ignorance and independence of one another's enterprise. While the cultivation of wheat did diffuse across Eurasia by contact of cultures, the rice revolution appears to have been *de novo*. There could have been no diffusion to speed the maize revolution. The three revo-lutions must be seen therefore as the outcome of the social evolution of humankind in regenerative feedback with the concurrent accumulation of objective knowledge.

All three revolutions, accelerating the growth of the populations they engaged, entrained their further social evolution on parallel cours-es. People settled in villages of a hundred or a few hundred inhabitants, depending upon the yield from the land within walking distance. The agricultural villages came to support up to 20 percent of the population off the land in other, new occupations. By hierarchically organized enterprise, the human species undertook to adapt the environment to its needs and pleasures. In time, that 20 percent concerted in the creation of agricultural civilization, its monuments, cities and empires and the making and writing of history.

Society's "normal structure"

Among the new occupations, from the beginning, was the exercise by religious and secular authorities of the moral and physical coercion that secured the living of people not bound to the land. The priests, often themselves the rulers, supplied the sanction of the gods to the disparate distribution of resources and product. They anchored what Alfred North Whitehead, in his *emeritus* professorship at Harvard University, called "the normal structure of society" in agricultural civilization: the better-off few and the semidestitute many.

Life expectancy in agricultural civilization remained at the 25 years of past experience. At that expectancy, the median age of a society is around 10 years. At the 10-year median, 60 percent of a population are children. For the many—the 80 percent—life expectancy and the median age were lower still. The 80 percent made possible for the 20 percent a life expectancy and median age often comparable to the norms of contemporary industrial society.

The privileged 20 percent, then, governed societies of children. The *proles,* the bearers and parents of all those children, were the readier to submit to authority, embrace the comfort of religion and accept the divinity of kings to which their children were hostage. They accepted the world they were born into. Some few found their way into the 20 percent. From time to time, others—the gladiators in Rome, peasants in Austria, slaves in Alabama—disturbed the peace.

The many were to remain in destitution as long as increase in the productivity of their labor made no substantial difference in a human lifetime. For nearly the duration of recorded history, increase in production could be secured only by increase in the land under cultivation. In accordance with what T. R. Malthus in 1798 called "natural law," the means of subsistence increased arithmetically while population increased in equilibrium with misery. In agricultural civilization, the few could improve their well-being only at the expense of the many.

To Asia Minor the final waning of the Wisconsin glaciation brought an age of warming and rainier climate. The Mediterranean forest spread from its narrow range on the coast eastward across the Anatolian highlands, the watershed of the increasing flow of the Tigris and Euphrates Rivers. On the piedmont, open forests and meadows spread. On the steppe of the Levant, once again barren and divided among Israel, Lebanon, Jordan, Syria, Iraq and Iran, a diverse community of plants and animals came to flourish.

The wheat revolution

The archaeologist James Breasted of the University of Chicago in the 1930s called this region the Fertile Crescent. Its desert landscape featured uncounted large and some larger mounds called, in Arabic, "tells." Archaeological digs here during that last 50 years confirm ancient fertility. Excavation of the Tell Abu Hureyra, in the steppe land of Syria in the early 1970s, proved especially rewarding. It uncovered at one site the debris of human occupation through the first millennia of the agricultural revolution. At the lowest level, dated to perhaps 12,000 years ago, "living into the environment" had already brought people into settled occupation in villages. Families lived in pit dwellings; holes around the pits show that posts held up walls and roof, probably of reed and grass thatch. Kitchen middens yielded bones of wild goats, sheep and pigs. Thin, sharp-edged "lunate" blades of flint, the most common of the many types of tools, probably tipped the arrows and the spears that brought in this prey.

The same blades, set in wood handles, were used in sickles for gathering grains. The kitchen middens yielded in some abundance grains of the primitive, wild einkorn wheat, as well as lesser quantities of wild barley and rye. Flat stones with depressions worn by rubbing with "pestle" stones may have served for milling. Some evidence supports the possibility that these grains were even then under cultivation at Abu Hureyra. From the long-buried soil, fine screening and flotation sorted out "phytoliths," fossilized bits of weeds that comprise, with cultivated plants, the artificial ecosystem of agriculture in the Near East today.

For reasons unknown, after some hundreds of years, people abandoned the settlement at Abu Hureyra. Then, after a millennium, around 9,500 years ago, the site came to life again as a kind of regional capital for many smaller settlements, some under other tells in the region today. Agriculture supported increase in the population of the town to several thousand. A nearby dry riverbed, scoured by runoff from now-infrequent rainfall, holds evidence of irrigation practiced in the past. Animal husbandry contributed as well to the wealth of the community. Cattle and pigs had been domesticated along with sheep and goats.

People lived in mud-walled houses, arranged in orderly blocks, separated by narrow alleys, each house with two or more rooms. Built into houses as a structural feature was a plaster bin for storage for grain. More recognizable mortars and pestles milled the grain.

Bronze and then Iron Age

Settled agricultural village life brought the end of the Stone Age. Metal-rich ores of copper occur in Asia Minor; the island of Cyprus derives its name from the metal. Evidence at many sites shows smelting of the ore in practice around 5,000 years ago. Simple furnaces reached 1,100 degrees C, the temperature at which copper melts. With arsenic as a contaminant in the ore, the furnaces yielded "natural" bronze. Soon craftsmen found that they could make bronze by alloying copper with tin. By casting the metal, they could make bronze artifacts of many kinds, starting probably with sickles. Forging came next, of weapons that kept their edge, and of shields and armor that withstood the weapons. The Bronze Age saw the sack of Troy.

Some furnaces employed the iron ores, magnetite or hematite, as the flux material to carry off the gangue from the ore. A by-product of these furnaces was sponge iron, an iron-silica compound with a melting point of 1,200 degrees C. Throughout the Bronze Age sponge iron was used as an ornamental metal. In the process of remelting, with charcoal as fuel, craftsmen found that iron hardened to steel. From furnaces pumped by bellows to the 1,537-degree melting point of iron, they made iron and then steel as products, not by-products. By hot and cold forging, they hammered steel into weapons and armor with twice the strength of bronze. The Iron Age succeeded the Bronze around 3,000 years ago, and steel held military supremacy until the middle of the 20th century. It shares honors now with aluminum and uranium.

First urban civilization

By 6,000 years ago, irrigation brought agriculture down the Euphrates to the delta formed then by its convergence with the river Tigris. (The much-diminished present flow of the rivers reaches the Persian Gulf at two widely separated mouths.) Within a millennium, the harvest from the villages was sustaining cities there—Ur, of biblical memory, Lagash, Nippur, Shuruppak, Kish, Erech, Asmar and a dozen more—with populations of 10,000 to 50,000. In city centers, the ziggurat—the terraced tower of Babel was a ziggurat—declared the presence of the religious authority that organized the economy. Substantial dwelling places secured the status and comfort of the authorities. Outside the walls of the cities, faint traces mark the dwellings of the many. This was the world's first urban civilization. Beyond evidence that the people spoke a Semitic language, their identity is unknown.

Their civilization is known by the name of their conquerors, the Sumerians. Around 5,000 years ago, these strangers came in bronze on wheels. They came apparently from somewhere around the Caspian Sea, where the wheel and horses in harness first appear in petroglyphs.

The Sumerian conquest may be taken as the first in the long series of conquests of centers of agricultural civilization by "barbarians." Another subsistence technology, the pastoral nomad, was to keep agricultural civilizations under recurrent siege.

History has its literal beginning in the empire of Sumer. By 6,500 years ago the Sumerians were keeping their books, writing their laws and recording current events in the first written language. In the cuneiform they invented, hundreds of thousands of clay tablets have preserved the beginning of written history. Sumer endured until 3,500 years ago, when the empire was conquered by the Babylonians.

Samuel Noah Kramer of the University of Pennsylvania attributed to the Sumerians the invention of "the potter's wheel, metal casting (copper and bronze), riveting, soldering, engraving, cloth fulling, bleaching and dyeing ... manufactured paints, leather, cosmetics, perfumes" and "command of a large assortment of *materia medica,* prepared from plants, animals and inorganic sources." From the work of his colleagues and his own, Kramer could acclaim "their houses and palaces, their tools and weapons, their art and musical instruments, their jewels and ornaments, their skills and crafts, their industry and commerce, their *belles lettres* and government, their schools and temples, their loves and hates, their kings and history."

Such was the release of human capability, energy and aspiration enabled by agricultural revolution not much more than 5,000 years after its beginning.

From the Euphrates-Tigris delta the practice of agriculture made its way down the Persian Gulf into the Indus valley in today's Pakistan. On Bahrain Island it reared a great city. In the Indus valley, new centers of urban civilization came to flourish at Mohenjo-Daro and farther upstream at Harappa. They vanished after 1,000 years, overrun by invaders and replaced by scattered villages. The cultivation of wheat spread throughout northern India and sustained a village culture there before the arrival of the rice revolution from China.

From the Levant, wheat cultivation spread to North Africa. Semitic farmers may have carried it to Egypt as early as 7,000 years ago. The biblical story of the Jewish enslavement may recall that time.

Indo-European languages

In the area from the south shore of the Black Sea eastward to the Caspian Sea, now the territory of Georgia and Azerbaijan, the wheat revolution supported another developing civilization. Some scholars argue that the agricultural revolution had its origin here; the evidence may have drowned in the flooding of the Black Sea around 6,000 years ago, remembered in the Old Testament deluge. Quite apart from that question, Thomas V. Gamkrelidze of the University of Tbilisi and V. V. Ivanov of the University of Moscow have shown that the Indo-European languages originated in this region. Those languages are now spoken by more than half the world population. The tracing of the root words of the ancestral language into the European languages has traced simultaneously the spread of the wheat revolution into the European landscape. In linguistics, the identification of the common and relatively common words in a family of languages reaches, by back-convergence, the proto-language. This procedure is analogous to those that establish the family trees of living organisms. Proto-Indo-European has words for "wheel," (a word like *roto*) and "horse" (a word like *equus*) and words for "axle" "yoke" and "foal" that sound like them in English and in German. In one assortment or another, these words appear in all the Indo-European languages. The ancestral language describes, in the roots of its daughter languages, the landscape at the foot of the Caucasus Mountains, its fauna and flora, and has terminology for agriculture.

Here, it appears, the horse was domesticated and the wheel invented. Petroglyphs of wheeled wagons and horses in harness, found in Uzbekistan on the far side of the Caspian from the Transcaucasus, are dated to around 5,000 years ago. From this general region, according to Kramer, the Sumerians descended on the Euphrates-Tigris delta.

Slow diffusion, not conquest, brought the spread of agriculture. The success compelled the growing agricultural populations to find new land. The spread went principally by cultivation of land at the margin of land already under cultivation. Indigenous peoples in lands on which they thus encroached would assimilate the new technology. Along with it, they would assimilate into their language the Indo-European vocabulary for agriculture and then for other aspects of the new way of life.

The people who introduced Indo-European root words into the Sanskrit of India eventually overran the urban civilization of the Indus valley. In the hearing of Portuguese navigators early in the 16th century, coincidences in vocabulary inspired the notion of the superfamily since confirmed by linguists.

Agriculture arrived in Mediterranean Europe deep in pre-Homeric times, 7,000 to 6,000 years ago. It brought the introduction of Indo-European words, especially those for the new technology, into what is now the Romance languages. While the wheel undoubtedly played its part, water transport must have hastened the passage. It left milestones in the Scythian cultures on the north shore of the Black Sea. The Hermitage Museum in St. Petersburg treasures the laurel leaves of gold that crowned Scythian kings.

Indo-European reached northern Europe by roundabout diffusion of agriculture eastward of the Caspian Sea—an Indo-European language is spoken in Chinese Turkestan—then northward around the sea into the steppes of today's Russia and Ukraine. There it left root words in the Slavic languages. Across the northern European plain and around the Baltic Sea, the contentment of the forager settlements delayed the penetration of Indo-European. The technology of agriculture did not take hold until around 3,000 years ago. Its success is then recorded in the Baltic, Nordic and Germanic, and the English and Celtic languages.

The rice revolution

The search for the site of earliest rice cultivation turned first to the peninsula of Indochina. Farther from the glacier that covered the northern Eurasian landmass and generously watered by rivers draining the melt of the Himalayan snowpack, it seemed more hospitable than regions on the continent in China. The earliest sites of rice cultivation so far identified in Indochina, however, date to paddy cultivations in the 6,000 years ago. It now appears the rice revolution began in the Yangtze valley, upstream in Jiangxi province in the heart of China.

Work in the last decade has established there a complete sequence of human occupation and social evolution from at least 24,450 years ago through the first appearance of wild species of rice, to rice under cultivation in the field and then under paddy cultivation between 9,000 and 7,500 years ago. The principal instigator of this still-ongoing work was Richard S. MacNeish. It was he who fixed, in the 1960s, the site of the maize revolution in Mexico [see page 417].

MacNeish enlisted colleagues at Beijing University and Jiangxi Institute of Archaeology in a joint venture with the corporate entity—the Andover Foundation for Archaeological Research—that supported his enterprise. Review of the literature pointed them to two caves and a lake in the countryside of the Jiangxi province.

In Diaotonghuan Dong (Buckethandle Cave), they dug a sounding to a depth of 15 meters, still 5 or 6 meters from the rock floor. They identified 21 distinguishable layers of soil on the way. The bottom three layers held no cultural artifacts. The next above yielded two stone tools. In the layer above that, along with nine stone tools, charcoal in a hearth made it possible to fix the date of 24,450 years ago.

In the fifth layer farther up, dated to 18,000–17,000 years ago, the pollen and phytoliths of the wild *Oryza rufipogon* first appear. The presence of a new tool, found in larger numbers in the upper layers, shows that rice was now in the diet. These are shells of a freshwater clam or mussel with two holes drilled through them. A cord through these holes secured a shell to the middle fingers of the hand; held thus in the palm, concave face outward, the shell, its edge sharpened, made a knife or sickle to cut rice stems.

Phytoliths of the more readily cultivable *Oryza sativa* turn up with *rufipogon* around 11,000 years ago. This suggests a start on rice cultivation as a hedge, perhaps, against the yield from foraging. The relative percentages of the two kinds of rice reverse over the next several thousand years. Pottery sherds then appear in increasing quantity. The foragers are incipient farmers. Occupation of the cave, seasonal in any case, ceased around 6,000 years ago. By that time people had settled in villages, supplementing gathered food by planting.

Sounding of soil layers, signs of occupation in Xian Ren Dong (Benevolent Spirit Cave) and core sediments from PoYang Lake bearing rice pollen and phytoliths all yielded parallel findings and richer evidence about later millennia. No evidence of rice was found in bottom core sediments; wild strains of rice were detected in sediments from around 12,000 years ago; at 4,000 years ago, the pollen and phytoliths of cultivated rice abounded. The top layers at Xian Ren Dong, more recent than those at Diaotonghuan Dong, yielded tools associated with rice planting, notably stones shaped to doughnut configuration to weight the dibble sticks used for planting rice.

Assay of the collagen in bones from 10 different human skeletons told a parallel story. Isotopes of carbon and nitrogen appear in different ratios in the food people eat and so in the tissues of their bodies. Even the strains of rice may be distinguished from one another by such analysis, and so may the bones of consumers of the rice. The oldest bones show no evidence of rice in the diet. The isotopic ratios characteristic of the wild *rufipogon* strain appear in the next oldest bones; of that strain and the cultivated *sativa* strain, in the next oldest; and finally the isotopes of the *sativa* and *indica* strains from paddy cultivation.

From all the evidence, MacNeish and his colleagues conclude that rice domestication was under way in this region from 11,000 to 9,000 years ago. Paddy rice cultivation was developing in the shallows of Po Yang Lake between 9,000 and 7,500 years ago. By that time, the two major cultivated rice strains—*japonica* and *indica*—were well established and harvested alternately, yielding two crops a year. Rice cultivation had begun meanwhile in neighboring Hunan province.

From China it is thought rice cultivation spread to Indochina in the period 8,000–7,000 years ago. Today rice supplies the primary calories of more than a quarter of the world population.

Qin's ceramic army

By 2,300 years ago, when the emperor Qin brought the half dozen "contending states" together in a single empire, agriculture had supported an urban civilization in today's northern China for 1,000 years. Excavation of Qin's capital at Xian not long ago unearthed the spectacular ceramic army that now marches, life-sized, into the daylight in the archaeological dig at the foot of the tumulus in which the founding emperor Qin himself may be buried.

Rice was not the grain under cultivation in the early empire. By that time, diffusion had brought grain cultivation from Southeast Asia. Coarse-grained millet was grown on the deep, soft and productive loess soils south of the Yellow River. The uncertainty of rainfall in the region made intensive agriculture and irrigation works rewarding. Agricultural technology was well advanced in the empire assembled by Qin and taken over by the Han dynasty upon his death.

The extent of land under cultivation around a village and the size of the village were largely a function of the supply of night soil and the distance to which it could profitably be transported. The fading of greenness with distance from Chinese villages brings this economic gradient into visibility even today. Villages formed a cluster around the provincial town, where the landlord gentry resided; the size of the town reflected the agricultural output of the region and the range of profitable transport of the goods. The towns clustered, in turn, around a walled city, not usually more than a few days' march from the most distant village within its political-economic sway. Here, the Mandarin administrative class managed the logistics of public works and tax collection. A contending state would command a circle of walled cities. As Owen Lattimore, historian of the inner Asian frontier, observed, Chinese civilization had a modular structure.

The landlord induced the labor of the peasant most often by sharecropping. Opportunity to work the land would go, by competitive bidding, to the peasant willing to surrender the largest share of the crop. By this strategy, Chinese "feudalism" reaped bigger harvests than ever could be secured by coercion. Labor for larger undertakings of shorter duration was enlisted, on the other hand, by the corvée; that is, by force.

Chinese bronze castings from as long as 3,500 years ago are among the treasures of the world. Around 2,800 years ago, the technique of "stack-casting" mass-produced iron-harness fittings and axle bearings and like hardware. From China just 700 years ago, trade on the Silk Road brought the magnetic compass, gunpowder and the cannon into Europe. Although Chinese mathematics was weak in geometry, their sky maps located the supernova of A.D. 1054 so accurately that modern astronomy was able to find it and with time exposure reveal it as the beautiful Crab Nebula.

Early in the 15th century, a few decades before Vasco da Gama sailed around the heel of Africa, 19-masted sailing vessels—the biggest ships ever launched before the 19th century—went to sea under the command of the imperial eunuch He Zheng. A dozen of them, escorted by a flotilla of smaller ships each larger than any Spanish galleon, altogether bearing 67,000 men, explored the Southwest Pacific and the Indian Ocean; paid calls on the coasts of India, Africa and the Arabian peninsula; mapped the southern sky and, after three years, sailed home and never went to sea again.

This episode holds perhaps the symbolic answer to why China did not go on to industrial revolution. The Central Kingdom had satisfied its interest in the world beyond. The splendors concealed behind red walls, such as those that enclose the Forbidden City in Beijing today, represented all that its privileged families could think of to want.

Around 2,300 years ago, as Qin was uniting China, Ashoka was bringing imperial unity for the first time to India. The tax-collector landlord practiced the same sharecropping strategy as his Chinese counterpart. Hunter-gatherer tribes, settled in agricultural villages, evolved to the caste communities of India today. Some 10,000 castes still ensure identity and, by mutual self-help, security to the members of the Hindu population. Certain castes brought their hereditary stoneworking, metal and textile crafts to high art; folkcrafts are significant in India's exports today. Wheat on the interior plateau and rice in garden and paddy cultivation in the lowlands sustained the great temples and palaces of Indian civilization, again behind red walls.

Thus it was, in the summary judgment of Alfred North Whitehead, that "arrested technology" perpetuated in both China and India "the exact conditions for the [operation] of the Malthusian law." Life expectancy of villagers, 80 percent of the total population, remained at about 25 years. Those populations grew slowly but steadily for 2,000 years. By the 20th century, the red-walled civilizations held nearly half of the world population.

Inner frontier of Eurasia

Visible from Earth satellites, the Great Wall of China stands as the largest monument to agricultural civilization. It was a part of what Owen Lattimore described as the "linked chain of fortified northern frontiers of the ancient civilized world from the Pacific to the Atlantic." The earliest walled frontier, in Southwest Asia, defended the nascent urban civilization of Persia. Against barbarians, the Romans built walls across the waist of Great Britain and on the Rhine and Danube. The Great Wall, begun by the emperor Qin, served as much to wall in the frontier Chinese communities and minimize their transactions with the pastoral nomads beyond as it did to exclude those nomads.

The derogatory "barbarian" has placed the pastoral nomad at an inferior stop on the way to agricultural civilization. In truth, as Lattimore showed, the pastoral nomads of inner Asia, the Mongols in particular, created an enduring, mobile civilization. Pastoralism made perhaps more efficient use of natural resources than did agriculture. The vigorous nomad existence surely testifies to the efficiency with which cattle converted meager pasturage to human nutrition.

The nomads preyed on the frontier of settled civilization to obtain utensils, such as weapons, of use to their mobile existence. The politics that organized their often-warring clans for occasional deeper assaults on settled civilization was as complex and devious as any to be found in capital cities. Nomad conquerors installed their families, more than once, as Chinese dynasties. The most famous, Genghis Khan, carried his 13th-century conquests across Asia to the Danube. His grandsons broke off their siege of Baghdad to hasten home for the more important business of choosing a successor to Kublai Khan. They learned of his death within 24 hours by heliograph from Peking.

The maize revolution

The domestication of maize—"corn"—establishes the inevitability of the agricultural revolution. The reproduction in the New World, then, of Old World urban civilization certifies that members of the same species were proceeding on the same course of social evolution.

While it is not known just when people made their way from Asia into the Western Hemisphere, they were at home in Mesoamerica as long as 30,000 years ago. By 7,000 years ago they had lived in that environment long enough to begin to take charge of it. These were the first findings of the campaign undertaken by Richard MacNeish in the 1960s to uncover the maize revolution. Director then of the Robert S. Peabody Foundation for Archaeology at Phillips Academy in Andover, Massachusetts, he enlisted an interdisciplinary group of 50 counselors and collaborators, including botanists and geologists.

No plant growing in the wild bears much resemblance to the plant that is harvested all around the world as the now third largest field crop. Corn does not look like a grass; it is nonetheless of the same botanical superfamily as wheat, barley and oats. More than five millen-

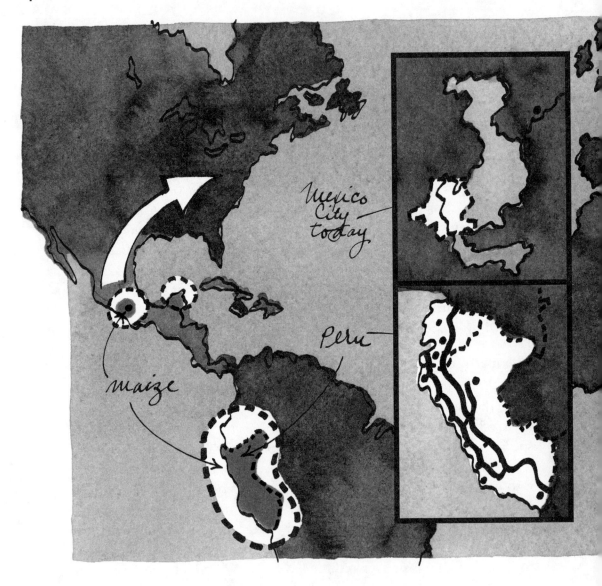

Mexico
City
today

Peru

maize

Three agricultural revolutions *began the increase by human beings of the planet's yield of sustenance to the support of their numbers. The revolutions happened in apparent entire independence of one another. Around 12,000 years ago in Asia Minor, people began to cultivate the wheat and other small grains they had been gathering [see page 407]. They soon settled in villages, and within 5,000 years, village agriculture was supporting the first urban civilization, in the Tigris-Euphrates delta. Wheat cultivation reached the Indus valley by sea*

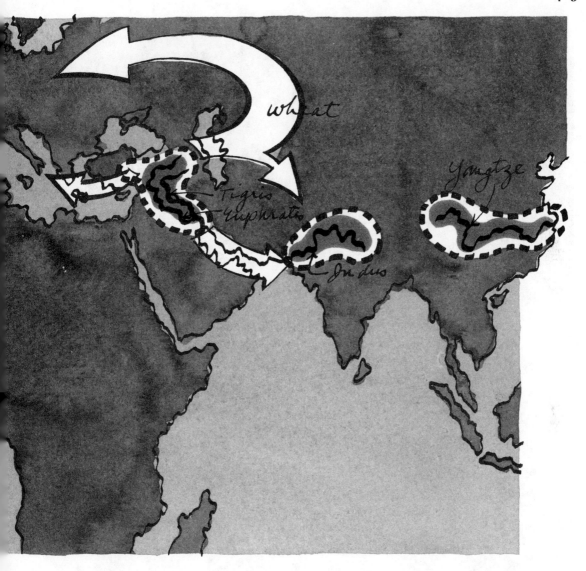

and northern India overland. It spread throughout the Mediterranean basin and later into northern Europe across the Russian steppes. Starting about the same time, rice cultivation was supporting high civilization in China by 5,000 years ago [see page 412]. Maize cultivation began in uplands south of the valley of Mexico about 7,000 years ago; again in 5,000 years, agriculture was supporting high civilization in the valley of Mexico and in Peru. Cultivation of corn had spread over most of North America before Europeans arrived [see page 417].

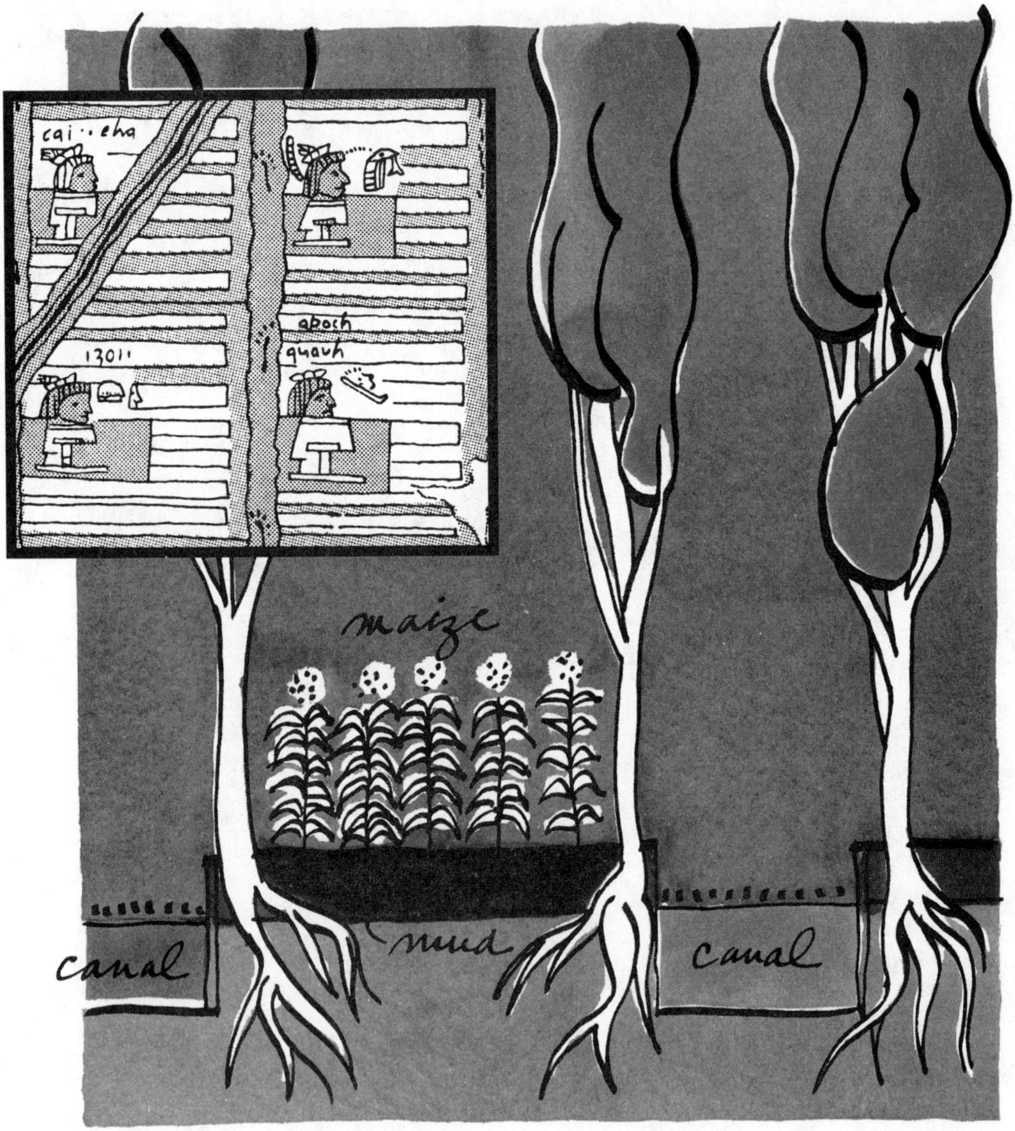

High-yielding chinampas ***gardens*** *in the fresh waters at the south end of the valley of Mexico supported the Aztec civilization. The layout of the gardens, resembling a many-piered marina, is shown in 16th-century Aztec drawing [inset]. Piers of earth, sustained by tree and bush planting, stood a foot or two above the water. To this labor-intensive horticulture, the bottom mud of the canals supplied annually renewed compost. Planting and harvesting went on the year round, yielding six or seven crops of varied produce* *[see page 424]**.*

nia of cultivation, hybridization and crossing of hybrids have obliterated all resemblance to the plant's wild ancestry. Corn is unsuited by its adaptation to agriculture for survival in the wild.

The favored ancestor was a wild grass teosinte, which bears its seeds in a spikelet and its pollen-bearing tassels at the tip of its stalk. George Beadle was convinced of this by his own backcrossing from modern cultivated corn. Paul Mangelsdorf at Harvard University argued that corn was its own ancestor and that wild corn had vanished in hybrids of its cultivated descendants.

In confirmation of the Mangelsdorf hunch, cores drilled in 1953 in ancient lakebeds in the landlocked Valley of Mexico—site of the Aztec civilization and of Mexico City today—yielded corn pollen dated to 8,000 years ago. On the consensus that wild corn had been a grass of dry uplands, MacNeish began the search for its early cultivation on the northern slopes of the valley. Finding none there, he turned to the uplands of Yucatán and Guatemala, where the Mayan civilization had preceded the Aztec. No sign of agriculture earlier than 3,000 years ago was found. Confident that they had bracketed their objective, MacNeish and his colleagues chose the highland valley of Tehuacán, on the border of the state of Oaxaca south of Mexico City. Here, in overlapping occupation levels in five different caves and on the valley floor, three years of excavation produced more than a million items of evidence of 12,000 years of human existence.

From 12,000 to 7,000 years ago, small bands of hunter-gatherer families roamed the Tehuacán valley. Bones of jackrabbits, rats and small birds, and the shells of turtles told of their day-to-day fare, along with plant foods in season. Bones of the then-not-yet-extinct New World horse and of antelopes signified occasional better days of hunting.

The lowest occupation level in one cave, dated to 7,000 years ago, yielded five miniature corn "ears" less than an inch long. Tiny kernels were paired in four to eight rows on the stem that was to become the cob. A tassel or the stub of a tassel tipped each stem. Here, almost without question, was the long-extinct wild corn.

On cultivated corn, of course, the tassels—the pollen-bearing male organ—grow at the tip of the stalk, separate from the seed-bearing female ear. To account for separation of the organs, Mangelsdorf now proposed that wild corn had crossed, by accident or design, with teosinte, the ancestor favored by Beadle. The crossing would have happened often in the wild. Corn and teosinte have homologous chromosomes: the corresponding chromosome carries, in each species, the genes for the corresponding features in the same sequence. Food-gatherers would have been attracted to the hybrid. It would be found with kernels well seated on the stem, whereas teosinte seeds would have scat-

tered from the brittle spikelets. Nor did the kernels, each seated in protective chaff, have the hard shells of the teosinte seeds.

In the caves of the Tehuacán valley, MacNeish and his colleagues found corn ears that reflected increasingly selective collecting and then the beginning of garden cultivation. At 5,400 years ago, 30 percent of the corn was cultivated; at 4,300 years ago, corn was being hybridized by its cultivators. At 3,500 years ago, excavation on the valley floor showed, village life had begun. A large settled population had succeeded the hunter-gatherer bands. Archaeologists have since unearthed nearly 25,000 ears of primitive corn undergoing transformation from the small, wizened, tasseled first specimens in the direction of the corn known to the world today.

The cultivation of corn spread across all of North America. The Pilgrims were glad to find it in Massachusetts in 1620. At the far other corner of the United States, the preliterate, near-urban Pueblo culture was thriving. On the Mississippi, the Mound Builders had already had their day. The hunter-gatherer way of life persisted, however, on the high prairies of the Great American Desert, west of the 20-inch rainfall line along the 100th meridian. There, nomad clans followed the migrations of the bison. They had a ready use for the horse, reintroduced in North America by the 16th-century Spanish *conquistadores.*

Along with the staple corn, the village gardens were producing the full variety of vegetables with which agricultural revolution in the Americas has endowed the world. Foremost are the plants of the remarkable nightshade family, the potato and the tomato; around the world, they come just after the grains from the grasses in production tonnage. There are then the squashes that Miles Standish found "like unto our melons, only lesse and worse." Tobacco must also be acknowledged. Amaranth, the first cereal grain not borne by a grass, may yet find its way to the world market.

It still remained to be shown that wild corn was not its own ancestor. In the early 1980s, Mary W. Eubanks at Duke University succeeded in crossing a perennial member of the teosinte family with Tripsacum. The fertile hybrid bore tiny tasseled ears, such as MacNeish had found in the Tehuacán valley cave three decades before. Her finding remained in controversy until in 1995, with DNA evidence, Eubanks established beyond dispute that Mangelsdorf's wild corn and its descendant from the second crossing with teosinte are true recombinants of the teosinte and Tripsacum families [opposite].

Teotihuacán

By 1,900 years ago—A.D. 100—the New World agricultural revolution had erected its first city. The ruins of Teotihuacán lie on the upland northeast of the Valley of Mexico. The city's main thoroughfare was 30 meters wide. It is aligned, perhaps by some forgotten astrological dictate, between 16 and 17 degrees east of due north. It stretches 4 kilometers from the Temple of the Moon, a large terraced pyramid at its north end, to a vast ceremonial and apparently administrative center. Just north of the center and east of the main thoroughfare stands the Temple of the Sun, a pyramid as large as those of Cheops at the base and fully half as high. This temple is a less elaborate structure than the Temple of the Moon, which is distinguished by bold sculptures adorning the face of each terraced level.

The minor axis of the city is another 30-meter-wide thoroughfare. It reaches, at a right angle to the main axis, 4 kilometers outward on either side of the ceremonial-administrative center. From the extent of the city and what can be made of digs on its outskirts, Teotihuacán must have had a population of 50,000 or perhaps more. The decline and fall of the city came, for reasons not disclosed in the unlettered ruins.

Aztec intensive agriculture

The subsistence of the large population of Teotihuacán carried the technology of New World agriculture to an intensity unmatched until the last century brought hydroponics. At the southern foot of the city, the ground bears the dry scars of what once were water gardens, called *chinampas.* Such gardens can be seen still under cultivation in the shallow lakes south of Mexico City. That some of these gardens—or their predecessors—must once also have supplied Teotihuacán is plainly suggested by the alignment of their principal canals between 16 and 17 degrees east of north, parallel to the major axis of the ancient city.

Chinampa gardening came to full bloom in support of the Aztec civilization, from around A.D. 1300 until the arrival of the *conquistadores* in 1532. The Lake of the Moon in the Valley of Mexico once covered 1,250 square kilometers of the 5,000 square kilometers of the land-locked watershed. In season, the rains filled the valley bottom with a single lake; in dry seasons, evaporation reduced the water level to expose bogs and marshland separating five lakes. Ages of evaporation concentrated salts and nitrates in the larger northern lakes. Springs still bring fresh water into the system along its southern and western margin. Here the *chinampa* technology flourished for 2,000 years. A few hectares still maintained as a tourist attraction provide live demonstration.

The *chinampas* are piers of earth, from 5 to 10 meters wide and 15 to 30 meters long, standing less than a meter above the surface of the lake. Separated by fingers of water, they reach out from long causeways to form a crowded green marina. Wattle held in place by wood pilings retains the narrower gardens; willow hedges hold the wider ones. Today, as for 2,000 years, hours of labor go to maintain their high fertility. The *chinamperos* boat away old topsoil and replace it with mud from the bottom of the canals enriched with compost of underwater plant life. A similar investment of labor goes to nursing individual seedlings in flats and then to planting them one by one. Such attention is lavished on all of the produce grown on the *chinampas* except corn and amaranth. Continuous gardening yields as many as seven crops a year. In Aztec times, the gardens yielded an abundance of flowers as well, marigolds and dahlias favored.

A 10-kilometer dike protected the fresh springwaters of the southern lakes from salination by the main body of water. Aqueducts 15 kilometers in length brought the springwaters to the *chinampas* around the gleaming capital city on the islands of Tlatelolco and Tenochtitlán—where Mexico City now stands. These engineering feats excited the admiration of the *conquistadores,* even as the *chinampa* canals and waterways impeded their conquest.

Absolute authority

The absolute authority asserted by human sacrifice on an annual cycle of ceremony ensured the supply of labor for the intensive *chinampa* agriculture and the engineering works that supplied its water. The chosen ones enjoyed a week or so of public celebration and, it is said, herbal tranquilizing of their apprehension. Conquest provided occasion and principals for such ceremony at other times. In redress of a calamitous engineering error on a major aqueduct, the emperor Ahuitzotl (1486–1502) had the hearts of the responsible officials cast into its waters.

At the peak of Aztec civilization, the population of the Valley of Mexico reached 500,000. Today it is the site of the world's most populous city, its new and old buildings sinking in the sediments of what was once the western shore of the Aztecs' Lake of the Moon.

As early as 3,000 years ago, village gardens supported the immense ceremonial centers of the Mayan culture in Guatemala and on the peninsula of Yucatán. Here, as later in the Valley of Mexico, human sacrifice disciplined the economic process. A 250,000-square-meter stone platform served this purpose at one center. From a wealth of pictorial and abstract glyphs carved in stone, scholars have teased out the history of the Mayans and their remarkable astronomical learning. Mayan astronomers kept the calendars of all the visible planets. The conjunctions in the travel of the planets on their orbits occasioned the cycle of larger and smaller sacrificial ceremonies. The calendars recorded also the accession and passing of divine rulers.

The Mayan is distinguished as a high civilization not organized around large cities. It vanished around 2,000 years ago. The pyramidal ceremonial centers today rear up in lonely immensity in the forests.

Village agriculture in the same millennia had spread to South America. The many streams flowing from the coastal mountains of Peru watered settlements strung along the Pacific coast for hundreds of kilometers. Canals on the upland contour lines spread water from the streams on the fields. With the growth of population, engineering of the canals became more ambitious. City centers brought the necessary organization of labor. Finally, under imperial authority, the grandest canal diverted water from one river valley to settlements 125 kilometers away. It was in the middens of these settlements that Junius Bird found the fragments of brocade he so admired [see page 401]. The Chimu civilization and empire, named for its substantial capital city, came under domination of the Andean Incas in the 14th century.

Inca civilization

The empire of the Incas in the 16th century reached from Colombia on the north to Argentina on the south and from the Pacific coast to the headwaters of the Amazon. To control this large territory, the Inca rulers decreed the building of 15,000 kilometers of highway. Two parallel highways in this system ran 2,500 kilometers north and south in the Andes on the east and in the coastal range on the west. Over most of their length they ran in straight lines up and down mountain, and across chasms on suspension bridges hung from many-stranded fiber cables 10 centimeters thick. The paved surface, 3 meters wide, was sheltered from blowing snow and sand by waist-high walls on either side. At longer and shorter intervals larger and smaller shelters eased the comfort of travelers. Over these highways—on which no wheel rolled—imperial couriers, in relays, ran the length of the empire in less than two weeks. Only the Roman highway system was comparable; by this time, the Roman highways long in disrepair, Europe had nothing like it.

In lofty sites above 10,000 meters, the Incas built their imperial cities of granite blocks, fitted without mortar. In witness to the power of manpower under Inca organization and discipline, 10-ton blocks, from quarries kilometers away and many meters lower in elevation, have been identified in those walls. The no more than semidomesticated, frail-boned llama could not have assisted in these feats of brute strength. In unexampled display of their power and wealth, the Inca rulers sheathed the facades of their palaces and temples in sheets of gold.

In the New World, accordingly, the 16th-century Spanish encountered what MacNeish has called "a series of cultures almost as advanced ... and quite as barbarous as their own." Compared to the conquest of any Old World civilization by barbarians from the inner frontier of Eurasia, theirs was absolute. At peak, the population of the extended Inca empire is estimated to have reached 500,000. The region did not see that population again until the 20th century. Prospecting in the early 1980s, MacNeish staked out sites of human presence and occupation reaching across more than 20,000 years, from traces of the first hunting and food-gathering arrivals to the ascendant urban civilization of the Incas. In no place in the Old World is a comparable record of human social evolution so open to inquiry. Human occupation there has been continuous and on the increase, obliterating the record.

Accounts between the New and the Old World were balanced, somewhat, in the melting down of Inca gold. The Spanish found their treasure made largely of copper. Inca metallurgists had mastered ways, by chemical treatment and hammering, to bring the gold in copper-gold alloys to the surface of objects made of them, such as the sheets of gold that sheathed the temples and palaces.

At A.D. 1600, the world population stood at 500 million. Another 30 billion, by the Deevey estimates, had lived and died. The land under cultivation sustained 100 persons per square kilometer. Worldwide, people occupied the Earth at the density of 3.7 per square kilometer. Life expectancy, with no improvement in the prospects of the 80 percent, still stood at 25 years. In the next period, the critical intervals of social evolution shrink to centuries and decades.

Life expectancy was lengthening in Europe. For the first time, the number of adults matched and then exceeded the number of children. Soon, the pennants of Portugal and Spain, then of the Netherlands and England, streaming from the masts of square-rigged merchant vessels and warships were carrying the first consequence of this change in the human condition worldwide. Over the next two centuries, as their populations exploded, the European countries brought the ancient civilizations of India and China and all the rest of the world, excepting only Japan, under occupation or political and economic domination.

Industrial revolution

The industrial revolution surely had roots in the classical civilizations of the Mediterranean. The scientific tradition of respect for experience reaches back to the Babylon of 3,000 years ago, where astronomers—for astrological purposes, to be sure—made the first precise observations of the naked-eye planets and established their orbital years. No red walls hid the temple of Athena; the democracy of the patriciate stood her temple on the Acropolis for all to see. The Europeans owed to Byzantium and Islam the preservation of the written record of the Classical civilizations through the Middle Ages, between the last barbarian sack of Rome and the awakening of urban civilization in the 15th-century Renaissance. Through those ages, the great monasteries maintained islands of literacy; there, scholastics were free to reckon with *probabilitas*—that which is subject to proof—even as they contemplated *veritas*—that which had been revealed.

Climate may also have played a role. Mutual interdependence compelled by single harvests and short growing seasons perhaps constrained the exercise of authority. In the interstices of the distribution of power among regional royalties and the elected emperor of the Holy Roman Empire, mercantile cities and the guilds of craft and learning claimed a measure of immunity. At Bologna, Paris, Prague, Vienna, Oxford and Cambridge, beginning in the 11th century, practitioners of the nascent professions of law and medicine joined their students in centers of teaching and learning that were to live on as universities. These seats of the independence of the learned professions of science and scholarship distinguish Western Civilization from all that came before.

The Black Death of the 14th century may also have had its role. Starting from Constantinople as a bubonic plague, it swept through the continent, turning lethally pneumonic. Raging through a decade, it reduced the European population by an appalling percentage, perhaps exceeding 50 percent. Thereby the population of the continent was reduced far below the carrying capacity at which temperate-zone agriculture had arrived. With reoccupation of abandoned farmland, more members of the better-fed next generation reached maturity. The higher value of scarcer labor compelled some equalization of shares between the 20 and the 80 percent. Concomitant improvement in the output of agriculture sustained the return of population growth to its prior trajectory in a better state of nutrition within 150 years.

Surely the conquest of the ocean had its role in igniting the revolution. The Portuguese navigators broke the Arab monopoly on trade with Cathay, at the far eastern end of the Silk Road. The collapse of prices on those imports ruined the Portuguese but excited trade in the rest of the continent. The pillage of the New World by the Spanish brought new liquid capital; English pirates, under crown license, secured its redistribution for productive investment. East India companies out of England and Holland began the plunder of China, India and the archipelagoes of the Southwest Pacific. Slaves from Africa, fed on Captain Bligh's breadfruit from Polynesia, opened the Caribbean islands to plantation. The yield from the new farmland under cultivation in North America ran ahead of the growth of population there and in England.

Mechanical energy

Whatever the enabling circumstances, the industrial revolution has one indisputable beginning: the invention of the steam engine in 1769. Without Watt's other invention, the flyball governor [see page 52], his steam engine would not have worked. Through a gear train, the output shaft of the engine sets the flyballs spinning. The angle at which they fly outward under centrifugal force adjusts the valve that admits steam from boiler to engine. Watt thereby established the principle of feedback control, which governs all the self-regulating machines and systems of industrial technology.

The evidence is good that Watt and his partner Matthew Boulton were conscious revolutionaries. With Erasmus Darwin, Joseph Priestley, Josiah Wedgwood (Charles Darwin's other grandfather) and a dozen others they joined in the monthly meetings of the Lunar Society. They kept no minutes of those meetings, but it is known that talk was not confined to machines and manufactures nor to common interest in natural philosophy. Discussion went on to consideration of what machines and mechanical energy might do to relieve human existence of want and toil. At century's end, the prospective consequence of what they were doing was in full-blown public controversy. Thomas Malthus published his *Essay on the Principle of Population* to rebut, he declared, speculation by certain contemporaries "on the perfectibility of man and society." They were "misled by great and unlooked-for discoveries in natural philosophy … into the opinion that we are touching on a period big with the most important changes that would in some measure be decisive for the future fate of mankind."

Malthus quoted the Marquis de Condorcet, one of the two authors he named: "A very small amount of ground will be able to produce greater variety of supplies of greater utility or higher quality; the manufacture of articles will be achieved with less wastage of raw materials and will make better use of them … each successive generation will have larger possessions." By way of impeachment, he cited Condorcet's declaration that "the principles of the French Constitution are already those of all enlightened men." Malthus dismissively quoted his countryman, William Godwin, in the wishful prophecy: "In a state of society where all men lived in the midst of plenty and all shared alike in the bounty of nature … the narrow principle of selfishness would vanish."

The Principle of Population

The Principle of Population, "the tendency of population to increase faster than food," put such fantasies down. A priest of the Church of England, Malthus held that the tendency served the Creator's purpose in peopling the Earth. "If the two tendencies [to increase] were balanced, I do not see what motive would be sufficiently strong to overcome the indolence of man and make him proceed to the cultivation of the soil. The population of any large territory, however fertile, would be as likely to stop at 500 or at 5,000 as at 5 millions or 50 millions." The "imperious necessity" of population growth ordained inequality in political economy and the "severe distress" of the unpropertied masses "who, in the great lottery of life, have drawn a blank."

The Principle of Population governs in popular opinion today as the unspoken statement of the ultimate economic problem. It accepts overpopulation as the terminal misery of humankind. The doubling of the population to 6 billion in the last 40 years and the misery of the 2 billion poorest of the poor affirm the prospect. In every foreign office, the misery and the prospect ratify the inexplicit premise of policy. The power of population authorizes the war of all against all.

Social evolution, it is apparent, has yet to transcend the heritage of the millennia of agricultural civilization. Change in human values nonetheless has attended increase in objective knowledge and its public demonstration in technologies that have so radically changed the material condition of an increasing percentage of the world population over the past three centuries.

Slavery becomes immoral

Output of mechanical energy began to exceed output of biological energy by men and beasts in the U.S. economy during the 1860s. That same decade saw the abolition of slavery in this country. Slavery—which attracted no mention in any ancient scripture—had become technologically obsolete. Economically noncompetitive in consequence, slavery could be abolished.

A man-year—the net output of mechanical energy that can be asked of a worker—is reckoned at 150 kilowatt-hours. From the country's central power stations U.S. citizens and the machines that do their work draw an average 30,000 kilowatt-hours per capita. The puissance of each citizen is multiplied therefore by 200 of what that moral prophet of industrialism, R. Buckminster Fuller, called "inanislaves."

Mechanical energy confers command of the other forces of nature engaged by industrial technology. It makes possible the making of the materials that contain, carry, sense and control those forces from the extremes of heat and pressure to the depths of cold and vacuum, from megavolts to nanovolts. It lifts any limit on the extraction of resources from the lithosphere: their economic availability is a function of the cost of energy. It makes limitless sources of energy itself accessible in the crust of the Earth and the light of the Sun. It puts the airfoil in competition with the wheel. To households in industrialized societies it supplies heat, light, most of the furnishings and a staff of mechanized and dutiful domestic servants.

The abundance generated by mechanical energy has brought the entire population of those societies out of want, even with little or no change in the shares of the 80 and 20 percent. No longer does scarcity impose poverty in those countries. Poverty persists stubbornly there as a socioeconomic institution.

The social revolutions brought on by industrial revolution in Europe in the late 18th and early 19th centuries celebrated the fresh experience of freedom of conscience, association and speech. Already suppressed in the social memory of the fortunate people of the industrial societies today is the experience, continuing still, of 80 percent of humankind. Again in the words of Alfred North Whitehead:

> The massive habits of physical nature, its iron laws, determine the scene of the sufferings of men. Birth and death, heat, cold, hunger, separation, disease, the general impracticability of purpose all bring their quota to imprison the souls of women and men.... The essence of freedom is the practicability of purpose.... Prometheus did not bring to mankind freedom of the press. He procured fire.

Not "the permanent problem"

In 1930, as the economies of the industrialized countries settled into the Great Depression, John Maynard Keynes called upon his anxious contemporaries to contemplate "the economic possibilities for our grandchildren." "Coal, steam, electricity, petrol, steel, rubber, cotton, the chemical industries, automatic machinery and the methods of mass production, wireless, printing, Newton, Darwin and Einstein" he said, had brought in sight the possibility that "assuming no important wars and no important increase in population, the *economic problem* may be solved, or at least be within sight of solution, within hundred years. This means that the economic problem is not ... *the permanent problem of the human race"* (Keynes's emphasis).

Since 1930, there has been at least one important war and the world population has more than doubled. The end of the economic problem has come nonetheless plainly in sight. Struck by the Principle of Population, Keynes proved less sure as prophet than the contemporaries whom Malthus confuted in his *Essay*. The Marquis de Condorcet and William Godwin each predicted the decline of fertility by the practice of contraception—which, to Malthus, was an abomination: "improper arts for the concealment of irregular connections"—that has come with the solving of the economic problem.

The popularization of abundance

The spread of abundance that began with the introduction of mechanical energy in the 18th century accelerated the lengthening of European life expectancy that had begun in the 16th century. Decline in the infant death rate brought Europeans in increasing numbers into the childbearing years of life. By the middle of the 19th century they were entrained everywhere in population explosion, with birthrate rising even with fertility declining. Foundling hospitals in the major cities of Europe and America accommodated the widespread resort to abandonment and infanticide, even as the population doubled between 1750 and 1850.

At mid-20th century, the Europeans—who numbered 50 million, one-tenth of the world population in 1600—had multiplied 15-fold to 750 million, fully one-third of the then world population. They had populated the New World and had carried industrialization onto all the continents and into all new and old cities around the world of more than 50,000 inhabitants.

Life expectancy in industrialized countries now exceeds 70 years. Not all of those countries are of European culture. They include Japan, South Korea and Taiwan and the metropolises of Hong Kong and Singapore. In all of the industrialized countries, fertility has fallen to or below, the population replacement rate of two children per childbearing

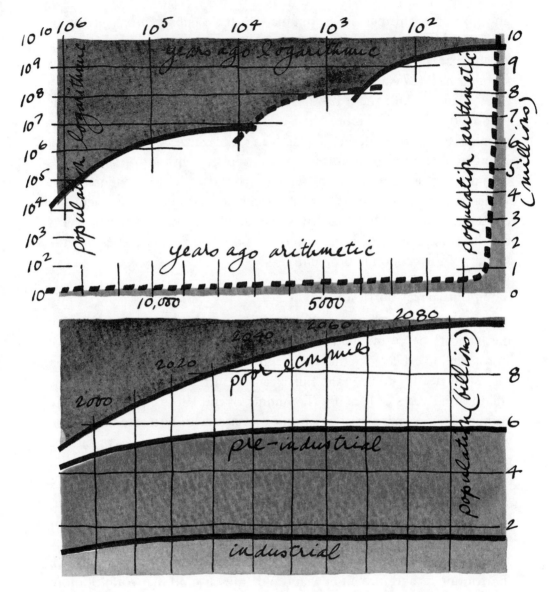

Three population explosions *have attended human biological and social evolution [top]. The cultural revolution brought the population to a million at 200,000 years ago. From 5 or 6 million at 10,000 years ago, the agricultural revolution carried population growth to 50 million at 1600. The industrial revolution has brought it to 6 billion. Future growth [bottom] will come principally from the now poorest billion. The advanced preindustrial countries will complete their demographic transition to zero growth in this century.*

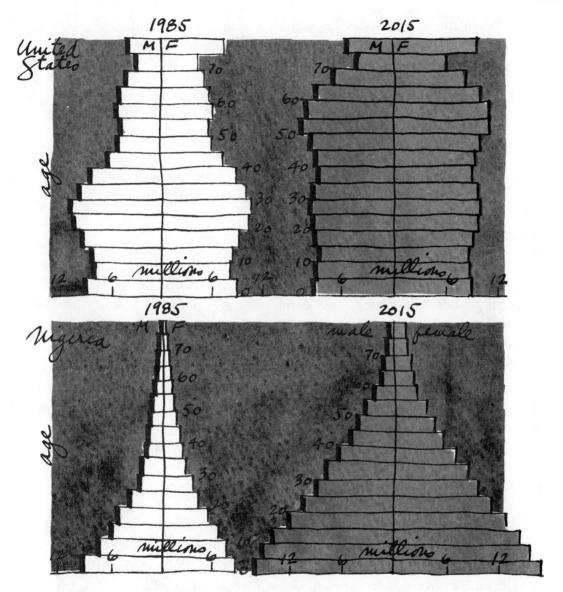

Demographic transition is illustrated in these graphs, by age and sex, of the populations of the United States and Nigeria for the years 1985 and 2015. With U.S. fertility at replacement rate in 1985, the child-bearing age-groups, offspring of the 1950s "baby boom," outnumber their offspring. Nigerians, in explosive phase of the transition with improving life-expectancy, see quadrupling of their cohorts in the child-bearing years in 2015; in evidence of declining fertility rate, the number of children under 14 increases by only a little more than 100 percent.

female lifetime. The population of the industrialized world has completed the demographic transition. This, as stated on page 41, is the transition of the world population:

a) from near-zero growth at high death rates and high birth rates and life expectancy of less than 30 years (as in all prior history of humankind)

b) through the population explosion

c) to near-zero growth again at low death rates and low birth rates and life expectancy exceeding 70 years (the experience of the living generation in increasing numbers)

Population has arrived at zero growth in all the industrialized countries. For the Europeans, engaged in mounting the industrial revolution, the transition required three centuries. With the great work accomplished, the Japanese made the transition in less than a century. The later arrivals have made it in still less time.

The European experience demonstrates that the transition has consequences that go well beyond the demographic. In the lengthening of life expectancy—in the deferral of death—people make what the demographer Jean-Claude Chesnais has called "the conquest of life." As the median age rises from 10 or 13 to 30 years and beyond, a population of children becomes a population of adults. Survival no longer preempts human consciousness. Future dwelling becomes possible. Future-dwelling parents make the calculation: the fewer the children, the more for each, and fertility declines. The population of the industrialized countries is the world's first adult population; its children constitute less than one-third of its number.

Living longer, people assert rights in their prolonged existence. The sanctity of the individual displaces the divine right of kings. It is no coincidence that the institutions of democratic self-government are coeval with industrial revolution. All of the industrialized countries today live under democratically elected representative governments. No government of any country, however dubious its legitimacy, fails to invoke the derivation of its power from the people.

Nations live, however, in the anarchy of national sovereignties that persists from agricultural civilization. The Thirty Years War of the 20th century, 1914–1945, carried off 10 percent of the people who lived in that time. Early in the War, in recoil from its horror and folly, the industrialized nations attempted to organize a world polity in the League of Nations. In 1945, they organized the United Nations. By its charter they charged themselves with two missions. The first was to keep the peace. This time, they had new compulsion. They could not—they cannot—protect their inhabitants from the Bomb that ended the War. In their second mission, the industrialized countries undertook to hasten the economic development of the new nations, the "underdeveloped countries" then emerging from their lapsed empires.

After half a century, it can be seen that the industrialized nations succeeded, partially at least, in the first mission. With deliberations in the Security Council to keep the parties talking, rational response to the assurance of mutual destruction postponed the next use of the Bomb. Nevertheless, at least 20 nations must now be counted in open or covert possession of the Bomb. In the drift of world affairs, this postwar era may any day become a prewar era.

Years ago, the overhanging menace to human existence prompted the mathematician Jacob Bronowski to this counsel:

> We have not let either the tolerance or the empiricism of science enter the parochial rules by which we still try to prescribe the behavior of nations....
>
> The body of technical science burdens and threatens us because we are trying to employ the body without the spirit; we are trying to buy the corpse of science. We are hagridden by the power of nature which we should command, because we think its command needs less devotion and understanding than its discovery.

Freedom from want

That the United Nations had a second mission is scarcely remembered today in the foreign offices of the nations that organized it. The war aim of "Freedom from Want" had nonetheless the full commitment of their postwar remorse and resolve. With recognition of the demographic transition, that aim came to comprehend a larger aim. Economic development would carry the rest of the world population through the demographic transition and bring population growth to a halt at a size that the Earth's resources might sustain in dignity and freedom.

With confidence, the not-yet-demobilized governments turned their still-enlarged authority to the task of economic development. In 1946, the United States called upon the UN General Assembly, convened in a former bombsight factory at Lake Success, New York, to organize an international conference on the conservation and utilization of resources. The preparatory sessions of that conference—where the business of such conferences gets done—supplied the inspiration and substance of Point 4 in Harry Truman's January 1949 inaugural address. Point 4 called on the United States to render financial and technical aid to the economic development of the underdeveloped countries, in extension of the program of massive economic assistance then running to Western Europe under the Marshall Plan.

One percent of GDP

Resolution 200 of the UN General Assembly later that year established the United Nations Fund for Economic Development (later called the Special fund, to raise the acronym from UNFED to SUNFED). It charged a "Group of Experts," including economists from India and the new Caribbean nation of Jamaica, to consider "measures for the economic development of the underdeveloped countries."

Their report constituted the first informed estimate of the dimensions of such an undertaking. It called for the extension of aid by the industrialized countries at the level of 1 percent of their GDP. "The amount which can be profitably invested at 4 percent," the authors explained, "depends upon the amount which is being spent at the same time on improving social capital, especially on public health, on education and on roads and communications."

Capital for social investment is usually generated by taxation. The poverty of the underdeveloped countries could not yield the necessary revenues. Given investment in the infrastructure, no one doubted that private enterprise would respond to the huge markets that would open up in the underdeveloped countries. Yet, no one expected that private capital would supply the entire investment.

Governments of the industrialized countries would accordingly contribute that 1 percent of their GDP by outright grant. Such assistance, equal to 10 percent or more of the GDP of the underdeveloped countries, would excite investment of 10 times its value from their underemployed manpower and unused resources in the infrastructure of industrial revolution. For the next level of financing, "soft" loans for long terms and low rates of interest, the then-to-be-developing countries would turn to the International Bank for Reconstruction and Development (i.e., the World Bank), which was to be financed by the same governments. On the infrastructure set by these investments, private capital and the market process would carry the industrial revolution worldwide. Then, at the dawn of the postwar era, it was expected that the underdeveloped countries would be launched on self-sustained, industrialized economic growth by 2000.

As late as 1962, John F. Kennedy affirmed the commitment of 1 percent of the U.S. GDP to economic assistance and called upon the General Assembly to declare the 1960s the "decade of development." The 1960s ended as the "decade of disappointment." The industrialized world divided into two ever more heavily armed camps. The economic assistance promised in 1945 to hasten economic development everywhere went instead to binding the loyalty of a few Cold War client states—giving "foreign aid" its bad name. Many times that promise went to fighting proxy wars among and in those client states.

Trade, not Aid

Counseled to seek development "by trade, not aid," the then-called "developing countries" made uneven progress, externally financed by debt and by the selective flow from the industrialized world of investment capital in search of oil and other natural resources and low-wage labor. The tiny trickle of funds through the multilateral channels of the UN Development Program and the UN technical agencies meanwhile introduced everywhere the most portable technologies of industrial revolution: public health and education. Statistics on the health and literacy of the people speak for the installation everywhere of the primary infrastructure of economic development.

Life expectancy has increased by at least 10 years on every continent. The infant death rate has fallen below 100 per 1,000 live births (from as high as 300). The fertility rate has followed; from an index of 6.5 it has declined to 4.2. The rate of world population growth peaked at just under 2 percent in 1970 and now approaches 1.5 percent. As do all averages, these conceal wide disparity between countries most advanced in development and those left behind.

The world's most populous country appears among the most advanced. The 1.3 billion population of China has a life expectancy of 70 years, up from 40 in 1950; an infant death rate of 32, down from 300; fertility at 2.3, down from 5.8 and on the verge of the replacement rate; and literacy at 98 percent in the population under age 60, the female rate within a fraction of the male. At the other end of the ranking in development are 30 countries in sub-Saharan Africa. Their combined population of 500 million presents correspondingly disheartening demographics: life expectancy at 50 years, infant death rate at 218, fertility rate at 6 and literacy at 27 percent—33 for males, 15 for females. These numbers reflect considerable improvement on conditions in 1950. Those numbers were down, however, from 1980. The escape, around 1980, of the HIV-AIDS virus from Africa and the ongoing worldwide pandemic ought to have taught, in the negative, that all must share the now common future of humankind.

To the statistics that describe the common present of humankind, the poorest nations have been making increasingly reliable contribution from their domestic bookkeeping—one measure of their progress in building the infrastructure of development. Their vital statistics assembled by UN population agencies constitutes a computer model of the world population. The model yielded in 1990 long-term projections of population growth that make it possible now to conjure with two questions: When might the world population, completing its passage through the demographic transition, arrive at zero growth? At what size might that population arrive?

In answer, the model was asked to project population growth at the present downward trends in fertility. On that hopeful assumption, it showed, the population would complete that passage and arrive at 11.5 billion and zero growth in 2100. Alternatively, the model projected population growth at present fertility rates. That run brought the population to 22 billion in 2100 and 28 billion in 2150 and still growing. To the 28 billion population, it showed the present poorest billion people contributing 14 billion descendants.

The development necessary to bring the 4.8 billion population of the developing countries through the demographic transition now hangs principally upon the flow of private-sector investment into what have come to be called "emerging markets." The 500 largest transnational corporations—conducting 30 percent of the global economic turnover and exceeding the combined GDP of the developing countries—make the principal connection between the industrialized and the preindustrial world. Through the 1990s, their annual direct investment has mounted to $150 billion a year, approaching the 1 percent of industrial world GDP once promised in aid.

For more than two decades, 80 percent of the private investment flow has gone to 10 of those markets. Until 1985, the 10 had a combined population of 673 million. With China now included in the capital flow, the population of the 10 countries totals 1.8 billion. (Once-favored Indonesia has fallen through the bottom of the first 10.) The low fertility rate of China brings the fertility rate of this population down to 2.6 and places it in the third phase of the demographic transition. For investment in the rest of the developing world, with 3 billion population, there remains $30 billion. The World Bank report for 2000 shows none of it going to sub-Saharan Africa and its 500 million of the world's poorest people, now in the second—explosion—phase of the transition.

Sustainable development

In UN deliberations on economic development, developed nations have yielded to the initiative seized in their own interest by developing countries. Over the half century, the General Assembly has convened a succession of UN conferences on human rights; on the welfare of children, women and the growing number of elderly; and on population, environment and economic development. The proceedings of these conferences and the body of world statistics kept by the secretariat and the technical agencies of the United Nations have now established the prospect and feasibility of a prolonged future for the human species.

Over the past half century, most developing countries have raised their own technically sophisticated, indigenous intelligentsia. There has been time for these people to put flesh on the bare bones of the 1950 formulation of the essential, catalytic role of foreign aid. Their engineering and economic studies of such undertakings as resource development; urban water supply and sanitation systems for the fast-growing cities of their countries; and transportation, communication and electric power networks have given tangible reality and specificity to the need for aid and for the infrastructure technology that it is to transfer.

A comprehensive sampling of this now-immense body of work was made by the secretariat of the UN Conference on Environment and Development held at Rio de Janeiro in 1992. The data bank underlying the estimates of the scale of overdue infrastructure investment is spelled out in Agenda 21, the principal work product of that conference. This agenda for "sustainable development" identifies and prices out the most urgent tasks in infrastructure building, including repair of environments damaged by poverty and heedless resource exploitation. These are the kinds of tasks for which foreign aid has always been intended. They require public financing. Paying no interest or dividends, they set the foundation for enterprises that will.

With these tasks undone, the world's poor will surely attract no gain-seeking interest from the globalized economy and will continue to live in poverty at the cost of environmental degradation. Agenda 21 embraces the ecosystems and populations presently at the margins of that economy or excluded from it. In the 40 poorest countries, in Africa and South Asia, projects of the highest priority await action, including halting of desertification, initiating reforestation, the capture and storage of uncertain monsoon rains, restoration of soils and education that will enable the next generation to shoulder these tasks and bring on the industrialization that will yield to their own people income from their countries' resources.

Agenda 21 projects the flow of aid at $125 billion per annum, or 0.7 percent of the 1992 GDP of industrialized countries. In the developing countries, Agenda 21 shows this transfer of technology generating a four-times-larger immediate investment from dormant human and physical resources. Sustained for 35 years, the combined investment by the industrialized and developing countries would trigger, for the rest of the 21st century, the fourfold expansion of the world economy that is needed to double well-being and eliminate the direst poverty in the doubling world population.

Ahead of any aid, developing countries look to the industrial powers to renew the growth of the world economy. It has been left to the market to determine the rate of investment. The 500 transnational corporations have been deciding what technologies to deploy and when and where and in whose interest to deploy them. Just as the market has no mechanism to internalize the cost of its operations to the environment, so it has no way to respond to human need, purpose or hope not qualified as economic demand. For that response, the self-governing citizenry of the industrialized countries must call upon other institutions.

The exercises that have been reviewed here—the projections of world population growth and of the economic development required to see the rest of the population through the demographic transition—show this next phase in the social evolution of humankind to be practicable. The more rapidly world economic development proceeds, the smaller the ultimate world population. If a prospective population of 11.5 billion seems overly large, the prescription will lie in measures that hasten development in the preindustrial countries.

It is possible to conceive such a mighty purpose because the technology to accomplish it is in hand. This finite end is in the reach of finite means. The success of the inquiry started by the first toolmakers challenges the humanity of H. *sapiens.*

NOTES

These end notes offer clues to, if not citations of, my sources. As will be seen, not many of them are primary and not all are here. The books and the SCIENTIFIC AMERICAN articles, in which I found pleasure and instruction, are commended to the reader whose interest I have stirred or baffled for dilation or clarification of my account.

CHAPTER 1

To the reaches of the observed universe there is no better introduction than: Morrison, Philip and Phylis and the Office of Charles and Ray Eames, *Powers of Ten*, SCIENTIFIC AMERICAN Library, W. H. Freeman and Company, New York, 1982. To the ground of knowledge: Hawkins, David, *The Language of Nature*, W. H. Freeman and Company, San Francisco, 1964; von Mises, Richard, *Positivism: A Study in Human Understanding*, Harvard University Press, Cambridge, 1951, and Bridgman, Percy, *Reflections of a Physicist*, Philosophical Library, New York, 1955. How science gets done: Merton, Robert K., *The Sociology of Science*, University of Chicago Press, 1973; and Kuhn, Thomas S., *The Structure of Scientific Revolutions*, University of Chicago Press, 1970.

In SCIENTIFIC AMERICAN: Gingerich, Owen, *The Galileo Affair*, April, 1986; Drake, Stillman, *Galileo's Discovery of the Law of Free Fall*, May 1973, and, with MacLachlan, James, Galileo's *Discovery of the Parabolic Trajectory*, March 1975; Zuckerman, Harriet, *The Sociology of Nobel Prizes*, Nov. 1967; Ingalls, Albert G., *Ruling Engines*, June 1952; Moore, A. D., *Henry A. Rowland*, Feb. 1982; Barnard, Chester I., *Arms Race v. Control*, Nov. 1949; Ridenour, Louis N.,

Bethe, Hans A., Bacher, Robert F. and Lapp, Ralph E., *The Hydrogen Bomb*, respectively March, April, May and June 1950; Newman, James R., review of *Herman Kahn, On Thermonuclear War*, March 1961; Blackett, P. M. S., *Steps Toward Disarmament*, April 1962; Bethe, Hans A. and Garwin, Richard, *Antiballistic Missile Systems*, July 1972; Scoville, Herbert, Jr., *Missile Submarines and National Security*, June 1972; von Hippel, Frank and Drell, Sidney, *Limited Nuclear War*, Nov. 1976; Forsberg, Randall, *A Bilateral Nuclear Weapons Freeze*, Nov. 1982; Briggs, Asa et al., *Technology and Economic Development*, a single-topic issue, Sept. 1963; Hutchinson, G. Evelyn, et al., *The Biosphere*, a single-topic issue, Sept. 1970; Dadzie, K. K. S. et al., *Economic Development*, a single-topic issue, Sept. 1980; Weaver, Warren et al., *Fundamental Questions in Science*, a single-topic issue, Sept. 1953.

CHAPTER 2

Accessible enlargement on much of the story told in this and the next two chapters is offered by George Gamow, himself a considerable contributor to the convergence of particle physic and cosmology that is the story of Chapter 4; see his *One Two Three...Infinity*, *The Great Physicists from Galileo to Einstein*, *Thirty Years That Shook Physics*, all available in reprint from Dover Publications. In James Clerk Maxwell, *A Dynamical Theory of the Electromagnetic Field*, Scottish Academic Press, with an appreciation by Albert Einstein, is a bridge from the classical to the new physics. *The Principle of Relativity*, A. Sommerfeld, Editor, Dover Publications, 1952, has the historic papers on Relativity, Special and General, and the reader will be enthralled to discover how far the arguments can be followed. And see Schilp, Paul Arthur, editor, *Albert Einstein: Philosopher Scientist*, Library of Living Philosophers, Evanston, 1949, for Einstein's autobiographical memoir.

In SCIENTIFIC AMERICAN: Einstein, Albert, *On the Generalized Theory of Gravitation*, April 1950; Cohen, I. Bernard, *Isaac Newton*, Dec. 1955; Wilson, Curtis, *How Did Kepler Discover His First Two Laws?* March 1972; Cohen, I. Bernard, *Newton's Discovery of Gravity*, March 1981; Wilson, Curtis, *Priestley*, Oct. 1954; Duveen, Denis I., *Lavoisier*, May 1956; Dyson, Freeman, J. et al., *Heat*, a single-topic issue Sept. 1954; Wilson, Mitchell, *Count Rumford*, Oct. 1960; Ehrenberg, W., *Maxwell's Demon*, Nov. 1987; Cohen, I. Bernard, *Benjamin Franklin*, Aug. 1948; Williams, L. Pearce, *Humphry Davy*, June 1960; Kondo, Herbert, *Michael Faraday*, Oct. 1953; Newman, James R., *James Clerk Maxwell*, June 1955; Morrison, Philip and Emily, *Heinrich Hertz*, Dec. 1957; Shankland, R. S., *The Michelson Morley Experiment*, Nov, 1954; Whittaker, Sir Edmund, *G. F. Fitzgerald*, Nov. 1953; Kline, Morris, *Geometry*, Sept. 1964; Dyson, Freeman, *Mathematics in the Physical Sciences*, Sept. 1964; Weisberg, Joel, et al., *Gravitational Waves from an Orbiting Pulsar*, Oct. 1981.

CHAPTER 3

Selective browsing through Feynman, R. P., et al., *Lectures on Physics*, Addison Wesley, Reading, Mass., 1963, and close study of Feynman, *QED*, Princeton University Press, 1985, together with Bohm, David, *Quantum Theory*, Dover Publications, New York, 1979, calmed and facilitated my apprehension of the quantum world.

In SCIENTIFIC AMERICAN: Weisskopf, Victor F., *How Light Interacts with Matter*, Sept. 1968; Lifshitz, Evgeny, *Superfluidity*, June 1958; Wilson, R. R., *The Batavia Accelerator*, Feb. 1974; Jackson, J. David, *The Super-conducting Super-Collider*, March 1986; Andrade, E. N. da C., *The Birth of the Nuclear Atom*, Nov. 1986; Feinberg, Gerald, *Ordinary Matter*, May 1967; Gamow, George, *The Exclusion Principle*, July 1959, and *The Principle of Uncertainty*, Jan. 1958; Schrödinger, Erwin, *What Is Matter?*, Sept. 1953; Dirac, P. A. M., *The Evolution of the Physicist's View of Nature*, May 8, 1963; Gell-Mann, Murray and Rosenbaum, E. P., *Elementary Particles*, July 8, 1957; Gell-Mann, Murray and Chew, Geoffrey, *Strongly Interacting Particles*, Feb. 1964; Morrison, Philip, *The Overthrow of Parity*, April 1957; Wigner, Eugene P., *Violations of Symmetry in Physics*, Dec. 1965; Lederman, Leon, *The Two Neutrino Experiment*, March 1963; Weinberg, Steven, *Unified Theories of Elementary Particles,* July 1974; Rubbia, Carlo, et al., *The Detection of Weak Neutral Currents*, Dec. 1974; Kendall, Henry W. and Panofsky, Wolfgang, *The Structure of the Proton and the Neutron*, June 1971; Glashow, Sheldon Lee, *Quarks with Color and Flavor*, Oct. 1975; Georgi, Howard A., *A Unified Theory of Particles and Forces*, April 1981; Freedman, Daniel Z. and van Nieuwenhuizen, Peter, *The Hidden Dimensions of Spacetime*, March 1985.

CHAPTER 4

Davies, Paul, editor, *The New Physics*, Cambridge University Press, Cambridge, 1992, brings together the authors who brought the convergence of particle physics and cosmology in a volume accessible, gradedly, to motivated curiosity. Mizner, Charles W., Thorne, Kip S. and Wheeler, John A., *Gravitation*, W. H. Freeman and Co., San Francisco, 1970, invites reading by "our fellow citizens who, for love of truth, take from their own wants by taxes and gifts … to forward the search into the mysteries and marvelous simplicities of this strange and beautiful universe, our home." Weinberg, Steven, *The First Three Minutes*, Basic Books, New York, 1988, relates the Big Bang standard model, and Guth Alan H.,

The Inflationary Universe, Addison-Wesley, Reading, 1997, amends it.

In SCIENTIFIC AMERICAN: Robertson, Harold P., et al., *The Universe*, a single-topic issue, Sept. 1956; Greenstein, Jesse, *Dying Stars*, Jan. 1959; Hewish, Anthony, *Pulsars*, Oct. 1968; Thorne, Kip S., *Gravitational Collapse*, November 1967; Le Corbeiller, Philippe, *The Curvature of Space*, Nov. 1954; Dicke, R. H., *The Eötvös Experiment*, Dec. 1961; Chaffee, Frederic H., *The Discovery of Gravitational Lens*, Nov. 1980; Weber, Joseph, *The Detection of Gravitational Waves*, May 1971; Gingerich, Owen, *Copernicus and Tycho*, Dec. 1972; Krauss, Lawrence M., *Dark Matter in the Universe*, Dec. 1986; DeWitt, Bryce, *Quantum Gravity*, Dec. 1983; Hoyle, Fred, *The Steady-State Universe*, and Gamow, George, *The Evolutionary Universe*, Sept. 1956; Webstger, Adrian, *The Cosmic Background Radiation*, Aug, 1974; Bok, Bart J., *A National Radio Observatory*, Oct. 1956, and *The Milky Way Galaxy*, March 1981; Whipple, Fred M., *The Dust Cloud Hypothesis*, May 1948; Sagan, Carl and Drake, Frank, *The Search for Extraterrestrial Intelligence*, May 1975; Osmer, Patrick S., *Quasars as Probes of the Early and Distant Universe*, Feb. 1982; Guth, Alan H. and Steinhardt, Paul J., *The Inflationary Universe*, May 1984; Green Michael B., *Superstrings*, Sept. 1986.

CHAPTER 5
Thompson, D'Arcy, *On Growth and Form*, Cambridge University Press, Cambridge, [1917] 1959, asked the questions central to the century's work. Schrödinger, Erwin, *What Is Life?*, Cambridge University Press, 1944, asks the naïve physicist's question. Margulis, Lynn, *Symbiosis in Cell Evolution*, W. H. Freeman and Co., New York, makes a breakthrough. De Duve, Christian, *Vital Dust*, Basic Books, New York, 1995, and Shapiro, Robert, *Origins: A Skeptics Guide to the Origin of Life*, Summit Books, New York, 1986, and Morowitz,

Harold J., *Beginnings of Cellular Life*, Yale University Press, New Haven, 1992, open that question.

In SCIENTIFIC AMERICAN: Brachet, Jean, et al., *The Living Cell*, a single-topic issue Sept. 1961; Lewontin, Richard C., *Adaptation*, Sept. 1978; Dobzhansky, Theodosius, *The Genetic Basis of Evolution*, Jan. 1950; Muller, Herman J., *Radiation and Human Mutation*, Nov. 1955; Mazia, Daniel, *The Cell Cycle*, Jan. 1974; Crick, Francis H. C., *The Structure of the Hereditary Material*, Oct. 1954; Mirsky, Alfred, *The Chemistry of Heredity*, Feb. 1953, and *The Discovery of DNA*, June 1968; Gamow, George, *Information Transfer in the Living Cell*, Oct. 1955; Fränkel-Conrat, Heinz, *Rebuilding a Virus*, June 1956; Stein, William H. and Moore, Stanford, *Chromatography*, March 1951; Pauling, Linus, et al., *The Structure of Protein Molecules*, July 1954; Kendrew, John, *The Three-Dimensional Structure of a Protein Molecule*, Dec. 1961; Cech, Thomas, *RNA as an Enzyme*, Nov. 1986; Lodish, Harvey, et al., *The Assembly of Cell Membranes*, Jan. 1979; Szent-György, Albert, *Muscle Research*, June 1949; Green, David E., *The Mitochondrion*, Jan. 1964; Arnon, Daniel, *The Role of Light in Photosynthesis*, Nov. 1969; Gehring, Walter, *The Molecular Basis of Development*, Oct. 1985; Wald, George, *The Origin of Life*, Aug. 1954; Mayr, Ernst, et al., *Evolution*, a single-topic issue, Sept. 1978; Margulis, Lynn, *Symbiosis in Evolution*, Jan. 1969; Woese, Carl R., *Archaebacteria*, June 1981.

CHAPTER 6
Cloud, Preston, *Oasis in Space*, W. W. Norton & Co., New York, 1988, and Schopf, J. William, et al., *Earth's Earliest Biosphere*, Princeton University Press, Princeton, 1993, and Stanley, Steven M., *Earth and Life Through Time*, W. H. Freeman and Co., New York, 1982, together recount the realization of the vision of Vernadsky, V. I. in, among other sources, *The Biosphere and the*

Noösphere, American Scientist, Jan. 1945. Simpson, George Gaylord, *The Meaning of Evolution*, Yale University Press, New Haven, 1952, states the "modern synthesis."

In SCIENTIFIC AMERICAN: Hutchinson, G. Evelyn, *The Biosphere*, a single-topic issue, Sept. 1970; Wilson, J. Tuzo, *Continental Drift*, April 1963; Dewey, John E., *Plate Tectonics*, May 1972; Dietz, Robert S. and Holden John C., *The Breakup of Pangaea*, Oct. 1970; Eiseley, Loren C., *Charles Lyell*, Aug. 1959, and *Charles Darwin*, Feb. 1956; Darlington, C. E., *The Origin of Darwinism*, May 1959; Siever, Raymond et al., *The Dynamic Earth*, a single-topic issue, Sept. 1983; Hurley, P. M., *Radioisotopes and Time*, Aug 1949 and *The Confirmation of Continental Drift*, Apr. 1968; Runcorn, S.K., *The Earth's Magnetism*, Sept. 1955; Orowan, Egon, *The Origin of the Ocean Ridges*, Nov. 1969; McMenamin, Mark A. S., *The Emergence of Animals*, Apr. 1987; Morris, Simon Conway and Whittington, H. B., *The Animals of the Burgess Shale*, July 1979.

CHAPTER 7
Whitehead, Alfred North, in *From Force to Persuasion in Adventures in Ideas*, The Free Press, New York, 1968, and Keynes, John Maynard, in *Economic Possibilities for our Grandchildren* in *Essays in Persuasion*, W. W. Norton, New York, 1963, describe "the normal mode of society" and the possibility of a different one. Chesnais, Jean-Claude, *The Demographic Transition*, Clarendon Press, Oxford, 1992, derives the new principle of population. Bronowski, Jacob, in *Science and Human Values*, Harper and Row, New York, 1965, makes the case for rescission of the Victorian Compromise [see page 324]. U. N. Conference on Environment and Development, *Agenda 21* sets the goal for the 21st century, and U. N. Development Program, *Human Development Report 2000*, and U. N. Conference on Trade and Development, *World Investment Report 2000*, all available at United Nations, New York, and show where the world is now. Rothschild, Emma, *Economic Sentiments: Smith, Condorcet and the Enlightenment*, Harvard University Press, Cambridge, 2001, recalls the origin of the science of economics in moral philosophy.

In SCIENTIFIC AMERICAN: Eiseley, Loren C., Alfred Russel Wallace. Feb. 1959; Stolarski, Richard A., *The Antarctic Ozone Hole*, Jan. 1988; Washburn, Sherwood, *Tools and Human Evolution*, Sept. 1960; Leakey, L. S. B., *Olduvai Gorge*, Jan. 1954; Leakey, Richard E. and Walker Alan, *The Hominids of East Turkana*, Aug. 1978; Howells, William, *Homo Erectus*, Nov. 1966; Deevey, *The Human Population*, Sept. 1960; Bickerton, Derek, *Creole Languages*, July 1983; Leroi-Gourhan, Arlette, *The Archeology of Lascaux Cave*, June 1982; Bosch, Perter W., *A Neolithic Flint Mine*, June 1979; Zvelebil, Marek, *Postglacial Foraging in the Forests of Europe*, May 1986; Moore, Andrew M. T., *A Pre-Neolithic Farmers' Village on the Euphrates*, Aug. 1979; Kramer, Samuel Noah, *The Sumerians*, Oct. 1957; Gamkrelidze, Thomas V. and Ivanov, V. V., *The Early History of the Indo-Europe Languages*, March 1990; Higham, C. C. W., *Prehistoric Rice Cultivation in Southeast Asia*, April 1984; Lattimore, Owen, *Chingis Khan and the Mongol Conquests*, Aug. 1963; Mangelsdorf, Paul, *The Mystery of Corn*, July 1950; MacNeish, Richard S., *The Origin of New World Civilization*, Nov. 1964; *Millon*, René, *Teotihuacan*, June 1967; Coe, Michael D, *The Chinampas of Mexico*, July 1964; Hammond, Norman, *The Emergence of Maya Civilization*, Aug. 1986; Protzen, Jean-Pierre, *Inca Stone Masonry*, Jan. 1986; von Hagen, Victor W., *America's Oldest Roads*, July 1952; Langer, William L., *The Black Death*, Feb. 1964; Ferguson, Eugene S., *The Origins of the Steam Engine*, Jan. 1964; Ritchie-Calder, Lord, *The Lunar Society of Birmingham*, June 1982; Leontief, W. W., *The Distribution of Work and Income*, Sept. 1982.

INDEX

(Italic numbers refer to art.)